C0-AWJ-523

PROCEEDINGS

OF THE

1995 INTERNATIONAL CONFERENCE

ON

PARALLEL PROCESSING

August 14 – 18, 1995

Vol. I Architecture
P. Banerjee, Editor
University of Illinois at Urbana-Champaign

Sponsored by

THE PENNSYLVANIA STATE UNIVERSITY

CRC Press
Boca Raton New York Tokyo London

The papers appearing in this book comprise the proceedings of the meeting mentioned on the cover and title page. They reflect the authors' opinions and are published as presented and without change in the interest of timely dissemination. Their inclusion in this publication does not necessarily constitute endorsement by the editors, CRC Press, or the Institute of Electrical and Electronics Engineers, Inc.

Neither this book nor any part may be reproduced or transmitted in any form or by any means, electronic or mechanical, including photocopying, microfilming, and recording, or by any information storage or retrieval system, without prior permission in writing from the publisher.

All rights reserved. Authorization to photocopy items for internal or personal use, or the personal or internal use of specific clients, may be granted by CRC Press, Inc., provided that $.50 per page photocopied is paid directly to Copyright Clearance Center, 27 Congress Street, Salem, MA 01970 USA. The fee code for users of the Transactional Reporting Service is ISBN 0-8493-2615-X/95/$0.00+$.50. The fee is subject to change without notice. For organizations that have been granted a photocopy license by the CCC, a separate system of payment has been arranged.

CRC Press, Inc.'s consent does not extend to copying for general distribution, for promotion, for creating new works, or for resale. Specific permission must be obtained in writing from CRC Press for such copying.

Direct all inquiries to CRC Press, Inc., 2000 Corporate Blvd., N.W., Boca Raton, Florida 33431.

Catalog record is available from the Library of Congress
ISSN 0190-3918
ISBN 0-8493-2619-2 (set)
ISBN 0-8493-2615-X (vol. I)
ISBN 0-8493-2616-8 (vol. II)
ISBN 0-8493-2617-6 (vol. III)
ISBN 0-8493-2618-4 (ICPP Workshop)
IEEE Computer Society Order Number RS00027

Copyright © 1995 by CRC Press, Inc.
All rights reserved
Printed in the United States of America 1 2 3 4 5 6 7 8 9 0
Printed on acid-free paper

Additional copies may be obtained from:

CRC Press, Inc.
2000 Corporate Blvd., N.W.
Boca Raton, Florida 33431

PREFACE

We are pleased to introduce you to the proceedings of the 24th International Conference on Parallel Processing to be held from August 14-18, 1995. The technical program consists of 27 sessions, organized as three technical tracks, one in Architecture, one in Software, and one in Algorithms and Architectures. We have also put together three panels, and three keynote speeches. A workshop on "Challenges for Parallel Processing" has been put together by Prof. Dharma Agrawal to be held before the conference. Two tutorials will be held after the conference and have been organized by Prof. Mike Liu.

There will be three keynote speeches at the conference to start each day of the meeting. The first keynote speech will be entitled, "Future Directions of Parallel Processing", by Dr. David Kuck of Kuck and Associates. Prof. Hidehiko Tanaka from the University of Tokyo will present the second keynote speech on "High Performance Computing in Japan". The third keynote speech will be given by Prof. John Rice from Purdue University on "Problem Solving Environments for Scientific Computing".

There are three panels with the conference to end each day. The first panel is entitled, "Heterogeneous Computing" and has been organized by Mary Eshaghian. The second panel "SPMD: On a Collision Course with Portability", has been organized by Tom Casavant and Balkrishna Ramkumar. The third panel on "Industrial Perspective of Parallel Processing", has been organized by A. L. Narasimha Reddy and Alok Choudhary.

The technical program was put together by a distinguished program committee. Each paper was assigned to two program committee members and two external reviewers. Reviews of each paper were handled using an electronic review process. The program committee met on March 17, 1995 in Urbana, IL, and decided on the final program. The decision on acceptance and rejection of each paper was made on the basis of originality, technical content, quality of presentation and the relevance to the theme of ICPP. A summary of the disposition of papers by area is presented in the following table:

Area	Submitted	Accepted	
		Regular	Concise
Architecture	160	14(8.75%)	27(16.8%)
Software	90	18(20.0%)	15(16.7%)
Algorithms	115	16(13.9%)	16(13.9%)
TOTAL	365	48(13.2%)	58(15.9%)

All papers submitted from the University of Illinois were processed by Prof. Chitta Ranjan Das of Penn State University.

We would like to thank all the people responsible for the success of the conference. First, we would like to thank the University of Illinois for providing the infrastructure necessary for preparing the program for the conference including the support staff and the mailing facilities. We would like to express our gratitude to Carolin Tschopp and Donna Guzy for managing the processing of all the papers, for preparing the proceedings, and for arranging the program committee meeting. We would like to thank the entire program committee for doing such a diligent job in reviewing so many papers in such a fine manner. Finally, we are grateful to Prof. Tse Feng for providing the guidance and wisdom for running this conference.

Prith Banerjee, Program Chair
Constantine Polychronopoulos, Program Co-Chair
Kyle Gallivan, Program Co-Chair
University of Illinois
Urbana, IL-61801

Program Committee Members

Chair: Prithviraj Banerjee — *University of Illinois*
Co-Chair: Kyle Gallivan — *University of Illinois*
Co-Chair: Constantine Polychronopoulos — *University of Illinois*

J-L. Baer	*University of Washington*	V. Kumar	*University of Minnesota*
U. Banerjee	*Intel Corporation*	S-Y. Kuo	*National Taiwan U.*
M. Berry	*University of Tennessee*	K. Li	*Princeton University*
Nawaf Bitar	*Silicon Graphics, Inc.*	A. Malony	*University of Oregon*
R. Bramley	*Indiana University*	P. Mehrotra	*NASA Langley Res. Ctr.*
T. L. Casavant	*University of Iowa*	D. Nassimi	*New Jersey Inst. of Tech.*
S. Chatterjee	*University of North Carolina*	L. M. Ni	*Michigan State University*
C. R. Das	*Pennsylvania State University*	A. Nicolau	*UC-Irvine*
S. K. Das	*University of North Texas*	U. Ramachandran	*Georgia Inst. of Tech.*
E. J. Davidson	*Michigan State University*	B. Ramkumar	*University of Iowa*
J. A. B. Fortes	*Purdue University*	S. Ranka	*Syracuse University*
W. K. Fuchs	*University of Illinois*	P. Sadayappan	*Ohio State University*
M. M. Eshaghian	*New Jersey Inst. of Tech.*	S. Sahni	*University of Florida*
D. Gannon	*University of Indiana*	D. Schneider	*Cornell University*
A. Gerasoulis	*Rutgers University*	J. Smith	*University of Wisconsin*
J. Goodman	*University of Wisconsin*	V. Sunderam	*Emory University*
M. Gupta	*IBM- T. J. Watson Res. Ctr.*	Y. Tamir	*UCLA*
M. Girkar	*Sun Microsystems*	N. H. Vaidya	*Texas A&M University*
W. Jalby	*Laboratoire MASI*	H. Wijshoff	*University of Leiden*
S. Kleiman	*Sun Microsystems*	K. Yelick	*UC-Berkeley*

Tutorial Chair: M. Liu, *Ohio State University*

Keynote Speakers

Speaker: **David Kuck, Kuck and Associates**
Topic: Future Directions of Parallel Processing

Speaker: **Hidehiko Tanaka, Univ. of Tokyo**
Topic: High Performance Computing in Japan

Speaker: **John Rice, Purdue University**
Topic: Problem Solving Environments for Scientific Computing

Panel Sessions

Panel I: Heterogeneous Computing
Moderator: **Mary M. Eshaghian, New Jersey Institute of Technology**

<u>Panelists:</u>

Gul Agha	Univ. of Illinois at Urbana-Champaign
Ishfaq Ahmad	The Hong Kong University of Science and Technology
Song Chen	New Jersey Institute of Technology
Arif Ghafoor	Purdue University
Emile Haddad	Virginia Polytechnic Institute and State University
Salim Hariri	Syracuse University
Alice C. Parker	University of Southern California
Jerry L. Potter	Kent State University
Arnold L. Rosenberg	University of Massachusetts
Assaf Schuster	Technion-Israel Institute of Technology
Muhammad E. Shaaban	University of Southern California
Howard J. Siegel	Purdue University
Charles C. Weems	University of Massachusetts
Sudhakar Yalamanchili	Georgia Institute of Technology

Panel II: SPMD: on a Collision Course with Portability?
Moderator: **Tom Casavant, University of Iowa**

<u>Panelists:</u>

Andrew Chien	University of Illinois at Urbana-Champaign
Alex Nicolau	University of California at Irvine
Balkrishna Ramkumar	University of Iowa
Sanjay Ranka	Syracuse University
David Walker	Oakridge National Laboratories

Panel III: Industrial Perspective of Parallel Processing
Moderator: **A. L. N. Reddy, IBM Almaden**

<u>Panelists:</u>

Tilak Agerwala	IBM
Ken Jacobsen	SGI
Bruce Knobe	Siemens-Nixdorf
Stan Vestal	Thinking Machines Corp.

Conference Awards

Daniel L. Slotnick Best Paper Award

A. Nowatzyk, G. Aybay, M. Browne, E. Kelly, M. Parkin, B. Radke, and S. Vishin, "The S3.mp Scalable Memory Multiprocessor"

Outstanding Paper Awards

S. P. Midkiff, "Local Iteration Set Computation for Block-Cyclic Distributions"
A. Heirich and S. Taylor, "A Parabolic Load Balancing Method"

List of Referees- Full Proceedings

Abdelrahman, T.
Abraham, S.
Adve, S.
Adve, V.
Ahamad, M.
Alexander, W. E.
Almquist, K.
Al-yami, A. M.
Armstrong, J. B.
Arrouye, Y.
Babbar, D.
Bagherzadeh, N.
Baker, J. W.
Baruah, S.
Bayoumi, M.
Beckman, C.
Beguelin, A.
Bekerle, M.
Bhuyan, L.
Bianchini, R.
Blough, D.
Boppana, R.
Bose, B.
Bruck, J.
Buddihikot, M. M.
Burr, J.
Carver, D. L.
Casavant, T.
Celenk, M.
Chae, S-H.
Chalasani, S.
Chalmers, A.
Chamberlain, R. D.
Chandy, J.
Chao, L-F.
Chatterjee, A.
Chatterjee, S.
Chaudhary, A.
Chaudhary, V.
Chen, P.

Chen, Y-L.
Cheng, A.M.K.
Cheng, K-H.
Chern, M-Y.
Chien, A.
Choudhary, A.
Clarke, E.
Conley, W.
Conte, T.
Cormen, T.H.
Craig, D.
Culler, D.
Cuny, J.E.
Das, C.R.
Das, S. K.
Das Sharma, D.
Davis, J.A.
Davis, T.
DeRose, L.
Deshmuth, R. G.
Dewan, G.
Dhagat, M.
Dhall, S.K.
Dietz, H. G.
Dincer, K.
Dowd, P.
Duato, J.
Dubois, M.
Dutt, S.
Dykes, S. G.
Efe, K.
El-Amawy, A.
Enbody, R.
Felten, E.
Feng, M. D.
Foster, I.
Fraigniaud, P.
Franklin, M.
Fu, J.
Fujimura, K.

Gallagher, D.
Ganapathy, K.
Gao, G.
Ge, Y.
Ghose, K.
Ghosh, J.
Ghosh, K.
Ghozati, S.
Gibson, G.
Gorda, B.
Grunwald, D.
Gupta, A.
Gupta, R.
Gupta, S.
Gyllenhal, J.
Haddad, E.
Haghighat, M.
Hanawa, T.
Hank, R.
Harper, M.
Harper, III, D.
Hassen, S. B.
Heath, M.
Herbordt, M.
Hermenegildo, M.
Hill, M.
Hirano, S.
Ho, C-T.
Holm, J.
Hou, R.
Hsu, D. F.
Huang, S.
Huang, Y-M.I.
Hwang, K.
Ibel, M.
Iwashita, H.
Jacob, J. C.
Jayasimha, D. N.
Jha, N. K.
Joe, K.

Johnson, D.
Jordan, H.
Joshi, B. S.
Kale, L.
Kavianpour, A.
Killeen, T.
Kim, H.
Kim, J-Y.
Kothari, S. C.
Krantz, D.
Krishnaswamy, D.
Krishnamoorthy, M. S.
Lai, A. I-C.
Latifi, S.
Lau, F.
Leathrum, J.
LeBlanc, T. J.
Lee, C-H.
Lee, C.
Lee, G.
Lee, J. D.
Lee, S.
Levy, H.
Lew, A.
Li, Q.
Lilja, D.
Lin, R.
Lin, W.
Lin, W-M.
Livingston, M.
Lo, V.
Lombardi, F.
Long, J.
Lough, I.
Loui, M.
Lu, M.
Lu, Y-W.
Maeng, S.
Mahgoub, I.
Mahmood, A.
Makki, K.
Marsolf, B.
Martel, C. U.
Mazumder, P.

McKinley, P.
McMillan, K.
Menon, J.
Michallon, P
Midkiff, S. P.
Mohapatra, P.
Moore, L.
Moreira, J.
Mouftah, H. T.
Mowry, T.
Mudge, T.
Mukherjee, A.
Mukherjee, B.
Munson, D.
Mutka, M.
Netto, M. L
Ngo, V. N
Nowatzyk, A.
Olariu, S.
Oruc, Y.
Padmanabhan, K.
Pai, M. A.
Palermo, D.
Palis, M. A.
Pan, Y.
Panda, D.
Park, J. S.
Parkes, S.
Passos, N. L.
Patil, S.
Patt, Y.
Pic, M. M.
Pinkston, T. M.
Pradhan, D.
Prasad, S. K.
Prasanna, V. K.
Quinn, M.
Radiyam, A.
Raghavendra, C. S.
Ramaswamy, S.
Ramkumar, B.
Rau, B.
Reeves, A.
Robertazzi, T.

Rodriguez, B.
Rogers, A.
RoyChowdhury, V.
Saha, A.
Saletore, V.
Samet, H.
Sarkar, V.
Scherson, I. D.
Schimmel, D.
Schouten, D.
Schwabe, E. J.
Schwan, K.
Sen, A.
Sengupta, A.
Seo, S-W.
Seznec, A.
Sha, E. H-M.
Shang, W.
Shi, H.
Shi, W.
Shih, C. J.
Shin, K.
Shoari, S.
Shu, W. W.
Siegel, H. J.
Singh, A.
Singh, J.
Singhal, M.
Sinha, B.
Siu, K.-Y.
Sohi, G.
Somani, A.
Srimani, P.
Stasko, J. T.
Stavrakos, N.
Stunkel, C.
Sun, T.
Sunderam, V.
Sundaresan, N.
Surma, D. R.
Suzaki, K.
Szymanski, T.
Taylor, S.
Teng, S-H.

Thakur, R.
Thapar, M.
Theel, O.
Theobald, K.B.
Thirumalai, S.
Tripathi, A.
Trivedi, K.
Tseng, Y-C.
Tzeng, N-F.
Ulm, D. R.
Vaidya, N.
Varma, A.
Varvarigos, E.
Vetter, J.
Wah, B.
Wang, F.
Wang, P. Y.
Wang, Y-F.
Warren, D. H. D.
Watson, D. W.
Weems, C.
Wen, C-P.
Wilsey, P.
Wittie, L.
Wojciechowski, I.
Wu, J.
Wu, K-L.
Wu, M-Y.
Wyllie, J.
Yalamanchili, S.
Yan, J.
Yang, Q.
Yang, T.
Yang, Y.
Yap, T. K.
Yew, P.
Yoo, S-M.
Youn, H-Y.
Yousif, M.
Zhang, X.
Zheng, S. Q.
Zwaenepoel, W.

Author Index - Full Proceedings

I = Architecture
II = Software
III = Algorithms and Applications

TABLE OF CONTENTS
VOLUME I - ARCHITECTURE

(R): Regular Papers
(C): Concise Papers

Session 1A: Practical Multiprocessors
Chair: Prof. J. Torrellas, University of Illinois

Session 2A: Networks
Chair: Prof. C. S. Ragavendra, Washington State University

The S3.mp Scalable Shared Memory Multiprocessor

Andreas Nowatzyk, Gunes Aybay, Michael Browne,

Edmund Kelly, Michael Parkin, Bill Radke, Sanjay Vishin

Sun Microsystems Computer Corporporation
2550 Garcia Ave., Mountain View, CA 94043
E-mail contact: agn@acm.org

Abstract

S3.mp (Sun's Scalable Shared memory MultiProcessor) is a research project to demonstrate a low overhead, high throughput communication system that is based on cache coherent distributed shared memory (DSM). S3.mp uses distributed directories and point-to-point messages that are sent over a packet switched interconnect fabric to achieve scalability over a wide range of configurations. S3.mp uses a new CMOS serial link technology that achieves transmission rates >1Gbit/sec and that is directly integrated into a packet router chip. Unlike other DSM systems, S3.mp can be spatially distributed over a local area via fiber optic links. This capability allows S3.mp to interconnect clusters of workstations to form multiprocessor workgroups that efficiently share memory, processors and I/O devices. Multichip module technology, the integrated arbitrary topology router, fast serial links, and a cache coherent DSM system that is integrated into the memory controller allow compact, massively parallel S3.mp systems.

1 Introduction

The S3.mp scalable multiprocessor system is an experimental research project that is being implemented by SMCC Technology Development group (TD) to demonstrate a low overhead, high throughput communication system that is based on cache coherent distributed shared memory (DSM). Advances in fiber optics and CMOS circuit technology allow the construction of low cost, high speed interconnect fabrics with bisection bandwidths in excess of 100 Gbytes/sec. Such systems cannot be used efficiently via conventional networking protocols. Instead S3.mp uses the notion of shared memory, where communication happens as a side-effect of accessing memory: a single store or load instruction is sufficient to send or receive data. The set of transactions that are required to support the DSM paradigm is small and well defined so that the S3.mp protocols were amenable to formal verification methods and could be implemented directly in hardware.

The S3.mp architecture is similar to ALEWILE, DASH, PLUS [2,3,5] and other cache coherent, nonuniform memory access (CC-NUMA) multiprocessors. How-

ever unlike these conventional CC-NUMA MP's, S3.mp is optimized for a large collection of independent and cooperating parallel applications that share common computing resources which may be spatially distributed. Consequently, S3.mp nodes may be separated by up to 200m, which means that a S3.mp system could be distributed over an entire building. S3.mp systems can be built by adding a specialized interconnect controller to the memory subsystem of an ordinary workstation.

FIGURE 1 : S3.mp System Overview

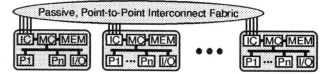

Workstations suitable for S3.mp interfaces must have a cache coherent processor / memory interconnect system. The first implementation of the S3.mp architecture is based on the *Mbus*, which is used in the SparcStation-10/20 series. Each node may have several processors, each equipped with caches that are kept consistent via snooping. When a processor accesses a remote memory location, the S3.mp memory controller translates the bus transaction into a message that is sent to the remote memory controller, which replies with the data and maintains a directory of blocks cached by external nodes.

S3.mp's interconnect controller (TIC) is part of a distributed switch that is optimized for the fine grained, irregular traffic of a DSM system. There is no need for a centralized switch, rather each node comes with a number of high speed links to connect to other nodes. Hence the interconnect fabric grows incrementally, as nodes are added to the system. Because the S3.mp architecture allows spatial distribution it must address a number of issues that are of lesser concern to conventional CC-NUMA machines:

- The interconnect topology is subject to external constraints, hence, the interconnect controller must be able to deal with arbitrary interconnect topologies.
- Fault tolerance is important to some S3.mp configurations. This requires hardware support to manage and protect multiple address spaces. This address translation mechanism

enables multiple, cooperating operating systems to run on one S3.mp system, each of which is isolated and can tolerate the loss of other OS domains.

- S3.mp must have facilities to dynamically change the system configuration, i.e. add or remove nodes.
- S3.mp components must be remotely controllable, so that the entire system can be maintained from a single point.

The S3.mp architecture supports real-time and multimedia applications by providing support for global synchronization, precise clock distribution, multicasting, and prioritized bandwidth allocation.

1.1 Motivation and project goals

S3.mp based multiprocessors will not compete with Supercomputers in terms of absolute performance. However the S3.mp project tries to build the most cost-effective and the easiest to use scalable computing environment.

The first goal, cost-effective computing, leads to resource sharing. Currently, most installed workstation class systems waste significant amounts of memory, disk space and other hardware components because they cannot be shared efficiently over conventional networks. The 64+ Mbytes of main memory in the idle system next-door cannot be used to aid a memory intensive application on another machine. Gbit/sec fiber technology provides the means to change this situation. Other ingredients for cost-effective computing include advances in caching, latency hiding, system integration, high density VLSI circuits, and new packaging methods.

Software for parallel systems does not come easy, but evidence is mounting that the shared memory paradigm leads to higher programmer productivity than explicit message passing. This comes at a modest increase in hardware complexity and some increase in bandwidth demand. This situation is not unlike virtual memory vs. program controlled overlays. While the latter can be more efficient in some situations and does require less hardware, virtual memory became ubiquitous because of its ease of use.

1.2 S3.mp system configurations

The S3.mp prototype system will use multichip modules (MCM) to integrate the interface, 64 Mbytes memory and two processors into one compact node (less than 0.9 in^3). Prototypes of this MCM technology have been successfully tested [16]. The processor memory bus (Mbus) is exposed so that the S3.mp processing element can be added to several existing workstation platforms. Interconnecting a number of workstations in this manner leads to a workgroup that can run shared memory applications.

Low end implementations may omit the processors and use the memory as a frame buffer that is integrated into the monitor. Such terminal nodes can be connected to work groups or centralized high end MPP servers. High

performance servers could be realized by densely packing many modules into one system.

FIGURE 2 : S3.mp Configuration Examples

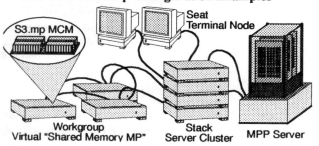

Workgroup Virtual "Shared Memory MP"   Stack Server Cluster   MPP Server

2 The S3.mp processing element

Each S3.mp node consists of 2 gate arrays that are part of one multi chip module (Figure 3). The memory controller (TMC, 172K gates) is in charge of maintaining memory coherency while the interconnect controller (TIC, 48K gates) is the building block for a scalable interconnect network. The TMC translates memory transactions from the processor bus (Mbus) to a set of messages exchanged among remote nodes. TMC can also initiate transactions on the Mbus as a result of messages coming from remote nodes. In the default configuration, two processors are connected to each TMC through the Mbus.

FIGURE 3 : A S3.mp Node

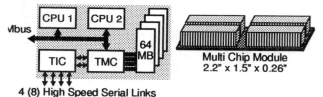

The TMC is designed to directly drive a 32-128 Mbyte memory array that uses 16 Mbit SDRAM devices with a high performance interface. Coherency is maintained on memory blocks of 32 bytes, which is the size of one cache line on Mbus based systems. The architecture is not tied to a particular cache line size because remote data is cached in the TMC and the TMC could maintain subblocks or multi-line objects.

TMC uses 18 bits of ECC overhead and 14 bits of directory overhead for every 32-bytes of memory. 14 bits provide sufficient storage to keep a 12-bit node pointer and 2-bit state for each cache line.

The Mbus of each S3.mp node can be connected to other devices, which is the primary facility to attach I/O devices to the system. In particular, it is possible to plug a S3.mp module into any Mbus slot (for example into a Sparc Station 10 or 20 Workstation).

FIGURE 4 : TIC Chip Microphotograph

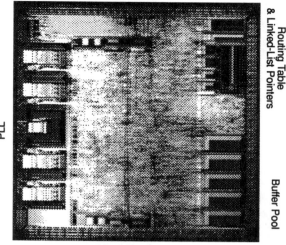

Vital Statistics

Chip Size:	11.8 x 11.8 mm
Technology:	0.65 µm CMOS, 2 Metal Layers
Power Supply:	4W @ 5V
# of Logic Gates:	48,000
Design Style:	Custom + Gate Array
On-chip Memory:	7520 bits
External Clock:	66 Mhz
Test Support:	Full ATPG Scan

3 The S3.mp interconnect system

The S3.mp interconnect system is built out of single chip packet switches (TICs, Figure 4). Each TIC is a six ported router that can receive and send one packet on each port during each routing cycle (= 6 system clock cycles). Fixed length packets are used, which allowed the use of a statically scheduled pipeline. Two of the TIC ports use a 16 bit wide, parallel, full duplex interface (Figure 5). The other four ports use high speed serializer and deserializer pairs that are an integral part of the TIC chip. Hence the TIC does not require any costly external serializers, such as Glink (HP), HotRod (Gazelle) or Taxi (AMD) chips.

Unlike the interconnect networks of other scalable systems that use a specific topology (meshes, rings, fat-trees, etc.), S3.mp has no preferred topology, rather it explores the actual interconnect network whenever nodes are added or removed from the system. An off-line process computes the routing tables which are then loaded into each TIC. It is possible to add or remove nodes and change the interconnect topology while other parts of the systems are operating normally.

The details of the TIC chip and its technology are described in [17, 18]. The TIC interconnect system has been operational since November 1994. For the remainder of this paper, the TIC based interconnect system should be regarded as a black box that provides reliable packet delivery between arbitrary nodes.

FIGURE 5 : The TIC chip[1]

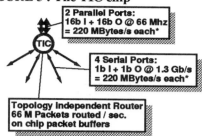

4 The Memory Controller

The TMC is responsible for handling accesses to local and remote memory and for implementing directory based cache coherence protocols. Besides acting as a normal memory controller, the TMC performs the following functions:

1. Maintaining the directory information for the local memory under its control

2. Constructing and sending messages to remote nodes on the network initiated by local Mbus transactions that require remote access or in response to messages received from other nodes on the network

3. Performing memory operations and Mbus cycles on the local node in response to messages received from remote nodes

4. Maintaining an Inter-Node Cache (INC) which is used to store a copy of every cache line retrieved from remote nodes

5. Sending and receiving diagnostic messages to/from other nodes to program configuration parameters, handle errors, etc.

From the CPU's point of view, the TMC operates as a virtual bus extender. Except for the extra latency in accessing remote locations, details of retrieving data from remote nodes and of maintaining a distributed cache coherence protocol, are completely hidden from these processors. Parallel application software targeted for a shared-bus cache-coherent multiprocessor system, written in accordance with SPARC memory models [7], executes correctly on a S3.mp system without any modification. However, performance tuning may be necessary to achieve good performance. This property of the S3.mp architecture is particularly important in being able to utilize existing SMP based applications. Besides supporting existing parallel code based on shared memory programming model, the TMC, acting as a virtual bus extender, enables single-threaded programs to utilize the collective physical memory in a network of workstations for memory intensive applications (e.g. CAD synthesis, formal verification, image processing, database applications).

---

1. The bandwidth figures are the usable packet throughput. The parallel ports transfer data only on 5 out of 6 cycles. The serial links carry an additional 16 CRC and handshake bits, that are not included.

The TMC is structurally divided into the following set of modules (Figure 7):

1. A bus controller to interface to the Mbus. Both passive (a local CPU access the TMC) and active (the TMC issues a bus transaction on behave of a remote node) are supported.

2. A memory controller that directly generates all signals necessary to drives multiple banks of synchronous dynamic memory (SDRAM). ECC is maintained over 128 bits, using 9 bits of redundancy codes. Given a standard 144 bit organization, this leaves 7 bits for each half of a cache line to store directory state information.

3. Two identical protocol engines to implement distributed cache coherency protocols (RAS and RMH).

4. A control unit to take care of configuration management, errors, diagnostics, timers, interrupts, etc.

5. Input and output queues for interfacing to the TIC (or to a general purpose point-to-point interconnect).

FIGURE 6 : TMC Structure

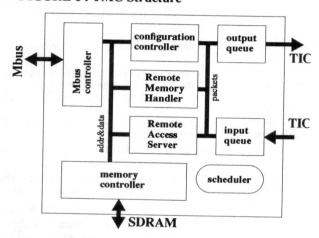

4.1 Modular Design Methodology

The TMC is designed as a system on a chip, where modules interact only through a small set of well defined transactions. Modules are designed to be as functionally self-sufficient as possible. The number of inter-module signals are kept to an absolute minimum. Functional units communicate with each other through an address bus, a data bus and a packet bus. Access to these busses are managed by a central controller, the *scheduler*. The scheduler receives dedicated request lines from each master module and ready lines from each slave module (Figure 7) and sends dedicated grant lines to each master and dedicated request lines to each slave. Requests requiring data or acknowledgments to be returned are handled as split transactions. Most slaves are required to be able to also function as a master and initiate transactions to return data and/ or acknowledgments. This methodology allows independent design and verification of the modules. It also aids design reuse: modules can be replaced easily to interface to different system bus, different memory devices, etc. On the downside, the inter-module interfaces require some extra pipeline stages that a monolithic design could avoid.

FIGURE 7 : TMC Resource Scheduler

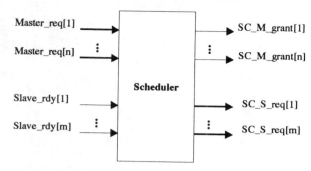

A typical transaction proceeds as shown in Figure 8:

1. The initiator asserts its dedicated request line (R_req) to the scheduler. R_req is a multi-bit signal including information about the resources to be used (i.e. A-bus, D-bus, P-bus), the number of cycles required for the transaction, module ID of the target unit and the type of the transaction. Each unit has its own dedicated R_req line going to the scheduler.

2. Slave modules assert a dedicated ready line (T_rdy) to the scheduler whenever they are ready to accept transactions.

3. Every cycle, the scheduler examines the request inputs from all master modules and from all slave unit ready lines. When the scheduler determines that all resources required for a request are available, it schedules a transaction for that request. Necessary communication resources are reserved for the duration of the transaction. A grant signal (SC_R_grant) is sent to the requestor and the request type and requestor ID is relayed to the target module (with SC_T_req signal).

4. Requestors are responsible for driving data on the bus as soon as they receive the grant signal from the scheduler. Typically, requestors latch the data into their output registers in the cycle they assert their request. The grant signal from the scheduler is used as an enable to drive data from these registers to the busses in the same cycle the transaction is scheduled.

FIGURE 8 : Internal Bus Protocol of TMC

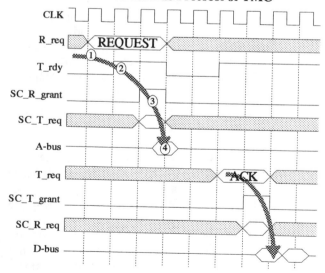

Three major transaction initiators on the TMC, *the Mbus Controller, the Remote Memory Handler* and *the Remote Access Server,* compete for access to the memory controller. The memory controller is designed to be able to handle two simultaneous transactions with a 4 or 8 bank memory subsystem. The split transaction communication protocol used in the TMC reduces the contention on internal busses by reserving the busses only when data transfer is taking place. It also introduces a level of parallelism *and* pipelining. It is possible to send the address for a transaction on the address bus while an independent data transfer is taking place on the data bus. Moreover, since acknowledges are generated by the target module as new transactions, communication does not depend on the number of cycles needed to process a request.

The concept is similar to object oriented programming. Each module has a small set of externally visible data structures and a set of transactions (methods) that operate on these structures. Modules can be tested independently of the whole system by constructing simple test environment that exercise all supported transactions. Retargeting the TMC design is important for this project so that the developed technology can be used in different platforms and/or technologies. For example, the memory controller module has a very simple functional definition: it either reads or writes a 32-byte block of memory (a cache line) of memory. This requires that some of the logic to handle byte insertion, etc. is moved into other modules, but it makes the memory controller extremely modular. A Rambus or E-DRAM based version of the memory controller can be integrated with a future version of the TMC design with minimal interface redesign overhead.

The Modular design methodology also decreases the complexity of the verification. Of particular importance is to formally verify the correctness of the inter-module communication protocol so that the problem of verifying the entire TMC decomposes into verifying each module independently. Simple module interfaces and the small set of transactions supported by each module allow to deal with the communication protocol at a high level of abstraction where formal verification tools like *Murφ* [13] are applicable. With the simple interface approach, integration of modules at different levels of abstraction is also possible. The TMC uses synthesizable Verilog HDL descriptions that are interchangeable with C++ objects that form the S3.mp architectural model.

4.2 TMC Directory Operation

TMC has a 128-bit datapath to the memory. It uses a SDRAM memory subsystem with 144-bit data interface. 9 bits are used to maintain ECC over 128 bit data words and 7 bits are left unused. This gives us 14-bits of extra storage for every 32-bytes of main memory. TMC uses these 14 bits to maintain the directory information for cache-coherency protocols. Two bits are used to maintain the state of the directory and 12 bits are used to maintain a pointer to a remote node or to a set of extension node pointers.

The TMC reserves a programmable fraction of the main memory and uses this storage as a cache for remote references. This internode cache (INC) is required to include all locally cached blocks of remote memory and is critical for an address compression scheme that conserves bandwidth for control messages. Since remote references have at least 3x higher latency than local memory references, even a relatively slow (= operating at main memory speed) INC is beneficial. The INC is programmable in size and may occupy up to 50% of the total memory. It is implemented as a 3-way set associative, write-allocate cache with LRU replacement policy. Although S3.mp in this configuration is still fundamentally a CC-NUMA architecture, it does have many of the properties of a cache only memory architecture (COMA). Hence S3.mp offers a variable degree of COMA behavior that can be used to fine tune the system for specific application.

Since the protocol engines are micro-programmable, it is possible to use a different cache-coherency protocol that supports Simple-COMA, which is a hybrid COMA scheme, where storage allocation is maintained in software. S-COMA microcode is being developed for S3.mp in collaboration with the Swedish Institute of Computer Science [19].

The S3.mp directory uses a multiple linked list scheme to keep track of nodes having copies of a cache line. We will use the following popular naming convention introduced in [2] to explain the basics of the S3.mp directory operation:

1. *Home* is the node which has the original copy of cache line in its main memory

2. *Local node* is the node which gets a remote copy of the cache line

3. *Remote nodes* are any other nodes having a copy of the same cache line

Initially, a cache line is only resident at the memory of the home node and directory is in the *resident* state. When a CPU on another node reads a cache line from a remote node for the first time, the following sequence of events takes place:

1. A CPU initiates a Coherent Read (CR) transaction on the Mbus.

2. The local TMC detects that this transaction is within its remote address range and it performs an address translation to convert the local address to a global address.

3. The Mbus controller checks the local INC to see if this cache line has been retrieved previously.

4. Upon missing in the INC, the Mbus controller sends a *LOAD_LINE* request to the RMH.

5. The RMH constructs a request packet for the cache line and sends it to the output queue. The process responsible for handling this transactions is suspended until a reply arrives.

6. The request packet traverses the output queue and is sent to the home node via the TIC interconnect system.

FIGURE 9 : Reading Data from a Remote Node

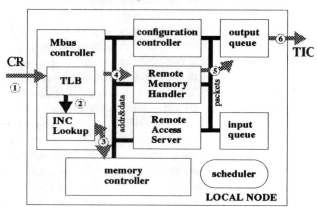

FIGURE 10 : Servicing Remote Read Request

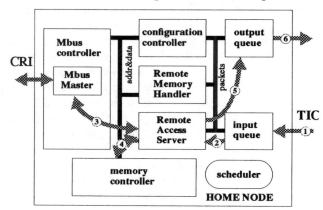

Figure 10 depicts the events of the TIC at the home node upon arrival of the request packet:

1. The request packet is removed from the interconnect by the input queue (like the TIC and the output queue, the input queue supports 4 priority levels that are guaranteed not to block each other in order to avoid deadlocks of the CC-protocol).

2. The input queue decodes the type of the packet to determine the destination unit and sends the packet to the RAS.

3. The RAS requests a snoop cycle on the home Mbus[1]. If one of the home caches have an exclusive copy of this cache line, data is read from this cache. This snoop action is necessary since the current S3.mp protocols do not distinguish between the resident directory state and states where a cache line is shared or owned by one of the CPUs at the home node.

4. If the snoop cycle has failed, the RAS reads the cache line and the associated directory information from the main memory.

5. If the directory is valid (i.e. the directory is not in *Exclusive_Remote* state), an acknowledge packet including the data is constructed and sent to the output queue. The RAS updates the directory information in the main memory if necessary. Assuming that no other node had a copy of the cache line involved in this transaction, directory state will be changed from *Resident* to *Shared_Remote_1* and the directory pointer will be updated to point to the local node.

6. The acknowledge packet including the data is sent back to the local node through the interconnect.

After the acknowledge packet is received by the remote node (Figure 11):

1. The acknowledge packet is removed from the interconnect by the input queue and sent to the RMH. This packet wakes up the sleeping RMH process which had initiated the request for this transaction.

2. The RMH reads the INC tags corresponding to the address of the cache line from the memory and allocates an INC entry. If there are no free INC entries available, RMH victimizes an existing INC entry in LRU fashion.

3. Data from the packet is written to the INC and INC tags are updated.

4. The Mbus controller is call to complete the transaction. This is done by enabling arbitration for the original requestor which will retry the transaction which will be served from a preload buffer.

FIGURE 11 : Completion of a Remote Read

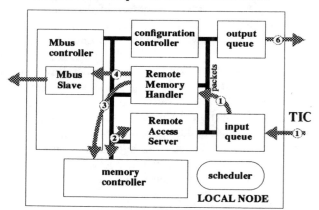

The basic flow of transactions outlined above illustrates the operation of the TMC and the interactions of its sub-component. The complete microcode for the RAS/RMH requires 400/432 instructions, most of which are needed to deal with corner cases. The details of this protocol exceeds the scope of this paper. This protocol was successfully verified in a collaboration with the University of Southern California [20].

---

1. The directory may store the shared local and the exclusive local states of the memory block, thus eliminating the need for snoop cycles on the home node. However, since the directory is kept in the main memory, this would convert many ordinary local memory read operations on the home node into Read/Modify/Write cycles. Being able to service read cycles going to local memory as fast as possible at the expense of having to do snoop cycles on the home Mbus for remote operations seemed to be a good trade-off for most applications. In future implementations of s3.MP systems, TMC will most likely be integrated into the CPU. In this case, TMC will be able to snoop the CPU cache much faster.

In the S3.mp architecture, the directory and the INC are closely related. Whenever a node accesses remote data, the INC will maintain a copy of it for the entire duration of the transaction. An active directory entry implies that there is at least one remote INC that has the corresponding data. Once an INC entry is reused, the directory entry corresponding to this block is updated. This relation allows the 64 bit global addresses to be abbreviated by only the node id and the INC index (Figure 12). As a result of this abbreviation scheme, S3.mp cache coherence protocols use very short messages (80-bits) for majority of the transactions.

FIGURE 12 : Address Abbreviation

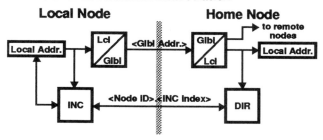

4.3 Protocol Engines - RAS and RMH

S3.mp directory protocols are implemented by two protocol engines on the TMC chip, the *Remote Memory Handler* and the *Remote Access Server*. The RMH is responsible for the transactions generated locally that refer to memory locations on a remote node. It also services invalidation and data forwarding requests originated by the home node and it maintains the INC. The RAS is responsible for all transactions that refer to local memory and that involve a remote cache. It services data request messages from remote nodes, maintains the directory and generates invalidation and data forwarding messages.

The structure of the RMH and RAS is shown in Figure 13:

FIGURE 13 : Protocol Engine Datapath

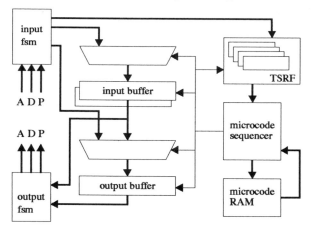

Directory protocols are separated into two parts, the server (RAS microcode) and the client (RMH microcode). Microcode is stored in on-chip RAM and appears as part of TMC.

A typical operation of the protocol engines is shown in Figure 14. Typically, there are two kinds of activity going on at the protocol engines:

1. Transactions that require an acknowledge. These kind of transactions typically have a short burst of local activity (i.e. Mbus cycles, memory cycles) which is terminated by sending a request packet. From this point on, the microcode waits for a reply from a remote node. This wait period is in the order of microseconds in the current implementation of the S3.mp system. When the reply is received, the transaction is terminated following another short burst of local activity.

2. Transactions that do not require an acknowledge. These are typically initiated by receiving a request packet and can be served using only the local resources of the receiving node. They are terminated by sending an acknowledge packet to the requestor.

FIGURE 14 : S3.mp Protocol Engine Activity

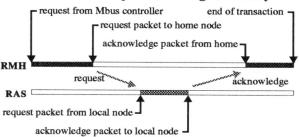

Due to the long latency associated with waiting for a reply, it would have been inefficient to run the protocol engines in a mode where the microcode sequencer was kept busy for the entire duration of the transaction while polling on the reply packet. To introduce a degree of parallelism in order to utilize the high bandwidth available from the interconnect network, the protocol engines were implemented as multithreaded state machines. RMH and RAS have Transaction Status Register Files (TSRF) with multiple contexts to keep track of multiple concurrent transactions. Each context includes all state that is needed to perform a memory operation (addresses, timers, state, pointers, etc.). Each TSRF entry consists of 121 bits, which exceeds the information that could be stored in either the directory or the INC. Hence the directory and INC can maintain only the stable states while active TSRF entries are required for all transient states that have outstanding messages.

The lowest 7 bits of the cache block address (bits [11:5] of the local address for 32-byte cache lines) are used as a tag for transactions. These 7 bits are preserved during address translation, thus, they effectively divide the global address space into 128 sets. Any protocol engine can be working on N of these sets concurrently where N is the number of register windows in the TSRF of that protocol engine. The current implementation of the TMC chip uses 4 TSRF windows. More register windows will be beneficial in hiding the latency of remote accesses when used in conjunction with processors that have non-blocking caches and compilers which utilize aggressive prefetching schemes.

Transactions received by the protocol engines can be of two types: request and acknowledge. Request transactions require a new microcode thread to be generated. For transactions received from the Mbus controller, the transaction type relayed by the scheduler is used as an entry point into the microcode. For packets received from the input queue, type information from the packet is used as the entry point. The next available TSRF window is assigned to this transaction and a new thread is generated and marked as ready to execute. When the microcode sequencer detects that this thread is ready to run, it starts executing it. The TSRF entry corresponding to this thread becomes part of the state space of the microcode sequencer for the duration of this transaction. Until this thread is completed, all other subsequent transactions with the same ID will be suspended.

When an acknowledge packet is received by the RAS or the RMH, the ID field of this packet is compared to the ID fields of threads that are currently sleeping. If a sleeping thread waiting for the particular acknowledge packet is found. This thread is woken up, i.e., marked as ready to run, otherwise a new thread is forked to deal with an unsolicited packet. The microcode sequencer resumes processing a thread whenever it is available.

A set of timers, one for each TSRF thread, are used to deal with the case where an acknowledge is expected but never received from a remote node. Whenever a thread is suspended, its timer is programmed to generate an error acknowledge if an reply is not received within a certain amount of time.

In addition to providing a way of matching sleeping threads and reply messages, the ID field provides a convenient way to lock the directory and/or the INC while these data structures are in a transient state. Each engine can potentially lock N out of 128 slices of the directory/INC independently, where N is the number of entries in the TSRF. ID-locking is utilized exclusively by S3.mp directory protocols to implement message-delay independent operation.

4.4 Address translation

The TMC uses a TLB based address translation mechanism. The size of the segments being mapped from the global dress space to the local address space can be as small as 4Kbytes or as large as 1Mbyte. TLB misses are handled by the operating system. The S3.mp architecture defines a 64-bit address space. In the current implementa-

tion, modules have a 36-bit local address space, therefore, each node can only see a limited portion of the global address space at any given time.

FIGURE 15 : TMC Address Translation

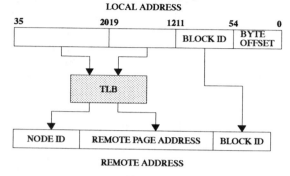

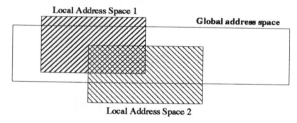

5 Performance

The first lot of TIC chips was received in November of 1994 and are used to interconnect a small network of workstation. The FPGA-based TIC demo boards only allow message passing and bulk memory-to-memory copy operations. In this capacity, the system is used to verify the correct operation of the TIC chip and to gain operational experience with the high-speed serial link technology. The performance of this system is largely limited by the Mbus-bandwidth, and the demonstrated transmission bandwidth was only slightly higher than 80 Mbytes/sec. Likewise, the rate at which messages were exchanged was limited by the software overhead to a sustained rate of 250K packets/sec with bursts to >1M packets/sec.

The TIC demo boards run at 51.840 Mhz (instead of 66 Mhz), which is due to some limitations in the analog section of the serial link core. The rest of the TIC chip has been tested at speed. The measured TIC traversal latency is 310ns, which includes 2m of cables and all serialization/ deserialization and synchronization overhead. The distributed clock synchronization provided by the TIC kept all clocks stable to within 100ps.

The TIC demo system (currently 5 nodes as of March, '95) is too small to stress-test the TIC. It has essentially infinite bandwidth compared to interface bandwidth, which is limited by the Mbus. However, the measured performance is in line with the simulation results. It does

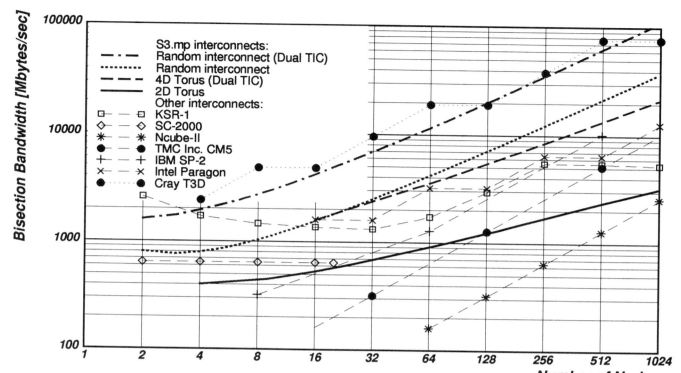

FIGURE 16 : Estimated Interconnect Performance.

however demonstrate the main point of the S3.mp effort: the software overhead of message passing dominates all hardware factors. Also, the internode bandwidth of a TMC based system is much higher because bulk data transfers do not need to traverse the Mbus.

Performance estimates for larger systems are based on simulations and analytical models. The TIC bisection bandwidth and latency for 2D meshes, 4D meshes and the average of random interconnect topologies with 1 and 2 TIC chips are given in Figure 14. The shown bandwidth includes only the data portion of single packets that are addressed to random destinations and does not include bandwidth used for flow control and CRC. To present a conservative estimate, the bandwidth is derated to 80%, to allow for congestion effects.

The reference points for other interconnect systems in Figure 16 use 100% of the raw physical bandwidth with no deductions for any communication overhead, congestion or router inefficiencies. They represent the upper performance limits which cannot be reached in real applications.

The raw bandwidth that is provided by the TIC interconnect system is not directly available to an application process, rather there is an average of 4 control messages for each data message. Since the control messages are 4 times smaller than the data message, this leaves still more than halve of the raw bandwidth.

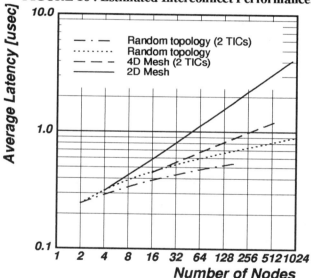

TABLE 1 : Interconnect System Performance.

Number of nodes	Bandwidth / node	Remote access latency
2	<406* (<812*) MB/s	0.84 (0.84) μs
8	112 (<313*) MB/s	1.8 (1.4) μs
32	63 (194) MB/s	2.8 (1.9) μs
128	43 (144) MB/s	3.8 (2.4) μs
512	32 (117) MB/s	5.0 (2.7) μs
1024	26 (96) MB/s	6.1 (3.1) μs

*: Limited by TMC performance

Table 1 summarizes the TIC performance for randomly generated topologies and random access to remote memory. In this case, the entire system of TMC and TIC is considered. However, the TMC capacity limits the realizable throughput for small numbers of nodes. In this case, the interconnect capacity is practically infinite. As system size grows, the usable bandwidth per node drops below the TMC capacity. The bandwidth and latency for single and dual TIC configurations below is the effective performance as seen by an idealized application. This includes the traffic that is used to maintain cache coherency, which is not included in the shown bandwidth

S3.mp nodes have bulk memory copy facilities. These operate cache coherently and can achieve peak transmission rates of up to 66% of the raw bandwidth in one direction.

6 Project Status

S3.mp is a experimental prototype project and not a currently planned Sun product.

The TIC chip has been fabricated and is used in a message interface board that provides message passing capabilities similar to those of the IBM SP-2 system or Myrinet.

The TMC design is in the final stages of verification and is expected to be submitted for fabrication in June. Functional TMC chips should become available in mid-August, leading to first operational S3.mp nodes in the late Summer of '95.

7 Summary

The main innovation of the S3.mp project is the support for spatially distributed shared memory multiprocessing. Unlike other cache-coherent DSM multiprocessors, S3.mp system can be distributed over a local area while still maintaining competitive performance. S3.mp remote miss latencies compare favorably [21] to the reported internode latency for the recently introduced Convex SPP-1000 Parallel Computer and S3.mp bisection bandwidth is close to the Cray T3D realm. This performance level is achieved at a relatively low system cost: all S3.mp components can be integrated in one modestly sized ASIC with standard CMOS technology. Other innovations of the S3.mp project include the use of DSM to reduce communication overhead. S3.mp allows spacial distribution of computational resources while preserving the convenience and efficiency of using memory semantics to access them. Additionally this allows workstations and I/O resources to be viewed as building blocks that make up a scalable system rather than a distributed collection of independent entities. Finally S3.mp hides the artifacts of adaptive routing, namely out of order delivery, by integrating it with a order insensitive coherency mechanism that is based on memory models defined for the Sparc architecture.

7.1 References

[1] Bell, G. *Ultracomputers, A Teraflop Before Its Time.* Communications of the ACM 35,8 (August 1992), 27-47.

[2] Lenoski, D. *The Design and Analysis of DASH: A Scalable Directory-Based Multiprocessor.* PhD Dissertation, Stanford University, December 1991.

[3] Bisiani, R., and Ravishankar, M. K. *Design and Implementation of The Plus Prototype.* Technical Report, Carnegie Mellon University, August 1990.

[4] Thapar, M., Delagi, B., and Flynn, M., *Linked List Cache Coherence for Scalable Shared Memory Multiprocessors,* Proceedings of the 1993 International Conference on Parallel Processing, pages 34-43.

[5] Agarwal, A.,Kubiatowicz J., Kranz, D., Lim, B., Yeung, D., D'Souza, G., Parkin, M. *Sparcle: An Evolutionary Processor Design for Large-Scale Multiprocessors.* IEEE Micro, June 1993, pages 48-61.

[6] Nowatzyk, A. *Communications Architecture for Multiprocessor Networks.* PhD Dissertation, Carnegie Mellon University, December 1989.

[7] SPARC International, *The SPARC Architecture Manual: Version 9,* Prentice-Hall, 1994

[8] Kendall Square Research, *Technical Summary.* 1992. 170 Tracer Lane, Waltham, MA 02154-1379

[9] Hagersten, E., Landin, A., and Haridi, S. *DDM - A Cache-Only Memory Architecture.* IEEE Computer 25,9 (September 1992), 44-54.

[10] Li, K. *Shared Virtual Memory on Loosely Coupled Multiprocessors.* PhD Dissertation, Yale University, September 1986.

[11] Bershad, B. and Zekauskas, M. *Midway: Shared Memory Parallel Programming with Entry Consistency for Distributed Memory Multiprocessors.* Technical Report CMU-CS-91-170, Carnegie Mellon University, September 1991.

[12] Bal, H., Jeffifer, S., Tanembaum, A. *Programming Languages for Distributed Computing Systems.* ACM Computing Surveys 21,3 (September 1989), 261-322.

[13] Herlihy, M. *Wait-Free Synchronization.* ACM Transactions on Programming Languages and Systems 11,1 (January 1991), 124-149.

[14] Dill, D., Drexler, D., Hu, A., Yang, C. *Protocol Verification as a Hardware Design Aid.* 1992 IEEE International Conference on Computer Design: VLSI in Computers and Processors, (October 1992), 522-52

[15] McMillan, K., *Symbolic Model Checking,* 1992, Carnegie Mellon University PhD Thesis, Pittsburgh, PA 15213

[16] Eichelberger, C.W., Davidson, H., *Viking Supersparc MCM Developement Program,* 1993 IEEE Multichip Module Conference, Santa Cruz, p. 28-32

[17] Nowatzyk, A., Parkin, M., *The S3.mp Interconnect System and TIC Chip,* Hot Interconnects'93, Stanford CA, Aug. 1993

[18] Nowatzyk, A., Browne, M., Kelly, E., Parkin, M., *S-Connect: from Network of Workstations to Supercomputer Performance,* Proceedings of the 22nd Symposium on Compter Architecture, June 1994

[19] Saulbury, A., Nowatzyk, A. *A Simple COMA Implementaion on the S3.mp Multprocessor,* 5th Workshop on scalable shared memory multiprocessors, June 1995

[20] Pong, F., Dubois, M., Nowatzyk, A., Aybay, G., *Verifying Distributed Directory-based Cache Coherency Protocols: S3.mp, a Case Study,* In proceedings of EuroPar'95, Stockholm/Sweden, August 1995

[21] Sterling, T., Savarese, D., Merkey, P., *Evaluation of the SCI based Convex SPP-1000 Parallel Computer,* SCI Workshop, Santa Clara, CA, 1995

BEOWULF:
A PARALLEL WORKSTATION FOR SCIENTIFIC COMPUTATION

Thomas Sterling   Donald J. Becker
Center of Excellence in Space Data
and Information Sciences
Code 930.5 NASA Goddard Space Flight Center
Greenbelt, MD 20771
{tron, becker}@cesdis.gsfc.nasa.gov

Daniel Savarese
Department of Computer Science
University of Maryland
College Park, MD 20742
dfs@cs.umd.edu

John E. Dorband
NASA Goddard Space Flight Center

Udaya A. Ranawake   Charles V. Packer
Hughes STX Corp.

Abstract – *Network-of-Workstations technology is applied to the challenge of implementing very high performance workstations for Earth and space science applications. The Beowulf parallel workstation employs 16 PC-based processing modules integrated with multiple Ethernet networks. Large disk capacity and high disk to memory bandwidth is achieved through the use of a hard disk and controller for each processing module supporting up to 16 way concurrent accesses. The paper presents results from a series of experiments that measure the scaling characteristics of Beowulf in terms of communication bandwidth, file transfer rates, and processing performance. The evaluation includes a computational fluid dynamics code and an N-body gravitational simulation program. It is shown that the Beowulf architecture provides a new operating point in performance to cost for high performance workstations, especially for file transfers under favorable conditions.*

1   INTRODUCTION

Networks Of Workstations, or NOW [4] technology, is emerging as a powerful resource capable of replacing conventional supercomputers for certain classes of applications requiring high performance computers, and at substantially lower cost. Another, less frequently considered, domain is the realization of the high performance workstations themselves from ensembles of less powerful microprocessors. While workstations incorporating between 2 and 4 high performance microprocessors are commercially available, the use of larger numbers (up to 16 processors) of lower cost commodity subsystems within a single workstation remains largely unexplored. The potential benefits in performance to cost are derived through the exploitation of commodity components while the performance gains are achieved through the concurrent application of multiple processors. The MIT Alewife project [1] seeks to provide a fully cache coherent multiprocessor workstation through modifications of the SPARC processor. The Princeton SHRIMP project [2] employs standard low cost Intel Pentium microprocessors in a distributed shared memory context through the addition of a custom communication chip. While both projects make heavy use of available VLSI components, they require some special purpose elements, extending development time and incurring increased cost. An alternative approach, adopted by the Beowulf parallel workstation project, recognizes the particular requirements of workstation oriented computation workloads and avoids the use of any custom components, choosing instead to leverage the performance to cost benefits not only of mass market chips but of manufactured subsystems as well. The resulting system structure yields a new operating point in performance to cost of multiple-processor workstations.

2   BEOWULF ARCHITECTURE

The Beowulf parallel workstation project is driven by a set of requirements for high performance scientific workstations in the Earth and space sciences community and the opportunity of low cost computing made available through the PC related mass market of commodity subsystems. This opportunity is also facilitated by the availability of the Linux operating system [7], a robust Unix-like system environment with source code that is targeted for the x86 family of microprocessors including the Intel Pentium. Rather than a single fixed system of devices, Beowulf represents a family of systems that tracks the evolution of commodity hardware as well as new ports of Linux to additional microprocessor architectures.

The Beowulf parallel workstation is a single user multiple computer with direct access keyboard and monitors. Beowulf comprises:

- 16 motherboards with Intel x86 processors or equivalent

- 256 Mbytes of DRAM, 16 MByte per processor board

- 16 hard disk drives and controllers, one per processor board

- 2 Ethernets (10baseT or 10base2) and controllers, 2 per processor

- 2 high resolution monitors with video controllers and 1 keyboard

The Beowulf prototype employs 100 MHz Intel DX4 microprocessors and a 500 MByte disk drive per processor. The resulting 8 GBytes of secondary storage available locally to Beowulf applications can substantially reduce LAN traffic to remote file servers in certain important cases such as dataset browsing. The DX4 delivers greater computational power than other members of the 486 family not only from its higher clock speed, but also from its 16 KByte primary cache (twice the size of other 486 primary caches) [6]. Each motherboard also contains a 256 KByte secondary cache. Two Ethernets running at peak bandwidths of 10 Mbits per second are used for internode communications, one a twisted pair 10baseT with hub and the other a multidrop 10Base2. Future Beowulf systems will employ more advanced versions of these component types but the basic configuration will remain the same. The Beowulf architecture has no custom components and is a fully COTS (Commodity Off The Shelf) configured system.

3 SCALING CHARACTERISTICS

3.1 Internode Communications

Communication between processors on Beowulf is achieved through standard Unix network protocols over Ethernet networks internal to Beowulf. Therefore the communication throughput of Beowulf is limited by the peformance characterisitics of the Ethernet and the system software managing message passing. However, Beowulf is capable of increasing communication bandwidth by routing packets over multiple Ethernets. This is made possible by a special device driver written by one of the authors which was facilitated by the free access to Linux kernel source code.

To evaluate the performance improvement derived from multiple networks, we measured the network throughput under a range of traffic demands using one, two, and three Ethernet networks. We assigned send/receive processes to pairs of processors which would exchange a fixed-sized token a particular number of times over the network. In this experiment, the load on the net was increased by exchanging larger tokens and also by increasing the number of tokens being exchanged. No processor was involved in the exchange of more than one token, i.e. each processor involved in the exchange of a token was assigned only one send/receive process. We used the BSD sockets interface and the User Datagram Protocol (UDP) to perform the token exchanges.

Figure 1 shows network throughput, measured in megabytes per second, as a function of the number of Ethernet channels available, token size, and number of tokens exchanged. At the time of the experiment, one of Beowulf's 16 processors was unavailable, allowing us to involve a maximum of only 7 pairs of processors in token exchanges. When performing the experiment using three channels, sufficient Ethernet cards were available to configure only 8 processors to use three channels. Hence a maximum of 4 tokens could be exchanged for that phase of the experiment.

It is evident that the small 64 byte tokens do not come anywhere near saturating the network for any number of channels. The 1024 byte tokens are able to saturate the 1 channel network with a throughput of about 1 MB/s, or 80% of the peak 1.25 MB/s possible on 10 Mbit/s Ethernet. Throughput for the 8192 byte tokens at 4 and 7 token exchanges is less than that for 1024 byte tokens

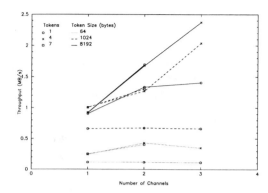

Figure 1: Beowulf Network Throughput

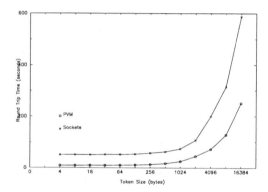

Figure 2: Beowulf Network Round Trip (1 channel, 16 processors)

because of additional network packet collisions. The minimum and maximum sizes of an Ethernet packet are 64 and 1536 bytes respectively [3]. Thus a 64 byte token and a 1024 byte token each require only one Ethernet packet for transmission. However, an 8192 byte token must be broken up into 6 Ethernet packets, increasing the likelihood of collisions on a 1 channel network. Figure 1 shows that multiple networks alleviate network contention, achieving throughputs of up to 1.7 MB/s (68% of peak) and 2.4 MB/s (64% of peak) respectively for 2 and 3 channel configurations.

System level applications like NFS are written making direct use of sockets, but most user level parallel programs use some higher level interface – usually PVM [10]. Figure 2 shows the overhead incurred by such high level message-passing interfaces. It shows the round trip time on one network channel across 16 processors for tokens of sizes ranging from 4 to 16384 bytes using PVM 3.3 versus BSD sockets and UDP. We define the round trip time as the time for a token to be sent from an initial processor, be received by a neighbor and be passed on to its neighbor, etc., visiting 15 intervening processors only once before finally returning to the initial processor. The PVM overhead is rather marked; the round trip time of a 256 byte token is 10 ms using sockets while the time us-

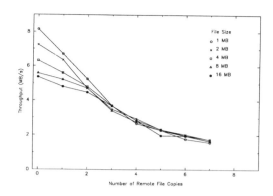

Figure 3: Beowulf File Transfers (2 channels, Total of 7 local and remote files)

ing PVM is 60 ms. The PVM experiment was run using the `PvmDataDefault` option for `pvm_initsend()` and the timing included the overhead of packing and unpacking a message.

3.2 Parallel Disk I/O

To ascertain the I/O performance of Beowulf, we measured the throughput of simultaneous file transfers across a mix of intraprocessor copies and interprocessor copies for a range of file sizes. No processor was involved in more than one file transfer, i.e. each processor involved in a file transfer was either performing a local disk file copy or was participating in a remote file transfer. Therefore there was no disk or processor contention caused by competing file transfers. File copies were performed using the Unix `read()` and `write()` system calls. In the case of remote transfers, BSD sockets and UDP were used to transmit a file between processors. At the time we performed this experiment, only 15 of Beowulf's processors were available to us. That meant we could only perform 7 simultaneous transfers instead of 8, because remote file transfers require the participation of 2 processors. Beowulf was configured to use 2 network channels for the experiment.

The empirical results of this key experiment are presented in Figure 3. The total number of file transfers is held constant while the ratio of local file transfers (those by a single processor to its local disk) versus remote file transfers (those between two disks and two processors over the network) is adjusted from 0.0 (all local) where there are 0 remote file copies to 1.0 (all remote) where there are 7 remote file copies. The data shows two dominant characteristics related to file transfer. Not surprisingly, file transfer throughput exceeds 8 MBytes per second (sustained) when all copies are local. It is seen that this value decreases for local copy only as the file size increases due to issues related to local buffering policies. As the number of remote file copies is increased to 1, overhead for the additional burden of moving over the network degrades overall throughput by about 15% in the worst case. Increasing number of remote copies beyond 1 adds an additional source of performance degradation, network contention. Two networks ameliorates

the degradation at two remote copies but between two and three remote copies, the rate of throughput degradation is seen to accelerate. Ultimately where all file copies are remote, throughput is entirely constrained by the network bandwidth and is approximately equal to the maximum network throughput measured for two networks as discussed in the previous subsection.

4 BENCHMARK EXPERIMENTAL RESULTS

Performance scalability of a multiple processor workstation is best characterized through the use of complete real-world applications rather than synthetic test programs. To this end, two full applications from the Earth and space sciences community were selected to bracket the dimension of communication and load balancing demands.

A 2-dimensional compressible fluid dynamics code, called Prometheus [5], has been implemented on a number of high performance computers including vector, shared memory, distributed memory, and SIMD architectures. The code solves Euler's equations for gas dynamics on a logically rectangular mesh using the Piecewise Parabolic Method (PPM). The message passing version of this code previously used on the IBM SP-1 and Intel Paragon was easily ported the the Beowulf parallel workstation and its PVM message passing environment. Parallelization was accomplished using a domain decomposition technique for which the computational grid was divided into 128x128 tiles. Communication between neighboring tiles is necessary only twice per time step. Because of the large number of floating-point operations required to update each grid point communication costs are relatively small.

A tree code for performing gravitational N-body simulations has been developed to reduce a classically $O(n^2)$ computation to $O(n \log n)$ and has been applied to shared memory [9], distributed memory, and SIMD parallel architectures [8]. The code is being used to study the structure of gravitating, star forming, interstellar clouds as well as to model the fragmentation of comet Shoemaker-Levy 9 in its close encounter with Jupiter. A range of number of particles were used from 32K to 256K. It is estimated that Beowulf can support a 1 million particle simulation for in-core computation and much larger if appropriate disk accessing can be coordinated.

Scaling characteristics of these two codes were evaluated on the Beowulf parallel workstation. The results are shown in Figure 4. The CFD code was executed on up to 16 processing modules and the tree code was performed on up to 8 processing modules. The CFD application showed good scaling characteristics with total degradation at 16 processors approximately 16% with respect to ideal. Single processor performance for this code is 4.5 MFLOPS. The full Beowulf delivered a sustained performance of 60 MFLOPS. This compared favorably with the Paragon of equivalent size as well as the TMC CM-5 (without vector chips). The CRI T3D performed less than 2.5 times better than Beowulf for the same number of processors. An additional experiment was performed to demonstrate the impact of multiple networks versus a single network in Beowulf. The dual network showed

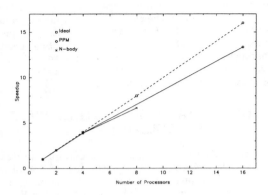

Figure 4: ESS Code Scaling

essentially no performance advantage over the single network case, indicating that communication bandwidth and contention is not an issue for this application.

The scaling characteristics for the tree code is somewhat poorer as should be expected. Much more communication per computation is involved with communication requirements being relatively global. Because the data structure is time varying, one part of the program does load balancing periodically (once every ten time steps) through a sorting strategy. Unfortunately, this problem has not been run on more than 8 processors. Performance degradation at 8 processors was observed to be 19% with respect to the ideal. Single processor performance was 1.9 MFLOPS with a total of 12.4 MFLOPS using 8 processing modules. The overall impact of multiple networks versus a single network was found to be only a few percent. But when the dynamic load balancing portion of the algorithm was tested, it was found to be network bandwidth constrained. Two networks provided a 50% improvement in performance for the sort algorithm versus a single network run.

5 DISCUSSION AND CONCLUSIONS

The Beowulf research project has been initiated to explore the opportunity of exploiting Network-of-Workstation concepts to provide high performance single user workstation performance at exceptional cost. The operating point targeted by the Beowulf approach is intended for scientific applications and users requiring repeated use of large data sets and large applications with easily delineated coarse grained parallelism.

Interprocessor communication proved to be the most interesting aspect of the Beowulf operation. Enhancements to the Linux kernel enabling multiple communications channels to be employed simultaneously showed excellent scaling factors. This new functionality alone will impact how Linux based PC's will be used in the future. But it was equally clear that the network, even in its dual configuration, is inadequate under certain loads. There was at least one instance where full application behavior was perturbed by network capacity. More importantly, parallel file transfers were seen to be limited by the network. It is clear from these results that higher

bandwidth networks are required. Fortunately, 100 Mbps Ethernet-like networks are now coming on the commodity market. The Beowulf project is beginning evaluation of this new technology and it is anticipated that dual 100baseTX or 100VG type networks will be incorporated in the new Beowulf demonstration unit being assembled in 1995.

Future work will focus primarily on advanced software technology that will make better use of the parallel computing resources. These include load balancing, parallel file distribution, global synchronization, parallel debugging and optimization, and distributed shared memory environments. But the lesson of this initial work is that a relatively simple capability as that offered by the Beowulf prototype can be of immediate value to real users in the arena of scientific computation

REFERENCES

[1] A. Agarwal, D. Chaiken, K. Johnson, et al. "The MIT Alewife Machine: A Large-Scale Distributed-Memory Multiprocessor," *M. Dubois and S.S. Thakkar, editors, Scalable Shared Memory Multiprocessors,* Kluwer Academic Publishers, 1992, pp. 239-261.

[2] M. Blumrich, K. Li, R. Alpert, C. Dubnicki, E. Felten, and J. Sandberg, "Virtual Memory Mapped Network Interface for the SHRIMP Multicomputer," *Proceedings of the Twenty-First International Symposium on Computer Architecture (ISCA),* Chicago, April 1994, pp. 142-153.

[3] D. Boggs, J. Mogul, and C. Kent, "Measured Capacity of an Ethernet: Myths and Reality," *WRL Research Report 88/4,* Western Research Laboratory, September 1988.

[4] K. Castagnera, D. Cheng, R. Fatoohi, et al. "Clustered Workstations and their Potential Role as High Speed Compute Processors," *NAS Computational Services Technical Report RNS-94-003,* NAS Systems Division, NASA Ames Research Center, April 1994.

[5] B. Fryxell and R. Taam, "Numerical Simulations of Non-Axisymmetric Accretion Flow," *Astrophysical Journal, 335,* 1988, pp. 862-880.

[6] Intel Corporation, "DX4 Processor Data Book," 1993.

[7] Linux Documentation Project, Accessible on the Internet at World Wide Web URL *http://sunsite.unc.edu/mdw/linux.html.*

[8] K. Olson and J. Dorband, "An Implementation of a Tree Code on a SIMD Parallel Computer," *Astrophysical Journal Supplement Series,* September 1994.

[9] T. Sterling, D. Savarese, P. Merkey, J. Gardner, "An Initial Evaluation of the Convex SPP-1000 for Earth and Space Science Applications," *Proceedings of the First International Symposium on High Performance Computing Architecture,* January 1995.

[10] V. Sunderam, "PVM: A Framework for Parallel Distributed Computing," *Concurrency: Practice and Experience,* December 1990, pp. 315-339.

PACE: FINE-GRAINED PARALLEL GRAPH REDUCTION

T.J.Reynolds, M.E.Waite and F.Z.Ieromnimon

Department of Computer Science, University of Essex,

Colchester CO4 3SQ, UK

email [reynt, waitm, ierof]@essex.ac.uk

Abstract -- *The PACE architecture is an extensible, distributed memory, multiprocessor designed specifically for graph reduction. It uses a specially designed processor, rather than currently available devices, as the basic replicable node. In this paper we give an overview of PACE together with results obtained by simulation of the hardware.*

INTRODUCTION

Graph reduction has long been advocated as an ideal model for parallel computation. However machine architectures that can exploit this potential have proved much harder to realise than first anticipated. We believe that this is due to a fundamental mismatch between the requirements of parallel graph reduction and the capabilities of currently available microprocessors. Over the past decade, advances in compiler technology have improved the performance of sequential implementations considerably. Most current attempts at parallel implementation seek to exploit these advances by employing similar techniques on MIMD architectures that incorporate fast serial processors as their basic processing nodes. However the requirements for sequential and parallel implementations are very different. The performance of a sequential graph reduction implementation depends solely upon the speed with which the required graph transformations can be executed. In a parallel implementation, however, performance is largely determined by how efficiently management tasks such as work allocation, interprocessor communication and distributed garbage collection can be implemented. These tasks have to be performed concurrently with the main graph rewriting activity, and, in a truly extensible design, by the same computing resource, namely the basic replicable node. This implies that each node has to be an efficient parallel processor in its own right if the overall system is to be successful. Conventional serial processors can be made to exhibit the required behaviour, but in order to maintain a viable ratio between useful computation and overhead, the grain size of tasks has to be quite coarse. This requirement means that opportunities for parallel execution are lost, and with fewer available tasks, scalability becomes restricted. Attempts to overcome the problem by means of program

annotations are far from satisfactory. It is not difficult to think of highly irregular or data-dependent examples where this approach does not work, and even where it does, the result is a machine-dependent program. At the very least it represents an unwelcome complication to the programmer who is required to think at two completely different levels of concern.

We therefore advocate a fine-grained approach to parallelism and special-purpose hardware to support it. This has long been the position of workers on various dataflow projects [1]-[4]. A much more detailed version of this paper is available in [5].

THE EVALUATION MODEL

Dataflow has the potential to use fine-grained multi-tasking to hide network communication latency. In PACE we combine this potential with the synchronisation properties enjoyed by graph reduction. Essentially, we treat a combinator graph as a dynamic form of dataflow graph, and then introduce a demand element to preserve the desirable aspects of a lazy evaluation strategy.

In dataflow architectures input data tokens are matched with process templates to form packets which are queued at processors for action. Since each packet contains all the necessary information to complete the task, the processor is able to perform completely independent tasks in rapid succession. In order to reproduce this situation in a combinator graph, we require that the processor is capable of performing a complete combinator rewrite as an atomic action. This is possible if we adopt a small fixed-size set of combinators such as that introduced by Turner in [6]. The rewrite rules associated with these combinators are very straightforward, usually involving simple rearrangements of their arguments. It is also necessary to present complete redexes to the processor so that it can perform the rewrite without the need to access memory. This is made possible by adopting a representation involving application sequences, rather than the simple binary applications used in Turner's original scheme.

Representation

Combinator expressions are represented by a

collection of labelled application sequences. For example, if we take the following program:

*DEF fac n = (n=1) -> 1; n * fac(n-1)*
fac 5 ?

then the definition of the function fac compiles into the following combinator expression:

*S (C' COND (C =1) 1) (S * (B fac (C - 1)))*

This is represented by the following set of sequences:

#L2: < S #L3 #L5>
#L3: < C' COND #L4 1>
#L4: < C = 1>
#L5: < S * #L6 >
#L6: < B fac #L7 >
#L7: < C - 1 >

The expression to be evaluated, fac 5, is represented by: #L1: < #L2 5 >

Classification, Reduction and Dereference

Sequences can be classified as being one of three types:

A *redex* is defined to be a sequence that consists of an operator (combinator or built-in primitive) followed by at least the required number of arguments. In the case of strict operators, the relevant arguments must be fully evaluated.

A *head-normal form* (HNF) can consist of a single data item, a constructed pair, or a combinator followed by fewer arguments than it requires.

An *intermediate sequence* (IS) cannot be classified as either redex or HNF. It can be a sequence representing an application of a strict operator to unevaluated arguments, or a function application where the function part is a reference, or just a simple indirection.

The classification of a sequence determines the operation that can be applied to it. There are just two basic operations, reduction and dereference.

Reduction is applied to redex sequences according to the rewrite rule for the leading combinator, e.g

#Li:<S a b c> =>   #Li:<a c #Lj>   +   #Lj:<b c>

Dereference inserts an evaluated HNF into a waiting IS. The dereference operation not only supplies arguments to strict operators, e.g.

#Li:< + 2 #Lj >   +   #Lj:< 4 >   =>   #Li:< + 2 4 >

but also supplies relevant parts of the compiled function definition code to the expressions undergoing evaluation. For example, the first step in the evaluation of our example expression <fac 5> is to dereference the leftmost label of the sequence, and thereby produce a redex sequence.

#L1:<#L2 5>+ #L2:<S #L3 #L5> => #L1:<S #L3 #L5 5>

The dereference operation also implements the elision of indirections, which in the distributed architecture provides a means for collecting remote values, e.g.

#Li:<#Lj>   +   #Lj:< 4 >   =>   #Li:< 4 >

The association of HNFs with the ISs that require them is straightforward, since an IS only has to be associated with one HNF (the one referred to by its leftmost label) for the operation to proceed.

Controlling the Evaluation

Evaluation proceeds by continually applying the operations of classification, dereference and reduction to the sequence representing the main expression, and to some subset of the sequences that become generated during the process. We employ a demand driven strategy by only including a sequence in the subset of active sequences if its label appears in a strict argument position within some already active sequence.

Parallelism is introduced by adding a control operator, PAR, into the demand driven model. PAR has rewrite rule

PAR f a => f a [make a active]

where <f a> is a function application in the ongoing evaluation and the argument a can safely be evaluated at the same time as <f a>. The evaluation of a is not activated if the processor is approaching its maximum capacity. At present we insert PAR by hand, but in principle the PAR combinator can be introduced as a result of strictness analysis. This has been a topic of much research [7][8]. PAR enables us to generate multi-threaded execution on one processor. Threads are spread between processors by a kind of diffusion scheduling: the decision to export or not is a simple heuristic based on the loading of the local processor and that of its immediate neighbours. It has proved efficient to restrict export to those sequences which represent the application of a defined function to some arguments. This is easily marked by the compiler.

THE PROTOTYPE ARCHITECTURE

Ultimately we wish to render each PACE processor in VLSI. But for now we plan a simple prototype which can be built from standard MSI components, to demonstrate PACE's viability on a wide range of example programs. We choose not to design the prototype around existing microprocessors for the reasons stated in the introduction. Rather we choose MSI circuits because we can precisely render the architecture we require. Furthermore it is easier to see how well such a design could fit onto silicon, and to judge what effects increased clock rates and improved architectural resources would have on performance.

Figure 1 shows a top-level view of the prototype distributed architecture. A simplified PACE processor is shown in figure 2, and figure 3 shows the internal organisation of the dereference and rewrite unit. A major architectural feature is a sequence-wide FIFO. It is the capacity of this FIFO which determines the number of active threads which can be accommodated on one PACE processor. The prototype design features a FIFO with a capacity for 32 entries. The rewrite unit also contains a wide crossbar/multiplexor which allows an output sequence to be produced in one step. During dereference the crossbar is also used to combine fields from the input sequence with fields from a stored sequence. Access to the sequence memory is as wide as the sequences themselves, as a high bandwidth for this path is vital.

For discussion of garbage collection and the

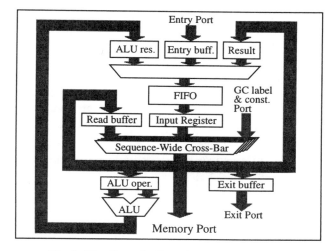

Fig 3: The Rewrite/Dereference Unit.

communications architecture the interested reader should see [5]

RESULTS AND CONCLUSIONS

We have estimated the performance of the PACE architecture using two simulators. The hardware simulation language verilog is being used to model the PACE architecture in considerable detail, in preparation for the construction of a prototype. We know the type of devices which will render each sub-system and have modelled the timing characteristics that we can expect. This modelling is accurate to the extent of determining the required clocking and hence telling us the absolute run-times for the whole machine. We clock our simulation at a conservative 25 MHz.

A C-simulator is used to model the functionality of the PACE architecture without attempting to model the timings of individual devices. The simulation cycle is the processing of a sequence at as many PACE processors that have sequences to process. The verilog simulation tells us that sequences an average of 5.7 clock cycles to process, and it is this average time which is used to convert C-simulator cycles into wall-clock times. This ratio is surprisingly constant over various programs and gives an average performance of 4.39M sequences processed/sec, of which around 1.4M/sec are combinator rewrites.

We show data here for four benchmark programs: nfib and queens are well-known, matm is a matrix multiply and cluster is a clustering algorithm taken from speech processing. Figures 4 and 5 give speedup graphs obtained from the calibrated C-simulator for the second two programs, employing progressively larger arguments to the functions. We can see extrapolation of the aligned results both to larger processor numbers and to larger

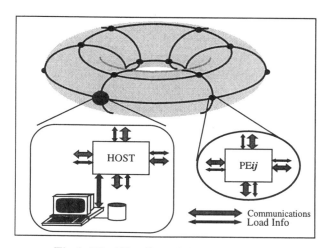

Fig 1: The Distributed Architecture

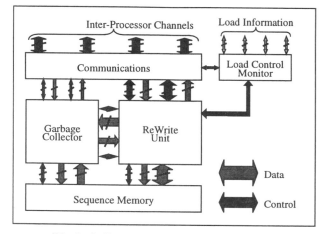

Fig 2: A Single PACE processor

arguments. As we hoped the speedups that can be obtained steadily improve as the size of the problem is scaled. This is encouraging. Figure 6 shows speedup graphs for all four programs. The poorer results for cluster are not intrinsic to the problem but rather arise because it is difficult, in slow simulations, to give some programs a large enough problem to really employ 49 processors.

We have shown that a fine-grained approach to parallelism is practical for SK-combinator graph reduction, provided we build a dedicated machine for the purpose. Our simulations show that our architecture can cope with all the problems inherent in massive parallelism. While running 10queens on 49 processors there are around 1000 threads in the machine for most of the computation. Many of these threads are short-lived, many make remote references and cause sequences to be exchanged between processors, much exporting of sequences occurs, and there is much garbage collection activity. Nevertheless good speedups can be obtained, because the architecture and the graph rewriting model are

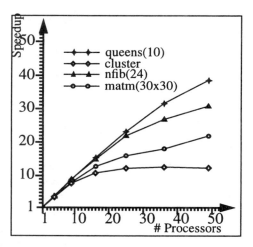

Fig 6: speedup graphs for several programs

in close harmony.

The programs we have run so far are small, but we see no problem in scaling to larger programs. A fine-grain approach thrives independently of the varied dynamic behaviour of larger applications by being maximally ready to exploit all types of parallelism as they arise.

REFERENCES

[1] A. Contessa, E.Cousin, C. Coustet, M. Cubero-Castan, G. Durrieu, B. Lecussan, M. Lemaitre, P. Ng, "MaRS, a combinator graph reduction multiprocessor", *LNCS 306,* (1987), pp.176-192.

[2] M.Parkin and D.Yeung, *The MIT Alewife Machine: A Large-Scale Distributed-Memory Multiprocessor,* MIT/LCS Technical Memo 454, (1991).

[3] R.Keller "Overview of Rediflow II Development", *Proceedings of the Workshop on Graph Reduction, New Mexico, LNCS 279 ,* (1986).

[4] K.R.Traub, G.M.Papadopoulos, M.J.Beckerle, J.E.-Hicks, "Overview of the Monsoon Project", *Int. Conf. on Computer Design,* IEEE (1991), pp.150-155.

[5] T.J.Reynolds, M.E.Waite and F.Z.Ieromnimon, "PACE: A prototype design", http:\\cswww.essex.-ac.uk, (1995).

[6] D.A.Turner, "A New Implementation Technique for Applicative Languages", *Software - Practice and Experience,* Vol 9, (1979) pp.31-49.

[7] E.G.Nocker, J.E.Smetsers, M.C.Eekelen and M.J.-Plasmeijer,"Concurrent Clean" *PARLE '91 , LNCS 506 ,* (June,1991) pp. 202-219.

[8] S.L.Peyton-Jones, "Parallel Implementations of Functional Languages", *The Computer Journal,* Vol32(2), (1989), pp. 175-186.

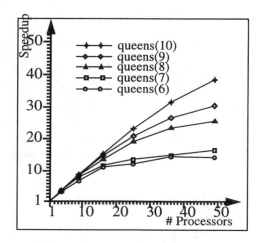

Fig 4: speedup graphs for queens

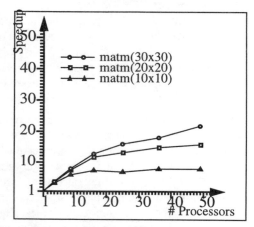

Fig 5: speedup graphs for matm

A Message-Coupled Architecture (MESCAR) for Distributed Shared-Memory Multiprocessors

Shigeki Yamada, Satoshi Tanaka*, and Katsumi Maruyama

NTT Network Service Systems Laboratories   *NTT Software Corporation

3-9-11 Midori-cho, Musashino-shi, Tokyo 180, JAPAN

shigeki@platina.nslab.ntt.jp

Abstract - *The proposed message-coupled architecture (MESCAR) for distributed shared-memory (DSM) multiprocessors reduces message latency and the kernel execution overhead needed for message passing. MESCAR uses duplicate DSMs located in sending and receiving processor modules (PMs). The DSMs are partitioned corresponding to pairs of sending and receiving PMs; messages and control data are automatically copied between corresponding DSMs. Each DSM has a FIFO memory structure that reduces the software overhead caused by DSM access contention and request polling. Programming experiments and latency calculations indicate that compared to a conventional DMA-based system, a MESCAR system reduces kernel overhead by 24% and 256-byte message latency by 50, 70, and 87% for inter-PM message passing for 10-, 50-, and 250-MIPS processors, respectively, with a 1.56-Gb/s bandwidth processor interconnect.*

1. INTRODUCTION

Efficient message passing is the key to implementing a message-passing programming model that allows concurrent execution of processes by exchanging messages. There are two underlying architectures that can be used to implement a message passing model on a multiprocessor system: a local memory architecture (e.g., [1]) and a shared memory architecture. A local memory architecture requires a significant amount of overhead for the direct memory access (DMA) control software used to transfer messages between local memories and to handle frequent DMA interrupts. This results in high message latency and low message passing throughput. The shared memory architecture, especially for massively parallel processors, often uses a distributed shared memory (DSM), which is classified into two types: hardware DSM (e.g., [2]) and software DSM (e.g., [3]). The primary drawback of a hardware DSM is its cost and complexity. A software DSM, on the other hand, has potential performance problems as its primary drawback.

To solve these problems, several DSM copy mechanisms have been proposed (Merlin[4], Memnet[5], SESAME[6], Galactica Net[7], and SHRIMP[8]) that automatically copy the data in the address location of one DSM into the same address location of another DSM. However, these proposed systems use the DSM copy mechanism mainly for the simple transfer of messages or shared variables between DSMs. They do not fully support the whole span of message passing processing, starting from allocating message buffers, writing to the buffers, sending and receiving messages, and ending at releasing the buffers.

This paper presents a message-coupled architecture (MESCAR) for distributed shared-memory multiprocessors. MESCAR efficiently supports the whole span of message passing processing; by balancing the software and hardware components, it reduces message latency and software overhead.

2. DESIGN APPROACH TO MESCAR

Asynchronous message passing, which is the target for MESCAR, can be broken down into the following processing primitives.

[A] Sender side
[1.1] A sending message buffer (MB) is allocated.
[1.2] A message is written into the sending MB by a sender object (lightweight process).
[1.3] The message in the sending MB is transferred to the receiver side.
[1.4] The sending MB is released.

[B] Receiver side
[2.1] A receiving MB is allocated.
[2.2] The message transferred from the sender side is stored in the receiving MB.
[2.3] A receiver object is notified of the message's arrival.
[2.4] The message in the receiving MB is read and processed by the receiver object.
[2.5] The receiving MB is released.
To efficiently implement these primitives on a multiprocessor system, we took the following systematic approach to designing MESCAR.

(1) Allocation of MBs and Control Data in Shared Memory
Allocation of the receiving MB in primitive [2.1] can be eliminated if the sending MB also serves as the receiving MB. We implemented this by using shared memory. The sending processor instructs the receiving processor of the sending MB address so that the receiving processor can read the message in the sending MB. The sending processor must be informed of the sending MB release timing by the receiving processor in primitive [2.5] so that the sending processor can release the sending MB for re-use. Primitives [2.1] and [2.5] thus require exchanging message control data between the sending and receiving processors. The control data area is also allocated in shared memory; this allows the sending and receiving processors to communicate by reading from and writing to the same area in the shared memory.

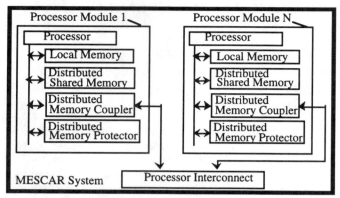

Fig. 1 MESCAR system configuration

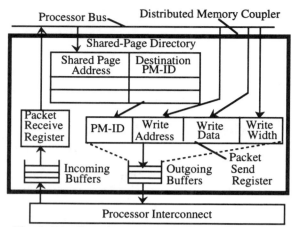

Fig. 2 Distributed memory coupler configuration

(2) Shared Memory Area Partition and Use of FIFO Memories

To avoid access contention for the MBs and control data area among sending processors, the shared memory area is logically partitioned, with each partition corresponding to a pair of sending and receiving processors. A sending processor can thus exclusively write messages and control data into its assigned area; the messages and control data can then be polled by the corresponding receiving processor. First-in first-out (FIFO) memories are used for some of the shared memory area to efficiently detect message arrival notices, which are needed for primitive [2.3]. This FIFO-based polling is described in detail in Section 3.2.

(3) DSM Duplication and Copying

Each logically partitioned area in the shared memory is physically duplicated in a sending processor's DSM (sending DSM) and the corresponding receiving processor's DSM (receiving DSM). The messages and control data in these duplicated DSMs must be coherent. Coherency control is very simple in the message-passing programming model because this model does not require a time-critical write-invalidate or write-update protocol to maintain coherency, as the shared-memory programming model does: This is because its message-passing semantics guarantee that an MB in the shared memory is accessed by the receiver only after the sender has written to the MB. Therefore, coherency control in MESCAR can be done by simple time-lenient copying between the sending and receiving DSMs. DSM copying is done by a distributed memory coupler.

3. MESCAR System Configuration

3.1 Overview of MESCAR System

In the MESCAR system (Fig. 1), each processor module (PM) consists of a processor, a local memory, a DSM, a distributed memory coupler, and a distributed memory protector. Each processor is allowed to access only its own DSM. The distributed memory protector protects the MBs in the DSM from illegal accesses, by using a lightweight capability-based protection mechanism[9]. The processor interconnect transfers messages and control data between the PMs. MESCAR uses a physical addressing scheme with physical addresses matched to virtual addresses. Virtual addressing can be easily implemented by simply adding logical-to-physical and physical-to-logical translation tables.

The distributed memory coupler (Fig. 2) copies the data written in an originating DSM into the destination DSM, which shares the same address space as the originating DSM. The distributed memory coupler has a shared-page directory, which stores the shared page addresses and the page-sharing PM-IDs. Every time the processor writes a message or control data into its own DSM, the address on the processor bus is compared with the shared page addresses in the distributed memory coupler. If the address matches any of the shared page addresses, the corresponding PM-ID is taken from the shared-page directory and added to the write-address, write-data, and write-width information (e.g., one, two, or four bytes) to form a copy packet. The copy packet is transferred through the processor interconnect to the receiving PM specified by the PM-ID. The distributed memory coupler in the receiving PM receives the copy packet and then copies the write-data into the specified address location of the receiving DSM. To allow bidirectional communication between sending and receiving PMs via the control data area in the DSM, the sending distributed memory coupler stores the PM-ID of the receiving PM for the shared control data pages, and the receiving distributed memory coupler stores the PM-ID of the sending PM for the same pages.

We assume write-through cache for simplicity. However, write-back cache can also be applied to MESCAR if a compiler generates machine instructions that push the dirty data in the write-back cache into the DSM at an appropriate time.

3.2 Data Allocation in DSM and DSM Structure

The shared memory space is partitioned into an inter-PM communication (IPC) area and a processing request

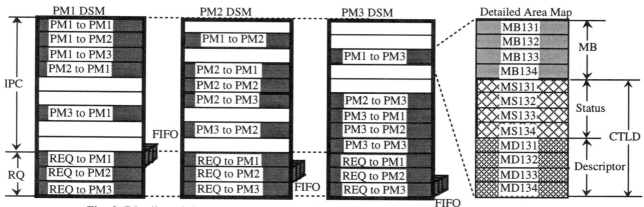

Fig. 3 Distributed shared memory map and memory structure

(RQ) area, as shown in Fig. 3. The IPC area is further partitioned corresponding to pairs of sending and receiving PMs. The partitioned IPC area is physically duplicated in the sending and receiving DSMs. Each partitioned IPC area consists of multiple MB areas and the associated control data (CTLD) areas. Each MB area has a fixed length of 256 bytes and can be accessed directly from the sending or receiving object in the user mode of a processor. This direct access to the MB from the application objects eliminates redundant copying between a user-space MB and a kernel-space MB. For messages larger than the MB size, two MB management schemes are available: the kernel can dynamically concatenate consecutive MB areas into a larger MB area or the system can provide several different MB pools, each with a different MB size.

The CTLD areas in Fig. 3 are used for bidirectional communication between the sending and receiving PMs. Each CTLD area is divided into MB status (MS) and MB descriptor (MD) areas. Each MS area contains a byte indicating whether the corresponding MB is allocated or released. The "allocated" status is written by the sending PM and the "released" status is written by the receiving PM. Each MD area holds a descriptor that stores message control information.

The processing request (RQ) area in Fig. 3 stores processing requests (or message arrival notices) and is used for unidirectional communication from sending PMs to receiving PMs. The RQ area is partitioned corresponding only to the receiving PM. Note that this partitioning is different from the partitioning of the IPC and CTLD areas, which are partitioned based on pairs of sending and receiving PMs. Although the RQ area in the sending DSM consists of conventional RAM chips, the RQ area in the receiving DSM uses FIFO memory chips.

Assume that two sending PMs simultaneously write their requests to the RQ area in their own DSM and both use the same address. While these two requests are simultaneously sent out from the corresponding sending distributed memory coupler, they are aligned in the order of arrival by the receiving distributed memory coupler. Because the RQ area in the receiving DSM consists of FIFO memories, the copying by the receiving distributed memory

coupler to the receiving DSM means that the newly received requests are accumulated in the RQ FIFO area while preserving any already stored requests. Thus, the FIFO memories in the receiving DSM automatically arbitrate any memory access contention among multiple requests from different sending PMs. A receiving PM can detect processing requests from other PMs by simply reading this single RQ FIFO area. This FIFO-based polling requires neither polling a large number of request-holding areas, each associated with a sending PM, nor large interrupt handling overhead.

3.3 Message Passing Implementation

The example of message passing from PM1 to PM3 shown in Fig. 4 uses message buffer MB134 (the fourth MB from PM1 to PM3), the corresponding message status MS134, the descriptor MD132 (the second descriptor from PM1 to PM3), and the request area RQ3 (to PM3). To distinguish the sending DSM areas from the receiving DSM areas, the receiving areas are named by adding a C to the sending area names.

In the "Sender Obj. Invoke" step, the sending kernel sets the sender object ID (SID) in the current capability register (CCR) in the sending distributed memory protector. Then, in the "MB Alloc." step called by the sender object, the sending kernel allocates free message buffer MB134 and updates status MS134 to "allocated". This update is propagated and copied to MS134C in the receiving DSM. The sending kernel then sets the SID in the memory capability register (MCR) in the distributed memory protector. Since both the CCR and MCR are assigned the same SID, only the sender object is allowed to access MB134.

In the "Write to MB" step, the sender object writes a message in 4-byte memory access units to MB134. This in turn causes the sending and receiving distributed memory couplers to copy data in 4-byte units from MB134 to MB134C. In the "SEND" step, the sending kernel clears the MCR and then writes control data to descriptor MD132 and a processing request to RQ3. MD132 and RQ3 are also copied respectively to MD132C and RQ3C in the receiving DSM. Then, in the "Sender Obj. Term." step, the sending

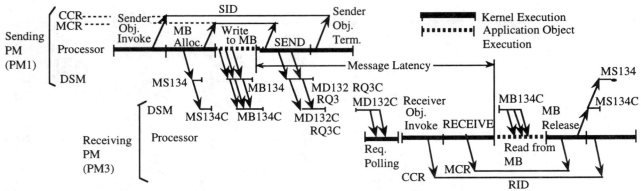

Fig. 4 Timing chart for inter-PM message passing

kernel terminates the execution of the sender object by clearing the CCR.

In the receiving PM, the receiving kernel periodically reads out processing requests from the FIFO memory of RQ3C in the "Req. Polling" step. When a request is found, the receiving kernel sequentially accesses descriptor MD132C, message buffer MB134C, and finally the receiver object ID (RID) in the message header. As a result, MB134C is enqueued to the message queue of the receiver object, specified by the RID. The "Receiver Obj. Invoke" and "RECEIVE" steps respectively set the RID in the receiving distributed memory protector to protect MB134C. In the "Read from MB" step, the receiver object directly reads the message from MB134C to execute the necessary processing. After the message is processed in the receiver object, the receiving kernel updates MS134C to "released" in the "MB Release" step. This update enables the sending and receiving distributed memory couplers to copy MS134C in the receiving DSM to MS134 in the sending DSM. The sending kernel can re-use the corresponding message buffer MB134 anytime after the copy to MS134 is completed.

4. Performance Evaluation

4.1 Kernel Execution Overhead

To evaluate the effectiveness of MESCAR, kernel overhead was compared between a traditional DMA-based system, based on the local memory architecture without distributed memory protectors, and a MESCAR system with distributed memory protectors.

The kernel execution overhead is defined as the sum of all the instruction counts of the sending and receiving kernels from sender object invocation to receiver object termination. These instruction counts were measured by programming and compiling a message passing kernel, based on the PLATINA kernel[10].

Table 1 shows the relative kernel overheads for the two systems. For intra-PM message passing, the kernel overhead of the MESCAR system is 17% larger than that of the DMA-based system. This is because the DMA-based system requires no physical message transfer, and therefore no DMA overhead for intra-PM message passing, while the

MESCAR system requires larger overhead for MB allocation to search for the appropriate MB corresponding to a pair of sending and receiving PMs. For inter-PM message passing, the kernel overhead of the MESCAR system is reduced by 24%, when compared with the DMA-based system. This is because the kernel overhead for sender and receiver DMA control in the DMA-based system significantly outweighs the kernel overhead for MB allocation in the MESCAR system.

4.2 Message Latency Evaluation

As shown in Fig. 4, message latency is defined as the minimum time from the initiation of SEND processing by the sending kernel to the termination of RECEIVE processing by the receiving kernel. Figure 5 compares the latencies for 10-, 50-, and 250-MIPS processors with a 156-Mb/s processor interconnect bandwidth, assuming the ATM AAL5 protocol. The DMA-based system message latencies are almost proportional to the message length because the DMA-transfer delay time is completely included in the message latency.

Table 1 Kernel overhead for message passing

Comm. Type	Functions	Relative Overhead	
		DMA-based	MESCAR
Intra-PM	MB Allocation	0.16	0.25
	SEND	0.25	0.27
	RECEIVE	0.20	0.23
	MB Release	0.16	0.14
	APL Obj. Invoke & Term.	0.23	0.28
	Total	1.00	1.17
Inter-PM	MB Allocation	0.16	0.25
	SEND	0.25	0.27
	Sender DMA control	0.35	-
	Request Polling	-	0.23
	Receiver DMA control	0.50	-
	RECEIVE	0.20	0.23
	MB Release	0.16	0.14
	APL Obj. Invoke & Term.	0.23	0.28
	Total	1.85(1.00)	1.40(0.76)

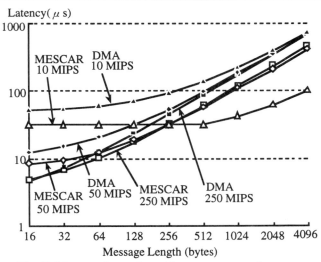

Fig. 5 Message latencies with processor performance as a parameter (with 156-Mb/s processor interconnect)

Table 2 Message latencies for a 256-byte message

Processor Interconnect Bandwidth (Mb/s)	Processor Performance (MIPS)	Latency in DMA-based System (μs) (a)	Latency in MESCAR System (μs) (b)	Reduction Ratio ((b)-(a))/(a)
15.6	10	484	437	0.10
	50	444	441	0.01
	250	437	442	-0.01
156	10	93	32	0.66
	50	53	32	0.40
	250	45	33	0.27
1560	10	62	31	0.50
	50	23	7	0.70
	250	15	2	0.87

The MESCAR system message latencies can be classified into two types: constant latency irrespective of message length (as shown by the line for 512-byte or shorter messages with a 10-MIPS processor) and scaling latency (in which the latency scales almost linearly with message length). The MESCAR system begins the look-ahead DSM copying in synchronization with writing to MBs before SEND primitive execution. Therefore, if the DSM copying has been completed by the time the SEND primitive execution is completed, the latency will be a constant value: (Ts+Tr), where Ts is the send-processing time and Tr is the receive-processing time, both determined by processor performance. If the DSM copying has not been completed by the end of the SEND primitive execution, the transfer delay associated with the remaining DSM copy operations are added to the constant latency of (Ts+Tr) and the total is used as the scaling latency.

Table 2 lists the 256-byte message latencies for combinations of 10-, 50-, and 250-MIPS processors and 15.6-Mb/s, 156-Mb/s, and 1.56-Gb/s processor interconnect bandwidths. The latencies for 10-, and 50-, and 250-MIPS processors with a 1.56-Gb/s bandwidth are reduced respectively by 50, 70, and 87%. The MESCAR architecture thus effectively lowers message latency by balancing processor performance and processor interconnect bandwidth.

5. Conclusion

MESCAR incorporates duplicate DSMs located in the sending and receiving PMs. Messages and control data in the DSMs, which are partitioned corresponding to pairs of sending and receiving PMs, are automatically copied between the corresponding DSMs. MESCAR efficiently supports the whole span of message passing processing; by balancing the software and hardware components, it reduces message latency and software overhead.

We are currently carrying out more detailed performance simulation for the MESCAR system.

References

[1] Palmer, J.: The NCUBE Family of High-Performance Parallel Computer Systems, *Proc. 3rd Conf. on Hypercube Concurrent Computers and Applications*, (1988), pp. 384-390.
[2] Lenoski, D., Laudon, J., Gharachorloo, K., Gupta, A., and Hennesy, J.: The Directory-Based Cache Coherence Protocol for the DASH Multiprocessor, *Proc. 17th Int. Symp. on Computer Architecture*, (1990), pp. 148-159.
[3] Li, K. and Hudak, P.: Memory Coherence in Shared Virtual Memory Systems, *ACM Trans. Computer Systems*, Vol. 7, No. 4, (1989), pp. 229-359.
[4] Wittie, L. and Maples, C.: Merlin: Massively Parallel Heterogeneous Computing, *Proc. 18th Int. Conf. on Parallel Processing*, Vol. 1, (1989), pp. 142-150.
[5] Delp, G. S., Farber, D. J., Minnich, R. G., Smith, J. M., and Tam, M.: Memory as a Network Abstraction, *IEEE Network Magazine*, Vol. 5, No. 4, (1991), pp. 34-41.
[6] Wittie, L. D., Hermannsson, G., and Li, A.: Eager Sharing for Efficient Massive Parallelism, *Proc. 21st Int. Conf. on Parallel Processing*, Vol. 2, (1992), pp. 251-255.
[7] Wilson Jr., A. W., LaRowe Jr., R. P., and Teller, M. J.: Hardware Assist for Distributed Shared Memory, *Proc. 13th Int. Conf. on Distributed Computing Systems*, (1993), pp. 246-255.
[8] Blumrich, M. A., Li, K., Alpert, R., Dubnicki, C., Felten, E. W., and Sandberg, J.: Virtual Memory Mapped Network Interface for the SHRIMP Multicomputer, *Proc. 21st Int. Symp. on Computer Architecture*, (1994), pp. 142-152.
[9] Yamada, S. and Maruyama, K.: A Lightweight Capability-Based Protection Mechanism for Object-Oriented Systems, *Trans. Information Processing Society of Japan*, Vol. 34, No. 9, (1993), pp. 2037-2047.
[10] Kubota, M., Maruyama, K., Tanaka, S., Osaki, K., and Yamada, S: Distributed Processing Platform for Switching Systems: PLATINA, *Proc. 14th Int. Switching Symposium*, Vol. 1, (1992), pp. 415-419.

AUTOMATIC SELF-ALLOCATING THREADS (ASAT) ON THE CONVEX EXEMPLAR

Charles Severance and Richard Enbody
Department of Computer Science
Michigan State University
East Lansing, MI 48824-1027
crs@msu.edu, enbody@cps.msu.edu

Steve Wallach and Brad Funkhouser
Convex Computer Corp.
Richardson, TX 75083-3851
wallach@convex.com, funkhous@convex.com

Parallel processing systems have an advantage over traditional supercomputers in price/performance, but traditional supercomputers retain a significant advantage over parallel processing systems in the area of flexibility. Traditional supercomputers can easily handle a mix of interactive, batch, scalar, vector, parallel, and large memory jobs simultaneously while maintaining high utilization. Often parallel processing systems do not effectively support dynamic sharing of the resources of a system which results in low overall utilization given a mix of jobs. This paper describes the implementation of thread balancing software intended to allow the sharing of a large parallel processing system under a variety of load conditions. The system under consideration for this study is the Convex Exemplar scalable, parallel-processing system. While this paper is focused on the performance of this technique on the Convex Exemplar, it is a general technique which should prove to be useful on many shared memory parallel processors.

INTRODUCTION

Parallel processing systems now have a significant price/performance advantage over traditional supercomputers. This advantage has led organizations to begin to replace their traditional supercomputers with these high-performance parallel processing systems. In many cases, these users are accepting a compromise when they install a parallel processing system. They are gaining peak performance at the cost of ease of use and ease of administration. Also, the new systems are more difficult to share among a large user population resulting in extra administrative effort and reduced flexibility.

This paper is an effort to make a large shared-memory parallel processing system perform as well as a traditional parallel/vector supercomputer under dynamic load conditions with many users. This work focuses on the performance of FORTRAN and C programs which have not been written to be explicitly parallel, but rather depend on parallelism identified by the compiler.

In our approach we attempt to balance the overall number of runnable threads with the number of processors in the system. The Convex C-Series performs this thread balancing using extensive hardware support called Automatic Self-Allocating Processors (ASAP). This work attempts to achieve similar overall balance using a software-only approach which we call Automatic Self-Allocating Threads (ASAT).

PREVIOUS WORK

The general topic of scheduling for parallel loops is one that is well studied. A number of scheduling techniques have been proposed and implemented. An excellent survey of these techniques is presented in [2]. These techniques include Pure Self Scheduling (PSS), Chunk Self Scheduling (CSS), Guided Self Scheduling (GSS) [4], Trapezoidal Self Scheduling (TSS) [7], and Safe Self Scheduling (SSS) [1].

The basic approach of these techniques is to partition the iterations of a parallel loop among a number of executing threads in a parallel process. The goal is to have balanced execution times on the processors while minimizing the overhead for partitioning the iterations.

The implementation of these techniques on most shared-memory parallel processors works with a fixed number of threads determined when the program is initially started. For the purpose of this paper, we call this technique Fixed Thread Scheduling (FTS). The FTS approach is reasonable for many of the existing parallel processing systems as long as each application has dedicated resources.

In [6] the problem of matching the overall system-wide number of threads to the number of processors was studied on an Encore Multimax. Their applications used explicit parallelism and a thread library. The thread library communicated with a central server to insure overall thread balance. They identified a number of the major problems with having too many threads including: 1) Preemption during spin-lock critical section, 2) Preemption of the wrong thread in a producer-consumer relationship, 3) Unnecessary context switch overhead, and 4) Corruption of caches due to context switches. In addition, they provide an excellent survey of related work.

As the speed of the CPU's has increased, the problem of a context switch corrupting cache has become an increasing performance impact. In [3], when a compute-bound process was context switched on a cache-based system, the performance of the application was significantly impacted for the next 100,000 cycles after the process regained the CPU. The context switch still had a small negative impact on performance up to 400,000 cycles after the context switch. In many situations, the cache impact dominated the overall cost of a context switch.

Other dynamic, run-time, thread management techniques which are geared toward compiler detected parallelism include: Automatic Self-Adjusting Processors (ASAP) from Convex [9] and Autotasking on Cray Research [10] computers. Convex ASAP is based on hardware extensions to the architecture and requires very little run-time library support. Cray's Autotasking is a software based approach and is supported in the run-time library and the operating system scheduler.

The Cray Autotasking product is similar to our work. The primary differences between Autotasking and ASAT are: 1) ASAT is targeted toward a lower cost environment with commodity processors using a coherent cache and 2) The initial version of ASAT requires no operating system or run-time library modifications and can be used by an end user or system administrator.

At some point ASAT may be more tightly integrated into the operating system and run-time library. Currently it is a research study into the potential benefit of ASAT. The concept and benefits of ASAT are not limited to the Convex Exemplar. Any cache-based parallel processor will have similar problems in dealing with thread imbalance. A previous study of the benefits of Automatic Self-Allocating Threads (ASAT) for the SGI Challenge was done in [5].

DYNAMIC LOAD ON THE C-240

Since our concept is based on ASAP, we begin by examining the Convex C-Series' [9] hardware support for dynamic thread management. As the load changes dynamically, the number of processors and threads assigned to an application changes. The Convex C-Series can adjust the number of threads within a very small number of hardware instructions because thread management is built into the ASAP hardware. As a result, thread assignment can be adjusted thousands of times per second. In addition, there is no special software required because the scheduler automatically removes processors and threads from a running application when appropriate due to changes in load. When ASAP is used on these systems, it is very easy to keep the system 100% utilized under dynamic load conditions. Parallel applications can get the best possible time to solution (under current load) while other programs continue to use the system effectively.

Consider the performance of a parallel application on a four-processor, C-Series system. In the Figure 1, the horizontal axis is time and the vertical axis is the percentage of the four CPUs in use. At the top of the figure, a parallel application uses all four CPUs because there is no other load on the system. The middle graph of the figure shows several unrelated serial jobs executing on an otherwise empty system. Each job has a different start time and duration. At times there are no CPUs in use and at other times, there are 3 CPUs in use. We call this collection of jobs the "load" that we put on the system. At the bottom of the figure, the parallel ASAP application is run at the same time as the serial load. The ASAP hardware automatically adjusts the usage of the parallel ASAP job to match the load on the rest of the system. The ASAP job "soaks-up" the available cycles resulting in high overall utilization under the dynamic load. In addition, the parallel job does not slow down the serial jobs.

Under light load, the parallel application grabs as many free processors as it needs. When there is a heavy load on the system, the scheduler does not assign more than one CPU to the ASAP application. In some sense, the parallel ASAP application politely allows the other applications to use the processors they need.

Table 1 shows the effect of ASAP under a similar load. For each test, the elapsed time to solution (wall time) and the total CPU time were measured for a four-CPU system. The CPU time is the sum of the time used on each CPU. For example, if a completely parallel job runs for 100 seconds on an empty system with four CPUs, the CPU time will be approximately

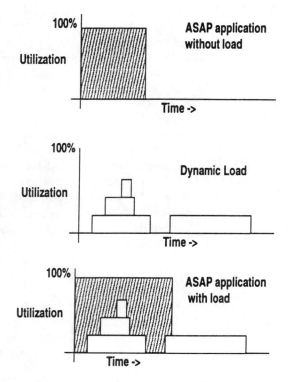

Figure 1: The Operation of ASAP Under Load

Table 1: Performance of ASAP on C-240

	Wall	CPU
Run Individually		
PARALLEL CODE (ASAP)	100.00	353.23
SERIAL LOAD	167.42	242.26
Total Time	267.42	595.49
Run Simultaneously		
PARALLEL CODE (ASAP)	176.88	387.20
SERIAL LOAD	220.75	243.01
Total Time	220.75	630.21

CPU time of the ASAP is due to the cost of the ASAP scheduling. The advantage of ASAP is that the parallel job soaks up excess cycles very efficiently whenever the cycles are available.

DYNAMIC LOAD ON THE EXEMPLAR

The target environment for Automatic Self-Allocating Threads (ASAT) in this study is the Convex Exemplar [8]. The Convex Exemplar provides a global cache-coherent memory which is distributed among several "nodes." Each node contains two to eight PA-RISC CPUs which access a common node memory via a crossbar switch. These nodes are connected together using a cache-coherent interconnect. Memory located on remote nodes can be directly accessed using load and store instructions, but with an increase in the latency of the load or store.

When a compiled parallel application is executed, the run-time environment starts up a number of threads[1]. The number of threads is the same as the number of available processors. These threads are scheduled by the operating system much like UNIX processes.

When the application starts, execution begins using one thread while the other threads wait for work. When a parallel loop is encountered, a routine is called to activate the waiting threads. The waiting threads join the computation and the threads work together for the duration of the loop to complete the work of

400 seconds.

The ASAP program in these experiments is a completely parallel application. The test "load" job contains two serial programs which are run together to simulate a random load on the system (much like part II of Figure 1 above). First, the ASAP and load jobs are run separately on an empty system. Then the jobs are run together on the system.

The numbers in the tables in this paper are not the actual time. They have all been scaled so the wall time of the parallel code run on an empty system is 100. This study is not intended to compare the absolute performance of the systems in this paper but rather their response to changes in load conditions.

The CPU time of the serial applications (load) did not change significantly when run in combination with the parallel application. The wall time of the load job increased about 32% when it ran with the ASAP job. This increase is primarily due to sharing resources with the ASAP job. The CPU time of the ASAP job increased only 10% when it was run in combination with the load job. However, the wall time of the ASAP application increased by 77% because it was sharing resources. Both applications experienced an increase in wall time, but the system was well utilized across the dynamic load. The 10% increase in

---

[1]In this paper, "threads" are the units of execution which are visible to the operating system or "kernel threads". It is quite possible to have a user-space thread management library in which logical threads are scheduled by library code. A parallel application can have some number of kernel threads and the same or different number of logical threads.

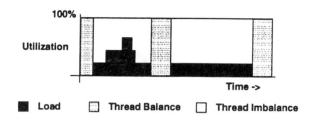

Figure 2: Thrashing Due to Thread Imbalance

the loop. When the loop completes, the threads execute a barrier and when all of the threads have completed the loop one thread continues executing in serial. The remaining threads wait for the next parallel loop.

To reduce the parallel-loop startup latency the routine which activates the waiting threads avoids using the operating system scheduler. The waiting threads execute a spin loop on a shared variable, waiting for the value of the variable to change. When an executing thread wants to activate a waiting thread, it changes this variable. The waiting thread notices the changed variable and joins the executing thread. A similar technique is used on other logically shared memory parallel processors such as the SGI.

If a waiting thread has been suspended by the operating system, the startup latency is much larger because the operating system must first re-schedule the waiting thread before it can execute and detect the changed variable. In the worst case, this latency can be on the order of milliseconds.

This spin-waiting design works well when there is a dedicated CPU available for each thread. However, as we see later, fewer CPUs than threads causes thrashing and severe degradation of performance.

The pseudo code for a parallel loop generated by the compiler is as follows:

```
Serial Code
nthreads = spawn(desired_number_of_threads)
Break up the work into nthreads pieces
Work on each piece
join_threads()
Serial Code
```

Note that the spawn routine could return a different number of threads each time it is called. However, in Fixed Thread Scheduling, the spawn routine always returns the same number of threads each time the parallel loop is executed.

EXEMPLAR EXPERIMENT

When the previous C-240 experiment (Table 1) is run on a four-CPU Exemplar system, we see significantly different results. On the Exemplar a fixed number of threads participates in parallel loops regardless of the system load throughout the duration of the application. When each thread has a dedicated processor, this system operates very well. However, when the number of threads exceeds the number of available processors, the threads must fight for processors. Significant thrashing results because caches become cold, and threads become suspended and must be restarted by the operating system. We call this state of having more threads than processors "thread imbalance."

The time profile of a fixed-thread application under varying load conditions is shown in Figure 2. As in Figure 1, we have a parallel application and some load on the system. At times, the application makes efficient use of the system resources but when other activity is present, the parallel application is in a state of "thread imbalance".

During the time marked "thread imbalance" in Figure 2, the parallel application is wasting a large amount of CPU time loading and flushing caches and scheduling four threads on three or fewer CPUs. In comparing iteration times under balanced conditions with iteration times in unbalanced conditions, the average wall and CPU time per iteration increased significantly during the periods of thread imbalance. It is important to note that this thrashing occurs even if there is only one excess thread.

One way to mitigate the performance impact of thread imbalance is to use gang scheduling. In gang scheduling, the parallel application is only scheduled when all the CPUs are available. This approach reduces the thrashing, but both the parallel load and non-parallel load experience over a 100% increase in time to solution.

Table 2 shows the performance results obtained when the experiment from the Convex C-240 is repeated on the Exemplar using fixed-thread scheduling. Again, both the parallel application and the serial jobs were run separately on an empty system and then the jobs were run in combination. Again, the numbers in the table have been scaled so the wall time of the parallel code run on an empty system is 100.

The impact of the thrashing can be easily seen. When the applications are run together, the load job experienced a 19% increase in wall time and a 4% increase in CPU time. The fixed-thread-scheduled parallel application experienced a 637% increase in wall time and a 397% increase in CPU time.

Table 2: Fixed Thread Scheduling On Exemplar

	Wall	CPU
Run Individually		
PARALLEL CODE (FIXED)	100.00	387.50
SERIAL LOAD	558.33	825.00
Total Time	658.33	1212.50
Run Simultaneously		
PARALLEL CODE (FIXED)	737.30	1924.87
SERIAL LOAD	662.50	862.50
Total Time	737.30	2787.37

An additional observation is that the time to solution for both applications in the combined run was 12% longer than if the parallel application simply suspended and waited until the other programs finished and then executed on an empty system using batch scheduling. This increase occurs because the thrashing of the fixed-thread parallel application consumes resources while making very little progress.

It is important to note that this phenomenon is not unique to the Exemplar architecture. This performance effect is due to the nature of cache-based parallel processors. It is well known that cache misses can dramatically affect the performance of today's high performance RISC processors [3]. In a multi-threaded, parallel processing system experiencing thread imbalance, the probability of cache misses is very high. Other RISC, cache-based parallel-processing systems exhibit the same behavior under thread imbalance [5].

Thread imbalance induced cache thrashing is a significant problem, and the typical solution is to tell users to simply stop using their equipment in a flexible manner, to purchase more CPUs or to partition CPUs into various "sub-complexes." Some sub-complexes are dedicated to gang scheduling in an attempt to minimize time-to-solution, while other sub-complexes are for interactive users. While these solutions may cleverly "avoid" thread imbalance, they limit flexibility and tend to make it difficult to maximize overall utilization of a system.

DESIGN GOALS OF ASAT

Convex C-series ASAP hardware would be prohibitively expensive on a hierarchical machine such as the Exemplar so we turn to software. The general goal of our Automatic Self-Allocating Threads (ASAT) is to eliminate thread imbalance by detecting thrashing

and then dynamically reducing the number of active threads to achieve balanced execution over the long term. In this way parallel applications will experience thread imbalance only during a small percentage of execution time of the application. To implement ASAT on a parallel processing system, there are a number of problems which must be solved:

1. Detecting if too many active threads exist.

2. Detecting if too few active threads exist.

3. Adjusting the number of threads.

ASAT takes advantage of the basic parallel loop structure shown earlier. Under Fixed Thread Scheduling the spawn routine returns the same number of threads each time it is called over the duration of application. When ASAT is used, the spawn call will return the number of threads based on the overall load on the system. The goal is to create the precise number of threads to match the available processors each time spawn is called.

At some point in their execution most parallel applications briefly run in serial to compute some global value or simply execute a barrier at the end of a time step. At the point when the applications are running single threaded, their thread count can be adjusted by ASAT. Given the current state of automatic parallelizing compilers, this parallel segment time duration tends to be shorter rather than longer (we assume between 0.001 seconds and 10 seconds). As compilers improve, the length of the parallel segments should increase.

It is acknowledged that some programs spawn once and run in parallel for very long periods of time. These applications will need to be modified to spawn more often to participate in ASAT. Such applications are often hand coded with explicit parallelism. Most applications which use compiler-generated parallelism will not exhibit this behavior.

EXEMPLAR IMPLEMENTATION

The primary feature of the Exemplar which allowed us to solve the implementation problems identified earlier on an SGI [5] is the hardware-based, real-time clock which is accessible in user space without a system call. With a high-resolution, low-overhead clock, ASAT on the Exemplar uses a timed barrier call to detect thread imbalance. Using the clock, the elapsed time between the first thread entering a barrier and the last thread leaving the barrier was measured using the following approach:

```
static double entering[MAX_THREADS];
static double leaving[MAX_THREADS];

double timed_barrier_test(int THREADS) {
  spawn(THREADS);
  barrier_code();
  first_in = min(entering);
  last_out = min(leaving);
  passage = last_out - first_in;
  return(passage);
}

barrier_code() {
  entering[MY_THREAD] = real_time();
  execute_barrier();
  leaving[MY_THREAD] = real_time();
}
```

The timed_barrier_test code must called while executing as a single thread.

During testing, there was a three-order magnitude difference between barrier passage times under thread balanced and thread imbalance conditions. With the directly-readable, real-time clock available, the balance could be checked with no system calls. In addition, the interval between barrier evaluations can be tuned. We set the ASAT software to only run the barrier test once every 0.5 seconds of elapsed time by default. The ASAT routine could then be called thousands of times per second, but most of the calls would return immediately because the time between ASAT barrier tests had not yet expired.

The number of spawned threads is decreased when the barrier transit time indicates a thread imbalance. ASAT has tunable values which determine the values for what is a "bad" transit time and the number of "bad" transit times necessary to trigger a drop in threads.

To determine whether or not to increase the number of threads, the ASAT barrier test is executed with one additional thread and the barrier transit time is measured. If the barrier transit time indicates that one more thread would execute effectively, the computation is attempted with one more thread. We call it "dipping your toe in the water". If the number of threads we are using has been working smoothly for a while, we test with more threads for a single barrier. If this barrier runs well, we dive in and run the whole application with more threads. Of course, if the increase in threads results in an imbalance, ASAT will drop the thread count at the next spawn opportunity.

The pseudo code for the ASAT thread adjustment heuristic is as follows:

```
/* The current number of threads */
```

```
static int ASAT_THREADS;

asat_adjust_threads() {
 Check Time
 if ASAT_EVAL_TIME has not expired, return
 BAR_TIME = timed_barrier_test(ASAT_THREADS)
 IF (BAR_TIME > ASAT_BAD_TIME) {
   ASAT_BAD_COUNT++;  ASAT_GOOD_COUNT = 0;
 } else {
   ASAT_GOOD_COUNT++;  ASAT_BAD_COUNT = 0;
 }
 IF (ASAT_BAD_COUNT >= ASAT_BAD_TRIG
     and ASAT_THREADS > 1) ASAT_THREADS--;
 IF (ASAT_GOOD_COUNT >= ASAT_GOOD_TRIG
     and ASAT_THREADS < MAX) {
   BAR_TIME =
       timed_barrier_test(ASAT_THREADS+1)
   IF (BAR_TIME < ASAT_EVAL_TIME)
   ASAT_THREADS++;
 }
}
```

The asat_spawn Routine

All the above concepts are integrated into the asat_spawn routine. This routine is a complete replacement for the CONVEX-provided spawn routine. The code is as follows:

```
asat_spawn(int desired_number_of_threads) {
  asat_adjust_threads();
  ithreads = spawn(ASAT_THREADS);
  return(ithreads);
}
```

The only difference between asat_spawn and spawn is that the number of threads that asat_spawn actually runs will change over time. The asat_spawn routine performs an asat_adjust_threads, and then calls the actual spawn with the proper number of threads.

The Exemplar does not support the dropping of hardware threads in a manner similar to the Convex C-Series and SGI Challenge. When the Exemplar version of ASAT detects thread imbalance and reduces the number of threads activated by the succeeding spawn calls, it results in one less hardware thread being used during the spawn. After the unused thread is idle for 100msec, the thread suspends itself and will not be re-scheduled until the number of active threads is increased by ASAT at some future asat_spawn call.

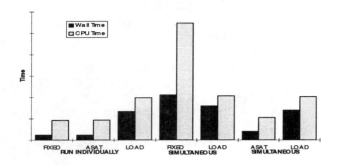

Figure 3: Fixed Thread Scheduling vs. ASAT

ASAT with Compiler-Generated Code

To use ASAT with compiler generated parallel code, the call to spawn is replaced by a call to asat_spawn as follows:

```
Serial Code
nthreads =
    asat_spawn(desired_number_of_threads)
Break up the work into nthreads pieces
Work on each piece
join_threads()
Serial Code
```

The asat_spawn call will spawn the appropriate number of threads based on the current system load.

EXEMPLAR PERFORMANCE

In this section, the previous experiments run on the Convex C-240 using ASAP and the Convex Exemplar using Fixed-Thread Scheduling are duplicated using an ASAT enabled application. Table 3 and Figure 3 summarize the performance results.

In Figure 3 the first three pairs of bars shown are for each job run individually on an empty system. Then the next two pairs of bars are for the fixed-thread job run together with the load job. The last two pairs of bars are the ASAT job in conjunction with the load.

When the execution time of the fixed-thread scheduled application is compared to the ASAT application on an empty system the execution time is effectively identical (about 1% different). This result shows that the ASAT evaluation routine has very low overhead.

When the ASAT and load jobs were run together, the load job experienced a 4% increase in wall time and a 3% increase in CPU time. The ASAT application experienced a 67% increase in wall time, but only

Table 3: Performance Summary on Exemplar

	Wall	CPU
Run Individually (earlier result)		
PARALLEL CODE (FIXED)	100.00	387.50
SERIAL LOAD	558.33	825.00
Total Time	658.33	1212.50
Run Individually		
PARALLEL CODE (ASAT)	100.00	391.67
SERIAL LOAD	558.33	825.00
Total Time	658.33	1216.67
Run Simultaneously		
PARALLEL CODE (ASAT)	166.67	441.67
SERIAL LOAD	583.33	845.83
Total Time	583.33	1287.50

a 13% increase in CPU time when run on a loaded system.

When these increases are compared to the increases experienced by the applications on the C-240, several things can be observed. 1) The time increases experienced by the ASAT application on the loaded Exemplar are nearly identical to the increases experienced by the ASAP application on the loaded C-240. 2) The time increases experienced by the load job on the loaded Exemplar are much smaller than those experienced by the load job running on the loaded C-240. Using ASAT on the Exemplar maintained the time to solution of the load job while making good use of the remaining cycles in the ASAT parallel application.

ASAT succeeds because it is very "polite" and will reduce threads when any other activity is encountered. As such, the load job operated as if it were essentially on an empty system. The increase in wall time of the ASAT application is due to the fact that it only used truly idle cycles.

In observing long runs of ASAT applications on systems with interactive usage, when a user logged in and began running compiles and other interactive commands, ASAT would often give up a thread. On an 8-CPU system, ASAT would use all the CPUs when the interactive users were idle. When the users began actively typing commands, it would drop to 7 threads. Once the interactive activity quieted down, ASAT would go back to 8 threads. This feature allows an ASAT application to soak up the excess cycles on an interactive system while having little or no impact on interactive response time.

FUTURE WORK

There is more work to be done. At some point, we will have access to an Exemplar with more CPUs. The current version of ASAT should be sufficient without modification as long as the well-balanced barrier times remain fast. If the barrier times increase by several orders of magnitude, the evaluation strategy used in ASAT may need some adjustment.

In addition, we need to further study how to best implement ASAT using compiler and operating system modifications. ASAT, as currently implemented, does not make or require any operating system changes. One operating system change we believe would be helpful to ASAT is to assign a lower priority to processes with more active threads. This modification would naturally encourage processes with the largest number of threads to give up their threads and balance overall usage in the long run.

Another area to study is the behavior of a system with several ASAT processes running at the same time. The goal should be to divide the available resources equally among processes. A low-priority ASAT job should yield threads to a higher priority ASAT or non-ASAT jobs. Preliminary work indicates that ASAT is quite stable in the case of two jobs running at the same time.

Another area of work is to do a long-term study of the overall effect of ASAT. This work would allow one to study the average time spent in a parallel section across a wide variety of applications. We hope to have a version of ASAT available via anonymous FTP. Please check the URL http://clunix.msu.edu/~crs/projects/asat for details on the availability of ASAT.

CONCLUSION

In this paper, we have taken the concept of Automatic Self-Allocated Threads (ASAT) and implemented it on a Convex Exemplar parallel processing system. Using available hardware features on the Exemplar, the performance was improved, and the overhead was reduced compared to the SGI implementation of ASAT. The ASAT software compares favorably with the ASAP hardware on the Convex C-240 in terms of flexibility and efficient use of resources. These results show that ASAT can be used to achieve nearly 100% utilization of a parallel processing system under dynamic load conditions without the use of expensive hardware support. Using ASAT, one can have the price/performance of the commodity-based parallel processor with the flexibility and efficiency of a supercomputer.

Thanks to: Duane Eitzen, Convex Corporation; Jerry McAllister, Michigan State University; Dave McWilliams, National Center for Supercomputing Applications and Lisa Krause, Cray Research.

References

[1] J. Liu, V. Saletore, T. Lewis, "Scheduling Parallel Loops with Variable Length Iteration Execution Times of Parallel Computers," *Proc. of ISMM 5th Int. Conf. on Parallel and Dist. Systems*, 1992.

[2] J. Liu, V. Saletore, "Self Scheduling on Distributed-Memory Machines," *IEEE Supercomputing'93*, pp. 814-823, 1993.

[3] J. C. Mogul and A. Borg, *The Effect of Context Switches on Cache Performance*, DEC Western Research Laboratory TN-16, Dec., 1990. http://www.research.digital.com/wrl/techreports /abstracts/TN-16.html

[4] C. Polychronopoulos, D. J. Kuck, "Guided Self Scheduling: A Practical Scheduling Scheme for Parallel Supercomputers," *IEEE Transactions on Computers*, Dec. 1987.

[5] C. Severance, R. Enbody, *Software-Based, Automatic, Self-Adjusting Threads (ASAT) for Parallel Supercomputers*, MSU Computer Science Dept. Tech. Report CPS-94-17, http://clunix.msu.edu/~crs/papers/asat_sgi.

[6] A. Tucker and A. Gupta , "Process Control and Scheduling Issues for Multiprogrammed Shared-Memory Multiprocessors," *ACM SOSP Conf.*, 1989, p. 159 - 166.

[7] T. Tzen and L. Ni, "Dynamic Loop Scheduling on Shared-Memory Multiprocessors," *Int. Conf. on Parallel Processing*, 1991, pp 247-250.

[8] Convex Computer Corporation, *Convex Exemplar Architecture*, Document DHW-014, Convex Press, Richardson, TX, Nov. 1993.

[9] Convex Computer Corp., *Convex Architecture Reference Manual (C-Series), Document DHW-300*, Convex Press, Richardson, TX, Apr. 1992.

[10] Cray Research, *CF77 Compiling System, Volume 4: Parallel Processing Guide.*

Performance Evaluation of Switch-Based Wormhole Networks*

Lionel M. Ni, Yadong Gui and Sherry Moore

Department of Computer Science
A714 Wells Hall
Michigan State Univercity
East Lansing, MI 48824-1027

Abstract

Multistage interconnection networks (MINs) are a popular class of switch-based network architectures for constructing scalable parallel computers. Four wormhole MINs built from $k \times k$ switches, where $k = 2^j$ for some j, are considered in this paper: traditional MINs (TMINs), dilated MINs (DMINs), MINs with virtual channels (VMINs), and bidirectional MINs (BMINs). The first three MINs are unidirectional networks, and we show that the cube interconnection pattern can provide contention-free and channel-balanced partitioning of binary cube clusters. BMINs based on butterfly interconnection are essentially a fat tree, and their routing properties are described. Both DMINs (dilation 2) and BMINs have a similar hardware complexity. We conclude that a 2-dilated MIN outperforms the corresponding BMIN (or fat tree) for most of the traffic conditions and is a better choice for the design of scalable parallel computers.

1 Introduction

Switch-based networks, or indirect networks, have emerged or resurged as another promising network architecture for constructing *scalable parallel computers* (SPCs). Most of the switch-based networks are based on some variations of *multistage interconnection networks* (MINs). An important metric used to evaluate a network is its communication latency. The *communication latency* equals the elapsed time after the head of a packet has entered the network at the source until the tail of the packet emerges from the network at the destination. The communication latency includes all possible delays encountered during the lifetime of a packet and is highly dependent on the switching technique used [1]. In order to offer low communication latency and reduce buffer requirements, wormhole switching [2] has been used in almost

* This work was supported in part by NSF grants CDA-9121641 and MIP-9204066, and DOE grant DE-FG02-93ER25167. (ICPP'95)

all new generation parallel computers and will be the only switching method considered in this paper. Interested readers may refer to [1] for a detailed survey of wormhole switching techniques for direct network architectures.

The basic component of those switch-based networks is the small-scale crossbar switches. A $k \times k$ switch has k ports at one side and k ports at the other side of the switch as shown in Fig. 1 with $k = 4$. Figure 1(a) is a traditional unidirectional switch with k input ports and k output ports, where each port has a single unidirectional communication channel. Figure 1(b) is a d-dilated ($d = 2$) unidirectional switch, where each port is associated with d unidirectional channels. Depending on the routing algorithm, an outgoing packet may use one of the d channels on a selected output port. With d channels, up to d different packets can be simultaneously transmitted over a port without blocking. Another approach to share the use of a port among multiple packets is to implement virtual channels. As shown in Fig. 1(c), two virtual channels share a physical channel in a time multiplexed manner.

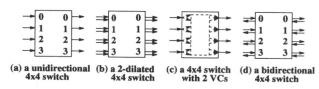

(a) a unidirectional 4x4 switch (b) a 2-dilated 4x4 switch (c) a 4x4 switch with 2 VCs (d) a bidirectional 4x4 switch

Figure 1. Four different types of 4×4 switches.

Unlike the above three unidirectional switches, Fig. 1(d) illustrates a bidirectional switch in which each port is associated with a pair of opposite unidirectional channels. This implies that two packets can be transmitted simultaneously in opposite directions between neighboring switches. For ease of explanation, it is assumed that processor nodes are on the left-hand side of the network, as shown in Figure 5. A bidirectional switch supports three types of connections: *forward*, *backward*, and *turnaround* (see Fig. 2). In forward connection, input port ℓ_i is connected to output port r_j, where $0 \leq i, j \leq k - 1$. In backward connection, input port r_i is connected to output port ℓ_j, where $0 \leq i, j \leq k - 1$. In turnaround connection, input port ℓ_i is connected to output port ℓ_j, where $0 \leq i \neq j \leq k - 1$. No connection is allowed from input port r_i to output r_j, where $0 \leq i \neq j \leq k - 1$. This property prevents the shortest-path routing from deadlock.

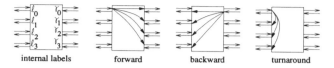

Figure 2. Possible connection patterns within a 4×4 bidirectional switch.

Note that in this paper a channel refers to a unidirectional communication channel. Strictly speaking, the switches in Fig. 1(b) and Fig. 1(d) should be considered as $2k \times 2k$ switches. However, in order to make our explanation easier, we will refer all those four types of switches in Fig. 1 as 4×4 switches with the understanding that some ports may have multiple communication channels.

Scalable parallel computers, typically based on distributed-memory architecture, are composed of a number of nodes. In this paper, it is assumed that there is exactly one pair of input channel and output channel connecting a node to the network, resulting in so-called "one-port communication architecture". This assumption, which is consistent with many existing parallel computers, implies that the local processor must transmit (receive) packets in sequential. The primary function of the network in a SPC is to route packets among those interconnected nodes. In this paper, four different switch-based networks based on those four different types of switches are considered as the network architecture for SPCs. MINs based on those unidirectional switches are *unidirectional MINs*, which imply a wraparound channel connecting the output of each MIN port to the input port of the corresponding processor. MINs based on bidirectional switches are *bidirectional MINs*. In order to make a fair comparison, we also assume that the input and output channel bandwidths are the same for all nodes in all network architectures.

The main objective of this paper is to evaluate and compare the performance of those four different MINs based on wormhole switching. Section 2 describes those unidirectional MINs using the three different unidirectional switches. Two known MIN topologies: butterfly MIN and cube MIN, will be considered. Section 3 discusses bidirectional MINs. Properties and routing algorithms of the butterfly bidirectional MIN will be detailed. Section 4 will address the network partitionability issue. Performance evaluation and comparison based on simulation experiments is reported in Section 5. Section 6 concludes the paper

and indicates future work. Due to space limitation, all proofs of theorems, additional simulation results, and some detailed descriptions are omitted here and can be found in [3].

2 Unidirectional MINs

An N-node MIN built with $k \times k$ switches can be represented as

$$C_0(N)G_0(N/k)C_1(N) \ldots C_{n-1}(N)G_{n-1}(N/k)C_n(N)$$

where G_i refers to the ith stage, C_i refers to the ith connection, and $N = k^n$. There are n stages. Each stage G_i consists of N/k identical $k \times k$ switches and thus is denoted as $G_i(N/k)$. Each connection C_i connects N right-hand side ports at stage G_{i-1} to N left-hand side ports at stage G_i and thus is denoted as $C_i(N)$. A connection pattern C_i defines the topology of the one-to-one correspondence between two adjacent stages, G_{i-1} and G_i, also known as *permutation*.

There are many ways to interconnect adjacent stages. An N-node ($N = k^n$) Delta network is a subclass of banyan networks, which is constructed from identical $k \times k$ switches in n stages, where each stage contains (N/k) switches. A unique property of Delta networks is their self-routing property [4]. Many of the known MINs, such as Omega, flip, cube, butterfly, and baseline, belong to the class of Delta networks [4] and have been shown to be topologically and functionally equivalent [5]. A good survey of those MINs can be found in [6].

This paper considers two popular interconnection patterns between adjacent stages: butterfly and perfect shuffle. Let β_i^k denote the ith k-ary butterfly permutation and σ denote the perfect k-shuffle connection. Two topologically equivalent MINs are considered below. The self-routing property of these two MINs allows the routing decision to be determined by the destination address. For a $k \times k$ switch, there are k output ports. If the value of the corresponding routing tag is i ($0 \leq i \leq k - 1$), the corresponding packet will be forwarded via port i. For an n-stage MIN, the routing tag is $T = t_0 t_1 \ldots t_{n-1}$, where t_i controls the switch at stage G_i.

Butterfly MINs. In a butterfly MIN, connection pattern C_i is described by the ith butterfly permutation β_i^k. Note that the ith butterfly permutation interchanges the 0th digit and the ith digit of the index. β_0^k is selected to be connection pattern C_n. For a given destination $d_{n-1}d_{n-2} \ldots d_0$, the routing tag is formed by having $t_i = d_{i+1}$ for $0 \leq i \leq n - 2$ and $t_{n-1} = d_0$.

Cube MINs. In a cube MIN (or multistage cube networks [6]), connection pattern C_i is described by the $(n-i)^{th}$ butterfly permutation β_{n-i}^k for $1 \leq i \leq n$. C_0 is selected to be σ. For a given destination $d_{n-1}d_{n-2} \ldots d_0$, the routing tag is formed by having $t_i = d_{n-i-1}$ for $0 \leq i \leq n - 1$.

Figure 3 illustrates two such 8-node MINs. We

shall refer to these traditional MINs as TMINs. Such TMINs have been extensively studied in the past and have been adopted in many research prototype parallel computers, such as the Illinois Cedar, the Purdue PASM, the IBM RP3, and the NYU Ultracomputer. Some commercial parallel computers have also adopted such networks, such as the BBN GP-1000 ($k = 4$), TC-2000 ($k = 8$), and the NEC Cenju-3 ($k = 4$). Both the GP-1000 and TC-2000 use circuit switching[1]. The NEC Cenju-3 adopts wormhole switching.

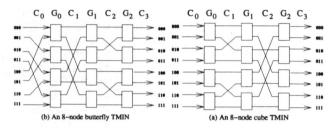

Figure 3. Two 8-node TMINs built with 2×2 switches.

An important characteristic which distinguishs between these two TMINs is their ability to be partitioned into different clusters without contention. This issue will be discussed in Section 4. It is known that both cube TMINs and butterfly TMINs are topologically and functionally equivalent [5]. However, we will show in Section 4 that the cube TMIN can more evenly utilize the communication channels than the butterfly TMIN when the network is partitioned and the network is unidirectional.

2.1 Dilated MINs (DMINs)

One of the nice features of the above TMINs is that there is a simple algorithm for finding a path of length $log_k N$ between any input and output pair. However, if a link becomes congested or fails, the unique path property can easily disrupt the communication between some input and output pairs. The congestion of packets over some channels causes the known *hot spot* problem [7]. Many solutions have been proposed to resolve the hot spot problem. A popular approach is to provide multiple routing paths between any source and destination pair so as to reduce network congestion as well as to achieve fault tolerance. These methods usually require additional hardware, such as extra stages or additional channels.

For ease of comparison purpose, the dilated MINs are considered in this paper [8]. In a d-dilated MIN (DMIN), each switch is replaced by a d-dilated switch (see Fig. 1(b)). By using replicated channels, DMINs offer substantial network throughput improvement [8]. Design of dilated networks has received much attention recently (e.g., [9]). Figure 4 shows a 2-dilated

cube MIN and a 2-dilated butterfly MIN with 8 nodes. Note that half of the input channels and half of the output channels to/from the network are not used in order to maintain the one-port communication architecture and to make a fair comparison with bidirectional MINs.

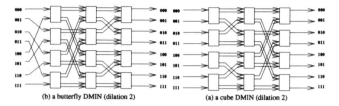

Figure 4. Two 2-dilated MINs using 2-dilated 2×2 switches.

The routing tag of a DMIN can be determined by the destination address as mentioned for TMINs. Within the network switches, packets destined for a particular output port are randomly distributed to one of the free channels of that port. If all channels are busy, the packet is blocked. The increase of throughput in dilated networks is obtained by adding redundancy to the network.

2.2 MINs with Virtual Channels (VMINs)

It is quite expensive to replicate each channel in a wormhole-switched network with its own unique set of physical wires. Furthermore, in most applications, the channel utilization is not high. A virtual channel is a logical channel with its own flit buffer, control, and data path [10]. A virtual channel may dynamically share a physical communication channel with other virtual channels.

Although the concept of virtual channels has been implemented in some direct networks, such as the Cray T3D and the MIT Reliable Router [11], to the best of our knowledge, this concept has not yet being implemented in any switch-based wormhole networks. Performance comparison between virtual channel MINs (VMINs) and other MINs will be studied in Section 5.

3 Bidirectional MINs

To allow for bidirectional communication, each port of the switch has dual channels as shown in Fig. 1(d). For ease of explanation, it is assumed that processor nodes are on the left-hand side of the network. The system architecture of an 8-node butterfly bidirectional MIN (BMIN) is illustrated in Figure 5. Note that due to turnaround connection, the cube interconnection is apparently not a good choice, which will become more clear in Section 3.3. Available ports on the right-hand side of the network are used to configure larger networks, which are not shown in the figure.

---

[1]With circuit switching, reply messages, such as acknowledgment, can be sent back via the same path.

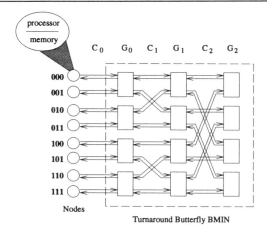

Figure 5. An 8-node bidirectional butterfly MIN

There are many commercial SPCs using BMINs with wormhole switching and turnaround routing including the TMC CM-5 [12], Meiko CS-2 ($k = 4$), and IBM SP-1/2 ($k = 4$) [13]. In the CM-5, the first two level stages use 4×2 switches, yielding a dual-port communication architecture. Although BMINs have been used in many commercial machines, to the best of our knowledge, the routing property of BMINs has not been formally described.

3.1 The Turnaround Routing

In a butterfly BMIN built with $k \times k$ switches, source address S and destination address D are represented by k-ary numbers $s_{n-1} \ldots s_1 s_0$ and $d_{n-1} \ldots d_1 d_0$, respectively. The function $FirstDifference(S, D)$ returns i, the position where the first (leftmost) different digit appears between $s_{n-1} \ldots s_1 s_0$ and $d_{n-1} \ldots d_1 d_0$. More formally, it can be defined as follows.

Definition 1 $FirstDifference(S,D)= t$ *if and only if* $s_t \neq d_t$ *and* $s_j = d_j$ *for* $t < j < n$.

A turnaround routing path between any source and destination pair is formally defined below.

Definition 2 *A turnaround path is a route from a source node to a destination node. The path must meet the following conditions:*

- *the path consists of some forward channel(s), some backward channel(s), and exactly one turnaround connection;*

- *the number of forward channels is equal to the number of backward channels; and*

- *no forward and backward channels along the path are the channel pair of the same port.*

Note that the last condition is to prevent redundant communication from occurring.

To route a message from source to destination, the message is first sent forward to stage G_t. It does

not matter which switch (at stage G_t) the message reaches. Then, the message is turned around and sent backward to destination. As it moves forward to stage G_t, a message may have multiple choices as to which forward output channel to take. The decision can be resolved by randomly selecting from among those forward output channels which are not blocked by other messages. After the message has attained a switch at stage G_t, it takes the unique path from that switch backward to its destination. The backward routing path can be determined by the "destination tag" routing: the message takes output channel ℓ_{d_j} on a switch at stage G_j.

3.2 Properties of the Turnaround Routing

Some properties of the turnaround routing in butterfly BMINs are described in this section.

Deadlock-free routing

A critical issue in the design of routing algorithms based on wormhole switching is to avoid deadlock [1]. Since a message only turns around once from a forward channel to a backward channel, the dependency graph for the routing paths selected in this way is free from cycles. Therefore, the turnaround routing is deadlock free.

Number of shortest paths

By the butterfly connection, any path connecting source S and destination D must pass through a switch at stage G_t where $t = FirstDifference(S, D)$. Therefore, the turnaround routing is the shortest-path routing. On the other hand, however, when it moves forward, a message can choose an arbitrary forward channel at a switch (there is no redundancy for backward channels). There are multiple choices of the shortest path which the turnaround routing may select between a source and a destination. This property can be formalized as follows:

Theorem 1 *In an N-node butterfly BMIN built with $k \times k$ switches ($N = k^n$), there are k^t shortest paths between source S and destination D, each of which can be generated by the turnaround routing, where $t = FirstDifference(S,D)$.*

Although there may exist multiple paths for a given source and destination pair, the butterfly BMIN with turnaround routing is a blocking network like other unidirectional MINs. In a blocking network, any source node may not be connected to any destination node without affecting the existing connections. For example, in Figure 6, the message sent from node 011 to node 111 and the message sent from node 001 to node 110 contend for a common channel, output channel ℓ_1 at the bottom switch in stage G_2. It is assumed that a source node has no knowledge of the traffic in the network. Under the circumstance that some (backward) channels are obstructed by other traffic, a message cannot predict a "correct" routing path to avoid channel collision. As a result, a "wrong" one might be chosen such that the message contends for backward channels with other messages.

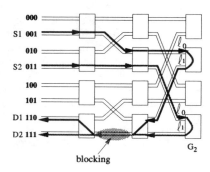

Figure 6. Blocking network

Path length

The path length is defined as the number of channels that a packet has to traverse in the network. For unidirectional MINs, the path length is a constant $n + 1$. For BMINs, the path length is dependent on the location where the packet makes a turn, which is $2(t + 1)$, where $t = FirstDifference$(S,D).

3.3 Analogy to Fat Tree

As shown in Figure 7, a butterfly BMIN with turnaround routing can be viewed as a fat tree [14]. In a fat tree, processors are located at leaves, and internal vertices are switches. When a message is routed from one processor to another, it is sent up (in forward direction) the tree to the least common ancestor of the two processors, and then sent down (in backward direction) to the destination. Such a tree routing well explains the turnaround routing as illustrated in Figure 7.

4 Network Partitionability

In a typical scalable parallel computer, the processors are allocated to different jobs, where each job (or application) usually has an exclusive subset of processors, called a *processor cluster*. If the system does not support contention-free processor allocation, network communication interference among processor clusters will affect the application performance. This section discusses the network partitionability and traffic localization issues. In this paper, we consider those processor clusters forming a cube defined below.

Definition 3 *In a MIN with $N = k^n$ nodes, a k-ary m-cube (cube) consists of k^m nodes which have the same $n - m$ radix-k digits (fixed variables) in their node addresses, where these same digits can be in any $n - m$ locations of the n possible locations. Two cubes are disjoint if they have different fixed variables and one is not a subset of the other.*

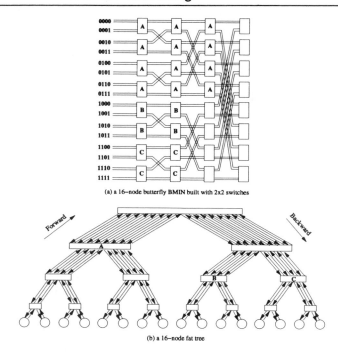

(a) a 16-node butterfly BMIN built with 2x2 switches

(b) a 16-node fat tree

Figure 7. Fat tree and butterfly BMIN

Definition 4 *A k-ary m-cube is referred to as a base k-ary m-cube (base cube) if these $n - m$ digits are in the most significant $n - m$ locations (or the remaining m digits are in the least significant m locations) of their node addresses.*

A base cube is a special case of a cube. Consider a system with $N = 4^4$ nodes. The cluster (21**) has 16 nodes ranging from (2100) to (2133) and is a base 4-ary 2-cube. The cluster (3*1*) has 16 nodes ranging from (3010) to (3313) and is a 4-ary 2-cube.

In addition to guaranteeing contention-free[2] network partitioning, it is important that the number of communication channels between two adjacent stages is the same as the number of nodes in the corresponding cluster. Thus, if a cluster has c nodes, the number of channels (or channel pairs) allocated to the cluster should be c between any two adjacent stages, and this is referred to as *channel-balanced allocation*. If the number of channels allocated is less than c, it implies the possibility of channel congestion within that cluster. If the number of channels allocated is greater than c, it implies that other clusters may be allocated with less number of channels or the channels have to be shared with other clusters.

Lemma 1 *A cube unidirectional MIN with $N = k^n$ nodes can be partitioned into contention-free and channel-balanced disjoint k-ary cubes.*

---

[2]Here we assume that there is no shared flit buffer in each switch; otherwise, the traffic in each cluster may affect other clusters.

In a more general case, when k is a power of 2, the restriction of k-ary cubes can be relaxed to binary cubes. From the proof of Lemma 1, we can immediately obtain the following result.

Theorem 2 *A cube unidirectional MIN with $N = k^n$ nodes, where $k = 2^j$ for some j, can be partitioned into contention-free and channel-balanced disjoint "binary" cubes.*

Figure 8 shows the partitioning of an 8-node cube MIN into three contention-free and channel-balanced binary cube clusters: 0XX, 1X0, and 1X1.

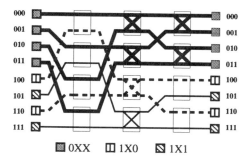

■ 0XX   ☐ 1X0   ◩ 1X1

Figure 8. An 8-node cube MINs is partitioned into three contention-free and channel-balanced binary cube clusters.

For the case of butterfly MINs, we have the following theorem.

Theorem 3 *A butterfly unidirectional MIN with $N = k^n$ nodes may not be partitioned into contention-free k-ary cubes.*

Figure 9 demonstrates the partitioning of an 8-node butterfly MIN into different binary cube clusters. In Fig. 9(a), there are three contention-free clusters: 0XX, 10X, and 11X. In all three clusters, the number of channels is reduced to half in some stages. In Fig. 9(b), there are two 4-node clusters: XX0 and XX1. Both clusters share the use of 8 channels.

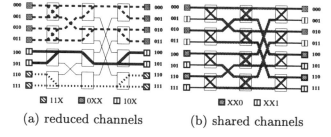

◩ 11X   ■ 0XX   ☐ 10X   ◩ XX0   ☐ XX1

(a) reduced channels   (b) shared channels

Figure 9. An 8-node butterfly MIN is partitioned into different binary clusters.

Due to the nature of butterfly interconnection in BMINs and the analogy to fat tree, we have the following theorem for the partitionability of BMINs. The proof is quite trivial and is ignored here.

Theorem 4 *A butterfly BMIN with $N = k^n$ nodes can be partitioned into contention-free and channel-balanced disjoint "base" k-ary cubes.*

5 Simulation Experiments

To study the performance of four variations of switch-based MINs under different network workloads, we simulate 64-node multistage networks consisting of 4×4 switches. Thus, each network consists of three stages with 16 switches per stage. For all four types of MINs, each node is connected to the network by a pair of unidirectional channels. All channels have the same bandwidth: 20 flits/μsec. Each input channel in a switch has a buffer the size of a single flit. The switches operate asynchronously, but synchronize to simultaneously transmit all of the flits in a *worm*, the portion of a packet in the network. Each node generates packets at time intervals chosen from a negative exponential distribution. Each message has an equal probability of being one packet between 8 to 1024 flits. Messages generated are queued at the source node and enter the network according to the FCFS policy. Messages arriving at a destination node are immediately consumed. The performance of the network under a given workload and routing algorithm is measured in terms of two main quantities: average communication latency and average sustainable network throughput.

5.1 Network Traffic Patterns

The pattern of message traffic, i.e., the destinations to which messages are sent, has a great impact on the performance of the network. Three message traffic patterns are considered in the simulation: *uniform, $x\%$ nonuniform, perfect k-shuffle*. In our simulation experiments, we consider two different clustering of nodes: global (one cluster with 64 nodes) and cluster-16 (four binary 4-cube clusters). For the case of cube networks, four clusters are 0XX, 1XX, 2XX, and 3XX to guarantee channel-balanced partitioning. For the case of butterfly networks, a *channel-reduced* clustering has four clusters: 0XX, 1XX, 2XX, and 3XX; and a *channel-shared* clustering has four clusters: XX0, XX1, XX2, and XX3. Note that in the channel-reduced clustering, the number of channels is reduced from 16 to 4, and in the channel-shared clustering, the number of channels is increased from 16 to 64. In non-uniform traffic, we consider the first node in each cluster as a hot node, which receive $x\%$ more packets than other nodes. The purpose of having such a non-uniform traffic pattern is to observe the effects due to hot spots [7]. The permutation pattern, perfect k-shuffle, is used to observe the network contention due to this special traffic pattern.

5.2 Cube MIN vs. Butterfly MIN

First, we compare the performance between cube unidirectional MINs and butterfly unidirectional MINs with the same workload. Figure 10 shows the performance between the 64-node cube TMIN and the

64-node butterfly TMIN under global uniform and cluster-16 uniform workloads. For the global uniform traffic, there is no difference between their performance as expected because the whole system is one partition. For the cluster-16 uniform traffic, the communication interference between four clusters in the butterfly TMIN degrades the system performance as shown in Fig. 10(b). As expected the channel-reduced clustering in the butterfly TMIN provides the worst performance. Simulation experiments for the cluster-32 uniform traffic and for DMINs and for VMINs were also conducted. The cube interconnection also showed performance improvement over the butterfly interconnection [3].

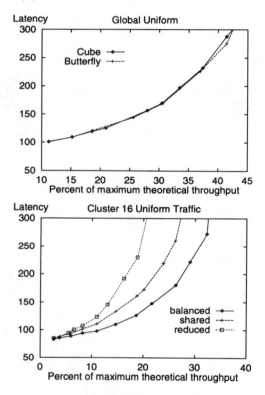

Figure 10. The performance of cube and butterfly TMINs under global uniform and cluster-16 uniform workloads.

We then try to understand the impact of channel-shared partitioning on butterfly networks. Intuitively, the channel-shared partitioning will be beneficial if some clusters generate more traffic than other clusters. Let a:b:c:d be the relative ratio of average network traffic generated from four 16-node clusters. Within each cluster, the network traffic still follows uniform distribution. Figure 11(a) shows the results of 4:1:1:1 ratio among four 16-node clusters. In this case, the channel-shared partitioning of the butterfly TMIN provides the best performance. The channel-reduced partitioning of the butterfly TMIN has the worst performance as expected. Figure 11(b) fur-

ther compares the channel-balanced partitioning of the cube TMIN with the channel-shared partitioning of the butterfly TMIN for two different ratios: 1:0:0:0 and 4:1:1:1. The channel-shared partitioning of the butterfly TMIN always has a better performance. The ratio 1:0:0:0 provides a smaller maximum network throughput because only one cluster of 16 nodes is able to generate network traffic.

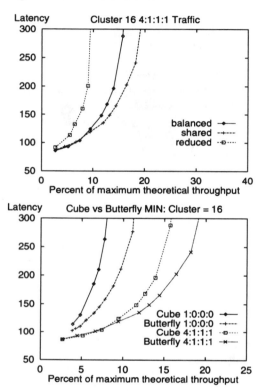

Figure 11. The performance of cube and butterfly TMINs with 4 16-node clusters: (a) ratio 4:1:1:1 and (b) ratios 1:0:0:0 and 4:1:1:1

The channel-shared partitioning of the butterfly TMIN will provide a better performance than the cube TMIN when the network traffic within each cluster is quite different. However, the channel-shared partitioning in butterfly MINs limits the partitionability of nodes than that of cube MINs. Furthermore, the channel interference among different clusters make the performance of applications in each cluster more difficult to predict. Thus, in the following discussions, we only consider cube interconnection for TMINs, DMINs, and VMINs.

5.3 Comparison of Different Networks

We then compare the performance of three unidirectional cube MINs: TMINs, DMINs, and VMINs, and the butterfly BMINs under different network workloads.

5.3.1 Uniform Traffic Pattern

Figure 12 shows the performance of four different networks with global uniform workload.

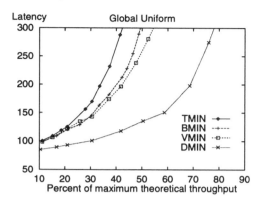

Figure 12. The performance of four networks under global uniform workload.

The TMIN performs the worst in both cases because it has only one path between each pair of source and destination. The DMIN performs consistently the best because the DMIN always has two choices at each outgoing port. The physical channel bandwidth available between stages in the DMIN always doubles than that of the TMIN and VMIN. Surprisingly, the performance of the VMIN is always slightly better than that of the BMIN. This can be explained that the channel blocking probability in the VMIN is reduced, while in the BMIN, there is only one path from the turnaround point to the destination. A similar relative performance difference was also observed for the cluster-32 and cluster-16 uniform workloads [3].

5.3.2 Hot Spot Traffic Pattern

Figure 13 shows the performance when there are hot spots in the network for the case of a global cluster. As shown in the figure, all four networks are congested as indicated by their reduced network throughput comparing with Fig. 12(a). The DMIN always has the best performance due to the same reason as in the uniform traffic case. The performance degradation for the DMIN is quite small (from 78% to 70%) as demonstrated in Fig. 12 and Fig. 13 when the hot node traffic is 5% more. When the hot node traffic is increased to 10% more, the maximum network throughput is reduced to 45% [3]. Since the TMIN does not provide any alternate routing path, it has the worst performance. Note that the performance difference between the TMIN and BMIN is quite small because the path down the tree is unique in the BMIN, while both the DMIN and VMIN have alternate paths all the way. The performance of the VMIN is expected to be better if there are additional virtual channels. The above relative performance difference among four different networks is the same for cluster-16 and cluster-32 partitionings [3].

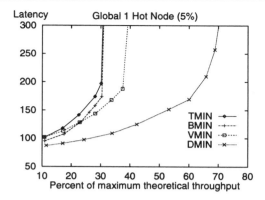

Figure 13. The performance of four networks under the global hot spot (5% more traffic) workload.

5.3.3 Permutation Traffic Pattern

Figure 14 shows the performance under the perfect 4-shuffle permutation traffic pattern. In this pattern, channel contention may occur in all four networks. Both the TMIN and the VMIN have a poor performance, because some channels have to be shared by four source and destinations pairs. The VMIN has the worst performance than that of the TMIN because the flit-level sharing of channels is based on round-robin scheduling, where the objective is to maintain fairness. Thus, all contending packets have about the same long delay. Note that the physical channel bandwidth is the same for both the TMIN and VMIN.

Both the DMIN and the BMIN demonstrate a better performance due to the existence of multiple routing paths. When the workload is heavy, the BMIN provides the best performance. In the DMIN (dilation 2), the channel contention is still possible due to the competition of four packets over some channels. In the BMIN, theoretically, all source and destination pairs can be transmitted simultaneously without contention if the forward channel is properly chosen. Even if not, the probability of contention is quite small due to the existence of multiple routing paths. The above relative performance difference among four different networks is the same for other permutation patterns, such as the 2-th butterfly pattern [3].

6 Conclusions and Future Work

This paper has described four switch-based wormhole networks. Some of these networks, such as TMINs and BMINs, have been adopted in many commercial parallel computers. This paper provides a comprehensive performance comparison among them. Among these four MINs, both DMINs (dilation 2) and BMINs have a similar hardware and packaging complexity. In terms of individual switch design, VMINs, DMINs, and BMINs have a similar hardware

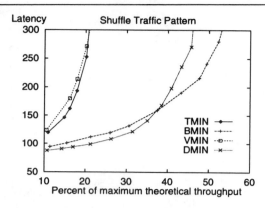

Figure 14. The performance of four networks under the perfect 4-shuffle permutation traffic pattern

complexity[3]. From our simulation results, we conclude that a DMIN (dilation 2) is the most cost effective design comparing with the corresponding popular BMIN (or fat tree) for most of the network traffic patterns. Note that our conclusion is based on the following conditions.

- Each node has one input port and one output port — the so-called one-port communication architecture.

- The network cycle time is the same for all network architectures.

- There is only one flit buffer for each channel.

The BMIN does not require wraparound connections and can provide a good performance if the application specific routing paths can be carefully scheduled. We have shown that although those Delta-class MINs are topologically and functionally equivalent, they are not equivalent in terms of network partitionability. Our additional work has shown that the Omega network and the cube network have the same network partitionability; while the baseline network and the butterfly network have a similar network partitionability.

Some may argue that our performance comparison among four different networks is unfair. However, it is difficult to make a truly fair comparison. We can double the channel bandwidth for both the TMIN and the VMIN, which becomes the same as the DMIN and the BMIN. Clearly, the performance of the TMIN and the VMIN will be significantly improved. However, this may not be fair to the other two networks because doubling the channel bandwidth implies that the input and output bandwidth to each node is also doubled. There are many directions for future research. Interested readers may refer to [3] for details.

---

[3]The hardware design complexity of switches in BMINs may be higher due to more choices of outgoing port for a given input packet.

References

1. L. M. Ni and P. K. McKinley, "A survey of wormhole routing techniques in direct networks," *IEEE Computer*, vol. 26, pp. 62 – 76, Feb. 1993.

2. W. J. Dally and C. L. Seitz, "The torus routing chip," *Journal of Distributed Computing*, vol. 1, no. 3, pp. 187–196, 1986.

3. L. M. Ni, Y. Gui, and S. Moore, "Performance evaluation of switch-based wormhole networks," Tech. Rep. MSU-CPS-ACS-96, Michigan State University, Department of Computer Science, July 1994. (ftp.cps.msu.edu:/pub/acs/msu-cps-acs-96.ps).

4. J. H. Patel, "Performance of processor-memory interconnections for multiprocessors," *IEEE Transactions on Computers*, vol. C-30, pp. 771–780, Oct. 1981.

5. C. L. Wu and T.-Y. Feng, "On a class of multistage interconnection networks," *IEEE Transactions on Computers*, vol. C-29, pp. 694–702, Aug. 1980.

6. H. J. Siegel, W. G. Nation, C. P. Kruskal, and L. M. Napolitano Jr., "Using the multistage cube network topology in parallel supercomputers," *Proceedings of the IEEE*, vol. 77, pp. 1932–1953, Dec. 1989.

7. G. Pfister and A. Norton, "Hot spot contention and combining in multistage interconnect networks," *IEEE Transactions on Computers*, vol. C-34, pp. 943 – 948, Oct. 1985.

8. C. Kruskal and M. Snir, "The performance of multistage interconnection networks for multiprocessors," *IEEE Transactions on Computers*, vol. C-32, pp. 1091–1098, Dec. 1983.

9. M. E. Becker and J. Thomas F. Knight, "Fast arbitration in dilated routers," in *Proc. of the First International Workshop on Parallel Computer Routing and Communication (PCRCW'94)* (K. Bolding and L. Snyder, Eds.), pp. 16–30, Springer-Verlag, May 1994.

10. W. J. Dally, "Virtual channel flow control," in *Proc. of the 17th International Symposium on Computer Architecture*, pp. 60–68, May 1990.

11. W. J. Dally, L. R. Dennison, D. Harris, K. Kan, and T. Xanthopoulus, "The reliable router: A reliable and high-performance communication substrate for parallel computers," in *Proc. of the First International Workshop on Parallel Computer Routing and Communication (PCRCW'94)* (K. Bolding and L. Snyder, Eds.), pp. 241–255, Springer-Verlag, May 1994.

12. C. E. Leiserson *et al.*, "The network architecture of the Connection Machine CM-5," in *Proceedings of the ACM Symposium on Parallel Algorithms and Architectures*, (San Diego, CA.), pp. 272–285, Association for Computing Machinery, 1992.

13. C. B. Stunkel *et al.*, "Architecture and implementation of Vulcan," in *Proc. of the 8th International Parallel Processing Symposium*, pp. 268–274, Apr. 1994.

14. C. E. Leiserson, "Fat-trees: Universal networks for hardware-efficient supercomputing," *IEEE Transactions on Computers*, vol. C-34, pp. 892 – 901, Oct. 1985.

SWITCH BOX ARCHITECTURE FOR SATURATION TREE EFFECT MINIMIZATION IN MULTISTAGE INTERCONNECTION NETWORKS*

M. Jurczyk and T. Schwederski

Institute for Microelectronics Stuttgart

Allmandring 30a

D-70569 Stuttgart, Germany

jurczyk@mikro.uni–stuttgart.de

Abstract -- *A switch box architecture for multistage interconnection networks is proposed that minimizes the degrading effects of saturation trees on the uniform background traffic under nonuniform traffic patterns that are known a priori. In this architecture, restricted buffer space is allocated for hot messages within a switch box, and switches can be implemented with little hardware overhead. Three transfer priority mechanisms are introduced and their effectiveness in saturation tree effect minimization under temporary hot-spot traffic is discussed. The proposed alternating transfer priority mechanism can minimize the uniform traffic delay and the length of the network overload phase under temporary nonuniform traffic scenarios, while it does not degrade performance under pure uniform traffic patterns.*

1. INTRODUCTION

In multiprocessor systems, multistage interconnection networks (MINs) are frequently used to interconnect the processors, or to connect the processors with memory modules [6]. Blocking MINs consist of stages of switch boxes and provide a unique path between any source and destination pair, and different source/destination paths might share common links and/or switch boxes. In this paper, one class of MINs, the **multistage cube network** [8] is considered that connects the processors with the memory modules in a shared memory MIMD parallel computer. Results can also be applied to message passing systems.

Hot-spot traffic, in which many processors send data (hot packets) to the same destination, can cause congestion within the MIN and degrade the overall performance of the MIN substantially [7]. Such traffic patterns occur in shared memory multiprocessor systems when, for example, processors access single shared variables. If the traffic rate to the hot memory exceeds a certain threshold, a saturation tree of full switch buffers builds up from the last network stage to the first one. The network is overloaded and even packets not destined to the hot-spot destination are delayed substantially [4, 7].

Several concepts to alleviate saturation tree effects on network performance have been proposed. In combining techniques [7], messages destined to the same destination

are combined into a single message in the switch boxes to reduce the number of hot messages within the network. This technique results in high hardware overhead, and fails if hot-spot messages are not combinable due to multiple hot locations within a module. Flow control techniques, like discarding networks [2], often result in decreased performance under uniform traffic.

In this paper, a switch box architecture with a priority-based flow control scheme is proposed in which hot packets have a negligible influence on the uniform background traffic during a hot-spot phase, and network performance does not degrade under pure uniform traffic. In the next section, the network and traffic models are described. The degrading effects of temporary hot-spot traffic in conventional networks are discussed in Section 3. In Section 4, a priority switch architecture is presented that minimizes saturation tree effects on the uniform traffic, while the performance of three transfer priority schemes are studied and compared in Section 5.

2. NETWORK AND TRAFFIC MODELS

The networks considered in this paper are synchronous, multi-buffered store-and-forward packet switching multistage cube networks [8], that connect $N=2^n$ processors with N memory modules in a shared memory MIMD parallel computer. The networks are constructed from $s = \log_B N$ stages of $B \times B$ switch boxes. Each stage consists of N/B switch boxes; two consecutive stages are connected via N network links. A multistage cube network with $N=8$ and $B=2$ is shown in Figure 1. It is assumed that a queue is associated with each processor that can buffer packets which cannot be injected into the network as a result of blocked network inputs. This way, no packets are lost during a temporary network overload. Furthermore, it is assumed that each memory access is acknowledged, as in the RP3 [6] and in other systems. Thus, each processor is able to detect and/or avoid out-of-order memory accesses, if necessary.

For **uniform traffic**, each processor generates λ fixed-sized packets per cycle ($0 \le \lambda \le 1$) under a Bernoulli process to be sent through the network; λ is the network load. The packet destinations are uniformly distributed.

If a set of processors accesses a single shared variable simultaneously, the processors send a message (hot packet) to the memory module the variable resides in, which results in a **temporary hot-spot traffic**. Because the pro-

*This research was supported by the German Science Foundation DFG scholarship program GK-PVS.

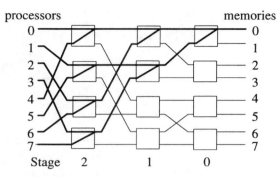

processors   memories

Stage   2   1   0

Figure 1: 8 × 8 multistage cube network with 2 × 2 switches;
bold lines illustrate hot-spot tree to output 0.

cessors in a MIMD system are independent, they send their
hot messages at different times. One way to model such hot
messages is a normal distribution with a mean μ and a standard deviation σ as proposed in [1]. This distribution is
illustrated in Figure 2. It is assumed that before and after

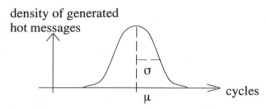

density of generated
hot messages

σ

μ cycles

Figure 2: Distribution of hot messages during a
temporary hot-spot traffic scenario.

a hot-spot access, processors generate uniform traffic with
load λ. The saturation tree that builds up under hot-spot
traffic to memory 0 is depicted in Figure 1 by bold lines.

The occurrence of a hot-spot is assumed to be known
a priori (as in [9]), e.g., determined by the compiler, so that
each processor can distinguish **hot** and **uniform packets**.
To determine the performance of the proposed architecture, a third class of packets will be considered, the **uniform-hot packets**. These packets belong to the uniform
traffic but are destined to the hot-spot.

3. TEMPORARY HOT-SPOT TRAFFIC IN MULTISTAGE CUBE NETWORKS

The effect of temporary hot-spots in networks with
conventional switches is studied in this section and serves
as a basis for comparison with the proposed box architecture. Consider a 1024×1024 multistage cube network
with 2×2 output buffered switch boxes (FIFO queue
length of $L=10$ packets). A hot-spot traffic with $\lambda=0.5$,
$\mu=1000$, and $\sigma=50$ was simulated with a parallel network
simulator [5] running on a MasPar MP-1 SIMD computer
with 16K nodes. Figure 3 shows the delay of the hot and
uniform packets. A simulation result shown at cycle t in
Figure 3 is the average of the results of the last 100 cycles
(from cycle t–99 to cycle t) of 10 independent simulation
runs.

The temporarily filled PE queues and tree buffers during the hot-spot phase result in a delay of the uniform packets of up to 800 network cycles. After most of the hot packets reached the hot destination, the buffers in the saturation
tree start to empty so that the average uniform packet delay
decreases until the tree has vanished. In this study, the
formation of a tree is detected when the uniform packet
delay is more than 5% higher than the delay under pure uniform traffic, while the **time point the saturation tree has
vanished** is determined by the first network cycle the uniform packet delay is less than 5% higher than the delay
under pure uniform traffic.

Thus, two overlapping phases can be defined during
temporary hot-spot traffic scenarios: 1) the **hot-spot
phase**, i.e., the time interval of length T_h from the injection
of the first hot packet into the network until all hot packets
left the network (see Figure 3a), and 2) the **overload
phase**, i.e., the time interval of length T_o from the first
detection of the saturation tree formation until the saturation tree has vanished (see Figure 3b).

To increase network performance under nonuniform
traffic, the packet delay of the uniform traffic and the
length of the overload phase have to be minimized. This
minimization might result in a longer hot-spot phase that
is in many cases preferable to a serious degradation of the
uniform traffic.

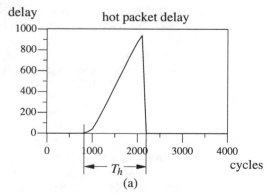

(a)

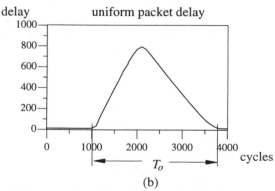

(b)

Figure 3: Delay of (a) hot and (b) uniform packets in a 1024 × 1024 network with 2 × 2 output buffered
switches under hot-spot traffic with $\lambda=0.5$, $\mu=1000$, and $\sigma=50$.

4. PRIORITY SWITCH BOX

To reduce the network performance degradation during the hot-spot phase, a priority switch box architecture is now proposed. As shown in Figure 4, in addition to the uniform buffers and FIFO queues for uniform packets, separate hot buffers of length one are added in which only hot packets are stored. Hot and uniform packets travel on the same inter-stage data lines. It is assumed that separate handshake lines for hot packets and uniform packets per switch box port are utilized. The handshake lines control data flow for the two packet classes, and determine the buffer type that is used to buffer data at the input ports, i.e., uniform or hot buffers. Other implementations of this scheme are possible. This architecture will not result in decreased performance under pure uniform traffic because under that traffic, the switch boxes are equivalent to conventional output buffered switches. Compared to conventional output buffered switch boxes, the proposed architecture results in a more complex switch hardware. However, one additional handshake line but no additional switch box data links are needed per switch box port so that the system complexity increases only slightly.

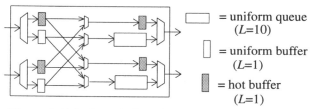

= uniform queue (L=10)

= uniform buffer (L=1)

= hot buffer (L=1)

Figure 4: Priority switch architecture.

5. TRANSFER PRIORITY MECHANISMS

In the proposed switch architecture, two output buffers, i.e., the uniform queue and the hot buffer, are competing for one output link. Thus, in each network cycle, a transfer priority mechanism has to determine which buffer will transfer a packet over the output link. In the next sections, three transfer priority schemes are proposed and their abilities to suppress saturation tree effects are discussed.

5.1 Low Transfer Priority of the Hot Packets

Assume switches with a **low transfer priority** of their hot buffers with respect to their uniform queues. Consider the transfer of a packet from a source switch in a stage to a destination switch in the following stage. In such switches, a hot packet in a source hot buffer is transferred over the link connecting the source and destination switches only if the destination hot buffer at the input of the destination switch has space available, and packets in the related source uniform queue cannot be forwarded, either because the source queue is empty or **blocked** (i.e., the corresponding uniform input buffer of the destination switch is full). Because the packets in the uniform queue always have priority over the hot packets, hot packets cannot

influence uniform packets, and a saturation tree cannot build up. However, under high uniform loads, each uniform queue will contain packets during each network cycle. Because these packets are prioritized over the hot packets at a port, the hot packets suffer from starvation during a high uniform load phase, which may result in an undesirably long hot-spot phase.

5.2 High Transfer Priority of the Hot Packets

Now consider switches with a **high transfer priority** of their hot buffers with respect to their uniform queues. In this scheme, a packet in a source uniform queue is transferred over the link connecting the source and the destination box only if the destination uniform buffer at the input of the destination switch has space available, and packets in the related source hot buffer cannot be forwarded, either because the source buffer is empty or blocked. Consider the hot output of a network constructed from such switches under hot-spot traffic. Because the hot output buffer has transfer priority over the uniform output queue, only hot packets are delivered to the hot destination during the hot-spot phase, while the uniform packets stay in the uniform queue. This way, a saturation tree of filled uniform buffers and queues builds up, resulting in undesirable network performance degradation during the hot-spot phase.

5.3 Alternating Transfer Priority of the Hot Packets

Now consider switch boxes with **alternating transfer priority**. Hot packets buffered in the hot buffer at a switch output port are given priority over the uniform packets in the related uniform queue only if at least K uniform packets were transferred following the transfer of the last hot packet. If only packets in one of the queues at a switch output port are present, then those packets are transferred, independent of K. This scheme can be implemented by employing an additional K-counter at each switch output port. The counter is incremented when a uniform packet is transferred over the output link, and it is reset to 0 when a hot packet leaves the port. A hot packet can leave a port only if either the uniform queue of this port is empty or blocked, or the counter is equal to K, while a uniform packet can leave a port only if either the hot buffer of this port is empty or blocked, or the counter is less than K.

To demonstrate the effects of different values of K on the network performance, the alternating transfer priority mechanism was simulated for a 1024×1024 multistage cube network under a temporary hot-spot traffic with λ=0.7, μ=1000, and σ=50. The packet delays for K=1, 2, and 3 are depicted in Figure 5. An increase of K results in a decrease in the delay of the uniform packets and in a longer hot-spot phase. Also, for K=3 and the network and buffer organization under consideration, the alternating priority scheme is able to fully suppress any saturation tree effects on the uniform background traffic during the entire hot-spot phase for uniform loads of $\lambda \leq 0.7$.

Assuming a maximum data link bandwidth of 1.0, the hot packets are assigned $1/(K+1)$, while the uniform pack-

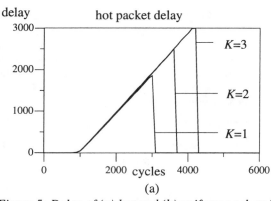

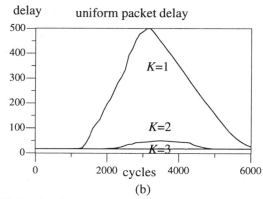

Figure 5: Delay of (a) hot and (b) uniform packets in a 1024 × 1024 network with 2 × 2 alternating priority switches under hot-spot traffic with λ=0.7, μ=1000, and σ=50.

ets obtain $K/(K+1)$ of the link bandwidth when packets are present in both queues of a switch output port due to the alternating priority. Because of the allocated bandwidth, on the average $K/(K+1)$ uniform packets are transferred over any network output link during the hot-spot phase. Thus, a saturation tree will not build up if the uniform traffic load is less than $K/(K+1)$. If a higher load is offered to the network, K has to be increased accordingly to avoid saturation trees. However, the increase of K results in a longer hot-spot phase because, in worst case, a hot packet is forwarded to the hot-spot every $(K+1)$th cycle only. Thus, the choice of K depends on the tradeoff between the maximum network load for which saturation trees will not develop and the length of the hot-spot phase. For $K \to \infty$, this alternating priority mechanism is equivalent to the low priority scheme, while, for $K=0$, it equals the high priority scheme.

Figures 6 and 7 illustrate the influence of the traffic load and K on the performance of a 1024 × 1024 alternating priority network under hot-spot traffic. For comparison reasons, the results for an output buffered network are shown as well. With increasing K, the hot-spot phase length increases, while the length of the overload phase, and the packet delays decrease. For $K>0$, a saturation tree can develop only if $\lambda>(K/K+1)$, while in the output buffered and in the $K=0$ cases, a saturation tree builds up even under low uniform loads, as was discussed earlier. This explains the different packet delay behavior of the networks shown in Figure 7. The high transfer priority of the hot packets in the $K=0$ network results in a minimal hot-spot phase length but in the longest overload phase length and the highest packet delays.

Comparing the output buffered and the $K=0$ networks, a crossover point of the uniform delays at a load of approx. 0.35 for the network organization under study can be observed that is due to the separate buffering of the packets and the high priority of the hot packets in the $K=0$ network. The saturation tree covers fewer network stages under low loads due to the separate packet buffering, while the high priority of the hot packets results in a higher delay of the uniform packets at higher loads in the $K=0$ network as compared to the output buffered network.

For $K=5$ and the range of traffic load depicted, the development of a saturation tree is fully suppressed so that the additional delay of the uniform traffic due to the saturation tree is minimal (recall that for $K=5$, a saturation tree does not exist as long as $\lambda \leq K/(K+1)=5/6=0.833$). The delay of the uniform-hot packets can be minimized for a wide range of loads as can be seen from Figure 7(b). Furthermore, at a load of $\lambda=0.8$, for example, the length of the overload phase can be decreased from 21000 cycles (output buffered switches) to 6000 cycles ($K=5$ switches).

Simulations with other traffic patterns in which saturation trees can build up also demonstrate the advantages of the alternating transfer priority. Example of such traffic patterns are transient partial hot-spot traffics, and bit-reverse permutation traffics [3] that result in **nonuniform traffic spots (NUTS)** within a multistage cube network. Thus, the proposed alternating transfer priority mechanism is able to minimize saturation tree effects, while it does not degrade network performance under pure uniform traffic patterns. To account for dynamic traffic load variations of hot-spot producing applications, an **adaptive transfer priority scheme** can be employed, in which the priority parameter is adapted to the traffic load behavior of applications during run-time by the parallel machine.

6. SUMMARY

A priority-based switch architecture was proposed that minimizes the saturation tree effects on the network performance under nonuniform traffic patterns. The switch contains separate buffers and queues for uniform and hot messages. Data links are shared by both packet classes that results in a hot-spot effect on the uniform traffic, which can be minimized by an alternating transfer priority mechanism. The length of the overload phase, the delay of the uniform traffic, and even the delay of uniform packets that are destined to the hot-spot can be minimized, while network performance under pure uniform traffic is not degraded. The proposed switch architecture requires limited additional hardware, both for the switch and the links between switches.

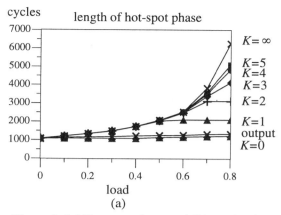 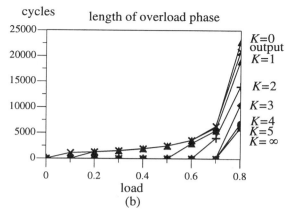

Figure 6: (a) Hot-spot phase and (b) overload phase length in a 1024 × 1024 network with 2 × 2 alternating priority switches under hot-spot traffic with $\mu=1000$ and $\sigma=50$.

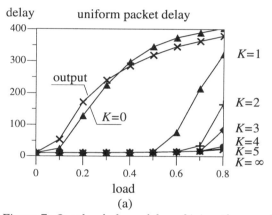 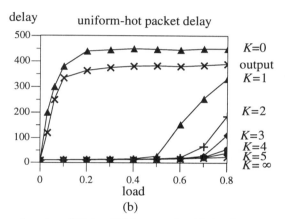

Figure 7: Overload phase delay of (a) uniform and (b) uniform-hot packets in a 1024 × 1024 network with 2 × 2 alternating priority switches under hot-spot traffic with $\mu=1000$ and $\sigma=50$.

REFERENCES

[1] S. Abraham and K. Padmanabhan, "Performance of the direct binary n-cube network for multiprocessors," *IEEE Transactions on Computers*, Vol. C-38, No. 7, July 1989, pp. 1000-1011.

[2] W.S. Ho and D.L. Eager, "A novel strategy for controlling hot spot congestion," *1989 International Conference on Parallel Processing*, August 1989, pp. 14-18.

[3] M. Jurczyk and T. Schwederski, "On partially dilated multistage interconnection networks with uniform traffic and nonuniform traffic spots," *5th IEEE Symposium on Parallel and Distributed Processing*, December 1993, pp. 788-795.

[4] M. Jurczyk and T. Schwederski, "Higher order head-of-line blocking in multistage interconnection networks," *7th ISCA International Conference on Parallel and Distributed Computing Systems*, October 1994, pp. 571-580.

[5] M. Jurczyk, T. Schwederski, H. J. Siegel, S. Abraham, and R. Born, "Strategies for the implementation of interconnection network simulators on parallel computers," *International Journal in Computer Simulation*, to appear, 1996.

[6] G. F. Pfister et al., "The IBM Research Parallel Processor Prototype (RP3): Introduction and architecture," *1985 International Conference on Parallel Processing*, August 1985, pp. 764-771.

[7] G. F. Pfister and V. A. Norton, "'Hot spot' contention and combining in multistage interconnection networks," *IEEE Transactions on Computers*, Vol. C-34, No. 10, October 1985, pp. 933-938.

[8] H. J. Siegel, W. G. Nation, C. P. Kruskal, and L. M. Napolitano, "Using the multistage cube network topology in parallel supercomputers," *Proceedings of the IEEE*, Vol. 77, No. 12, December 1989, pp. 1932-1953.

[9] M.-C. Wang, H. J. Siegel, M. A. Nichols, and S. Abraham, "Using a multipath network for reducing the effect of hot spots," *IEEE Transactions on Parallel and Distributed Systems*, to appear, 1995.

Design And Analysis of Hierarchical Ring Networks for Multiprocessors

Clement Lam †, Hong Jiang ‡
†Department of Electrical and
Computer Engineering
Queen's University
Kingston, Ontario, Canada K7L 3N6

V. Carl Hamacher †*
‡Department of Computer Science
and Engineering
University of Nebraska-Lincoln
Lincoln, Nebraska 68588-0115

Abstract

Hierarchical ring networks with 2 and 3 levels for multiprocessors are studied through simple analytical modeling and extensive simulations. Comparison of deflection routing and buffering shows that the transaction delays in systems using deflections increase faster than in systems with buffering with an increase in traffic intensity. However, the performance gain by reconfiguring from a 2-level deflection system to a 3-level deflection system is significant, and the gain can outperform buffering in a 2-level system. Non-contention optimal configurations are found by minimizing the maximum transaction delay and the average transaction delay. When contentions are considered, configurations that minimize the average non-contention delay perform worse than those which minimize the maximum non-contention delay.

1 Introduction

There has been strong interest in shared-memory multiprocessors from both commercial and research viewpoints. On the commercial side, the availability of very high performance single-chip processors has meant that powerful computing systems can be built by interconnecting a number of computing units (processor plus memory) containing such processors. The aggregate computational throughput of these multiprocessor systems can compete with that of conventional mainframe/supercomputer systems at significant cost/performance advantages.

One interconnection network (IN) form that has been used for shared-memory multiprocessors is a hierarchical ring structure. This is a natural extension of the single-ring multiprocessor system, typified early on by the MIT Concert multiprocessor and popularized recently by the SCI (*Scalable Coherent Interface*) IEEE standard, in countering the problem of poor scalability inherent in the single-ring system. The hierarchical structure takes full advantage of the spatial locality of communication often exhibited in multiprocessors. Spatial locality, which says a processor is likely to communicate with a physically near neighbor, is described as a key to size scalability. This paper focuses on a class of hierarchical ring structures as INs for shared-memory multiprocessors. Examples of multiprocessor

systems using such network structure as INs are Giga-Max, Paradigm, KSR-1, and Hector.

The purpose of this paper is to investigate important design issues pertaining to hierarchical ring networks These design issues include: (1) What is the optimal configuration of the hierarchy, given the total system size, with respect to message latency *or* system throughput? (2) How will traffic patterns affect optimal configurations of the hierarchy? (3) What is the optimal configuration under a given traffic condition? (4) Given the hierarchical nature of the network, what flow control (i.e., message routing) mechanism will be more effective and/or efficient?

The paper is organized as follows. Section 2 describes the hierarchical interconnection network model, including enough structural and operational detail for performance evaluation purposes. Section 3 presents optimal structure design subject to some constraints such as traffic conditions and the depth of the hierarchy. Two major flow control schemes, namely, the buffered approach and the deflection routing approach (including all four possible schemes), are investigated for their effectiveness through extensive simulations in Section 4. Also presented in this section are other simulation results comparing various design alternatives for the hierarchical ring structure. Finally, some concluding remarks on future work are made in Section 5.

2 Network Architecture
2.1 Structure and Operation

Hierarchical networks are a class of NUMA architectures in which memory access delay depends on the memory location. A general m-level hierarchical ring network logically resembles a tree in which every node represents a ring. The root of the tree, level m, is the *global ring*; and the nodes at level 1, the lowest level, to which the leaves are attached are *local rings*. The leaves of the tree are *processor clusters* or *stations* and they contain shared memory modules as well as processors. A balanced or symmetric tree structure is assumed throughout the paper. A 2-level hierarchical ring network with 12 processor clusters or stations is shown in Figure 1. Each station is a collection of processors (possibly one) and memory modules connected by a bus. Two types of interfaces which can be realized by simple logic are needed in the system. A *station interface*, which is basically a transceiver.

---
*Supported by an NSERC (Canada) Research Grant

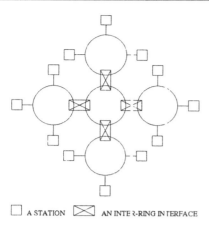

□ A STATION   ⊠ AN INTER-RING INTERFACE

Figure 1: 2-level Ring Network with 12 Stations.

connects a processor cluster to a local ring. Rings at different levels are interconnected by *inter-ring interfaces*, which are essentially 2×2 crossbar switches. Each ring in the network is unidirectional and is divided into fixed-size slots or segments, the size of a packet. The slots can be realized by associating a set of latches with each station interface and inter-ring interface on every ring. In one clock cycle, a packet can be transferred between two adjacent station interfaces on a local ring or two adjacent inter-ring interfaces on a higher level ring; between two adjacent inter-ring interfaces on different rings; from the latches in a station interface to one of its processors or memory modules; or from a processor or memory module on a station into the latches of the next station interface on the ring.

Conflicts may result during an attempted transfer at a station. When a new packet from a station requests a transfer from the station into a slot on the local ring, a conflict occurs if another packet on the local ring is to be transferred to the same slot in the next clock cycle. However, if a station is receiving a packet while there is a pending transfer of a new packet to the local ring, both transfers can complete successfully in the same clock cycle. Another type of conflict can occur at an inter-ring interface when two packets on different rings request the same output channel.

The general system assumption is that of a single address space in which all processors can access all memory locations transparently for memory read/write operations. Besides data, each packet contains the source and destination addresses for routing purposes. When a read or write request is introduced into the network from a station, the request packet will move around the rings until it reaches the destination station. At the destination station, if the target memory is free, the request is accepted and a positive acknowledgement is sent back in the same clock cycle if the request is a write. A write memory *transaction* is assumed to be completed when the acknowledgement packet is finally removed at the source station. For a read request that is accepted, the request packet is removed at the destination and a *response* packet will be sent back at a later time with the required data. A read transaction is completed when the response packet is removed by

the source station. If the memory is not free to accept the read request, a negative acknowledgement is sent back to the source immediately. The request will be resent by the source station at a later time. In what follows, we will define a transaction delay to be the time it takes a packet to move from the source to the destination and back again, i.e., the round-trip delay.

2.2 Flow Control Mechanisms

To resolve conflicts described earlier, some flow control mechanism must be employed. Here we describe two major schemes relevant to hierarchical ring networks, namely, the buffering scheme and the deflection routing scheme. In Section 4, the effectiveness and efficiency of these two schemes will be studied through extensive simulations.

2.2.1 Buffering Scheme

In Hector [4], the conflict is resolved based on a buffering scheme. FIFO buffers are associated with the outputs of the crossbar switches at the inter-ring interfaces. Priority is given to the packet on a higher level ring to complete the transfer while the other packet is buffered until the slot is free. The reason behind this priority scheme is to minimize the delay of the packets that are descending the hierarchy [3]. However, with the buffering scheme, a packet can be dropped when it finds the buffer full. To deal with dropped packets, a time-out mechanism can be used in which a source retransmits a request when no response is received within a time-out period. Although one solution to prevent packet loss is to increase the size of the buffers as suggested by Stumm et al. [2], it leads to higher cost. Furthermore, in some systems, such as optical networks which are becoming more and more popular in light of their extremely high transmission rates, even small buffers are not acceptable for their adverse effect on cost and performance. This is because pure optical buffers are still impractical while electrical buffers coupled with optical links incur o-e-o (optical to electrical or electrical to optical conversion) delays. Buffers can be eliminated by using a deflection scheme.

2.2.2 Deflection Routing Scheme

In this scheme, when two packets from two different ring levels arrive at an interface at the same clock cycle and request the same output link, one packet is allocated the requested link and the other can be deflected to the other output. With a 2×2 crossbar switch, four deflection routing schemes are possible to resolve the conflict.

The HRP (*Higher Ring Priority*) scheme grants the requested link to the packet coming from the higher level in the hierarchy. If the requested output link is at the higher level, the packet at the lower level is deflected into the same ring on which it arrived. When the requested link is at the lower level, the lower-level packet is deflected to the higher level ring. As an alternative to HRP, the requested link is granted to the packet coming from the lower level in LRP (*Lower Ring Priority*) scheme. Transaction delays are minimized for source and destination stations on the same local ring under LRP. The third possibility is the CRP

(*Current Ring Priority*) scheme in which the requested link is allocated to the packet coming from the same level as the requested link. The last possibility is the ORP (*Other Ring Priority*) scheme which gives priority to the packet which is crossing from one level to another.

It may appear that LRP and ORP lead to longer delays because some or all packets, which try to stay on the same ring, may potentially be deflected to another ring at every inter-ring interface. However, the effect on the average transaction delay over all packets is not clear without further quantitative studies. Due to the complexity of the system, an analytical model with simplifying assumptions may not accurately predict the performance of the schemes. As a result, simulation is used to study the four schemes in detail. The results and discussions, in relation to the buffering scheme, will be presented in Section 4.

3 Optimal Structure Design

We will first focus on optimal-structure designs based on minimizing transaction delays under no contention. (Simulations will be used in Section 4 to study the practical situation where contentions occur.) To obtain useful insight into the relationship between the system configuration and its performance, we first derive some simple performance bounds with simple traffic assumptions. Optimal design parameters under such traffic conditions are calculated. This is done in the first part of this section. The results here are taken from our earlier paper [1]. The second part of the section defines a practical traffic condition, or workload, and then derives optimal design parameters for the system subject to the workload. We only consider hierarchical ring networks with 1, 2, and 3 levels, denoted H1, H2, and H3, respectively.

3.1 Uniform Traffic Conditions

Maximum transaction delay, denoted MD_{Hi} for Hi ($1 \leq i \leq 3$), under non-blocking is also usually referred to as the diameter of the network. Let $N = n^2$ be the size of the system, i.e., the total number of local ring stations. The maximum transaction delays for H1, H2, and H3 are derived as follows.

$$MD_{H1} = n^2$$

$$MD_{H2} = \frac{2N}{l} + l = 2\sqrt{2}n$$

where l is the number of local rings, whose optimizing value for MD_{H2} is $\sqrt{2}n$. Similarly,

$$MD_{H3} = \frac{2N}{l_1 l_2} + 2l_1 + l_2 = 3\sqrt[3]{4n^2}$$

where l_1 and l_2 are the numbers of local rings and second level rings, respectively, whose optimizing values for MD_{H3} can be shown to be $l_1 = l_2/2 = \sqrt[3]{n^2/2}$.

3.2 Localized Traffic Conditions

Since most application programs exhibit properties of communication locality, delay measures for localized traffic become important. Here we adopt the

clusters of locality model by Holliday and Stumm [2], in the context of clustered communication patterns. In their model, The computational locality of an application, as opposed to physical locality of a parallel machine, is captured. This allows us to compare systems of different architectures under the same loading conditions. In the following analysis, we assume a 3-level clustered traffic, namely, $S = (1, x, N - x - 1)$ and $P = (P_1, P_2, 1.0)$, where N is the number of processes (or tasks), $x \in \{2, 3, ..., N - 2\}$. That is, each process communicates with itself (i.e., cluster 1) with probability P_1 and communicates with processes of its local cluster (cluster 2) of size x with probability P_2 given that it does not communicate with itself.

However, some simple mapping of the computational locality model, which is network topology independent, onto a given network topology is needed before the analysis of transaction delay can be done. For simplicity, clusters are evenly mapped with respect to a given processor such that it is the geometric center of its corresponding cluster. In the hierarchical ring networks, we consider a simple method, adopted from that in [2], where the locality is defined on a one-dimensional grid (or linear array). Thus, processors are numbered from left to right in the logically formed tree from the ring hierarchy, as described earlier in the paper, and the cluster set 2 for processor i, for example, would consist of processor $i - 2$, $i - 1$, $i + 1$, and $i + 2$ (modulo N). Clearly, the mapping is also direct.

The analysis is carried out for each of the networks under study. Let D_{Local_t} denote the average delay under localized traffic for network topology t, where $t \in \{H1, H2, H3\}$. Due to space constraints, detailed derivations are omitted and only the results are given.

H1: Since every processor connects to the same ring, message delay is independent of traffic locality. Thus,

$$D_{Local_{H1}} = n^2$$

H2: A message can experience two distinctive delays, N/l when travelling in local ring and $2N/l + l$ when travelling globally. Further, depending on the logical number of the source processor and the value of x relative to N/l, a message's target processor in a local cluster may or may not reside inside the same local ring. For simplicity, assume x to be an even number. Conditioning on a message not being home-bound (i.e., $1 - P_1$), we have the expression for $D_{Local_{H2}}$ as,

$$
\begin{cases}
\frac{P_2}{4}\left(\frac{(x+2)l^2}{N} + \frac{4N}{l} + x + 2\right) + \frac{(1-P_2)}{4(N-x-1)} \\
\quad \{\frac{2x(N-1)l^2}{N} + 4xl - x(x+2) \\
\quad + \frac{8N(N+1)}{l} - \frac{4N^2}{l^2}\} & x \leq \frac{N}{l} - 1 \\[2mm]
\frac{P_2}{x}[(x+1)l + \frac{N(2x+1)}{l} - \frac{N^2}{l^2} - N] \\
\quad + (1-P_2)(\frac{2N}{l} + l) & x \geq \frac{2N}{l} - 1 \\[2mm]
P_2[\frac{1+x}{x}\frac{l}{l} + \frac{x-2}{4N}l^2 + \frac{l}{4} + \frac{x}{4} - \frac{1}{2}] \\
\quad + (1-P_2)[((2 - \frac{3x}{2})l^2 - l - \frac{N}{l} \\
\quad + 4N - 2 - \frac{7x}{2})\frac{2N - l(x+2)}{N-x-1} \\
\quad + \frac{x}{N}l^2 - l - \frac{2N}{l} + 2x] & \text{otherwise}
\end{cases}
$$

H3: In a three-level network, a message may experience one of three delays, $N/(l_1 l_2)$, $2N/(l_1 l_2) + l_1$, or $2N_i'/(l_1 l_2) + 2l_1 + l_2$, depending on the destination, similar to the case of H2. The expression for $D_{Local_{H3}}$ [3], which is similar to but much more complicated than that for $D_{Local_{H3}}$, is omitted due space constraint.

With the estimates of the average delay in localized traffic for a given network configuration described above, the next logical question to ask is what would be the best configuration for the hierarchical networks that minimizes the average delay, given a localized traffic pattern. The answer to the question may be found by deriving optimal values for l in H2 and l_1 and l_2 in H3 that minimize $D_{Local_{H2}}$ and $D_{Local_{H3}}$, respectively, while keeping values P_1, P_2, and x constant. Unfortunately, closed form solutions for the optimal values for H2 and H3 don't seem to exist. Therefore, in the absence of a closed form solution for the optimal structures, we derive them through numerical methods with given values of P_1, P_2, and x. Some representative examples of optimal structures are presented in the next subsection.

4 Comparisons and Discussions

All the simulation results presented here have a 95% confidence interval. The results are organized into two parts. The first part focuses on the comparative performance of the different flow control strategies and the tradeoffs between the deflection routing scheme and the buffering scheme. The other part concentrates on the performance of different network topologies.

4.1 Flow Control Strategies

The performances of the four possible deflection routing schemes and the buffering scheme are studied via simulation. The experiments assume the optimal topology designed by minimizing the maximum transaction delay in the system described in the previous section. The physical locality-of-communication model, instead of the computational locality model, is used because the model has direct control of the percentages of transactions that are local, second level, or global. That is, in the case of 2-level ring network, a locality of, say, 80% implies that 80% of all transactions take place inside local rings. Similarly, in a 3-level ring network, a locality of $P = (0.9, 0.5, 1.0)$ means that 90% of the transactions are local, 50% of the rest are second level transactions, and the remainder are global with probability 1.0.

4.1.1 2-level Hierarchy Results

A system with $N = 512$ stations is chosen to study the different flow control schemes. Three sets of experiments were performed with request rates $\lambda = 0.002$. The locality of communication is varied in each set with 80%, 60%, 40%, and 20% local transactions. Although 40% and 20% local transactions may be unrealistic since spatial locality is not present, they serve to study the network behavior under various workloads. The average transaction delays of the different schemes are plotted against the fraction of remote transactions in Figure 2. The rates are chosen in the experiments such that the network experiences a range of loads from light to saturation. The results for the

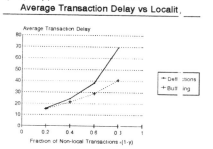

Figure 2: Avg. Transaction Delay in a 2 level System with $L = (16, 32)$, and $\lambda = 0.002$.

Figure 3: Avg. Transaction Delay in a 3 level System with $P_2 = (y, 0.2, 1.0)$.

2-level hierarchy show that the differences among deflection routing schemes are insignificant when communications are localized or when the rate of requests is low. LRP and ORP perform slightly better than HRP and CRP when the system becomes more heavily utilized. The performance between LRP and ORP or the performance between HRP and CRP cannot be differentiated.

When the performance of the buffering scheme is compared to the deflection schemes, the simulation results show average transaction delays of deflection routing schemes are comparable to the buffering scheme only when the traffic is highly localized or when the traffic intensity is low. The performance of deflection routing degrades at a faster rate than buffering with a decrease in locality of communication or an increase in traffic intensity.

4.1.2 3-level Hierarchy Results

The performance of the different deflection schemes is further studied in 3-level hierarchical networks. A 3-level hierarchy with $N = 504$ stations, $L = (7, 6, 12)$, and $\lambda = 0.0025$ is used in the experiments. A set of experiments were performed by varying the locality of communication. The experiments are performed with $P = (y, 0.2, 1.0)$ where y is varied from 0.8 to 0.2 in steps of 0.2. The simulation results, in terms of the average transaction delays, are plotted against the fraction of transactions with a station not within the locality set in Figure 3. Similar to the 2-level results, no significant differences can be found among the four deflection routing schemes. Under high traffic inter-

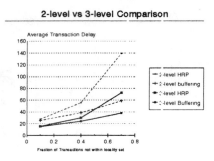

Figure 4: 2-level vs 3-level system avg. transaction delay with $N \approx 450$.

sity or non-localized communication patterns, the average transaction delay increases and Figure 3 shows the difference between deflection and buffering. The increase in transaction delays is, again, due to heavy contention at the global ring. A direct comparison between 2-level systems and 3-level systems is not possible because the physical locality-of-communication model, which is network configuration dependent, is used. The comparative performance of systems with different numbers of levels is studied next.

4.2 2-level vs. 3-level Systems

In section 3, the non-contention transaction delay is shown to be lower in a 3-level hierarchy than a 2-level hierarchy given the same traffic pattern. This section presents the simulation results to compare the performance of 2-level and 3-level hierarchies under varying traffic loads with contention occurring. In order to compare hierarchies with different numbers of levels, the computational locality-of-communication model is used because it is independent of the network configuration. The experiments also assume the optimal configuration designed by minimizing the maximum transaction delay as in the last section.

The 2-level topology with $N = 450$ and $L = (15, 30)$, and the 3-level topology with $N = 468$ and $L = (6, 6, 13)$, are used in the experiments. Communication locality with $x = 4$ and P_2 varied as 0.9, 0.6, and 0.3 is used. A request rate of $\lambda = 0.0025$ is chosen so that both configurations are below saturation. The topologies are chosen so that the number of stations in the 2-level and 3-level hierarchy are similar. Moreover, because results from the last subsection show no significant difference among the deflection schemes, only HRP and buffering scheme results are presented. Figure 4 shows the plot of the average transaction delays versus the fraction of transactions to a station outside the locality set. Simulation results show that the performance gains in 3-level structures are quite significant. In fact, in cases where the traffic exhibits localization, the performance of deflection routing in a 3-level hierarchy is better than in a 2-level hierarchy with buffers. Moreover, the degradation in performance from the buffering to deflection routing is less significant in a 3-level system than a 2-level system. In fact, Figure 4 shows that the average transaction delay of a 3-level system with deflection routing is comparable to the average transaction delay of a 2-level system

with buffering. Therefore, an alternative to replacing deflection routing with buffering in a 2-level system in order to improve performance is to reconfigure from 2 levels to 3 levels. From a hardware complexity standpoint, such a reconfiguration might be attractive, because buffer management at high clock rates may be more difficult to implement than deflection routing.

5 Concluding Remarks and Future Work

This paper has studied some important design issues, namely, optimal structures and flow control strategies, in hierarchical ring networks for multiprocessors. Extensive simulation studies are carried out to investigate the performances of various flow control strategies. The performance of the four possible deflection routing schemes are basically the same using network configurations which minimize the maximum transaction delay under no contention. The delay is comparable to buffering systems under low load. The performance of deflection routing degrades significantly with either an increase in traffic intensity or a decrease in the locality of communications because of heavy contention at the global ring in 2-level systems. However, the advantage of buffering in 3-level systems is not as significant as in 2-level systems. Since the performance of 3-level systems is significantly better than 2-level systems with deflection routing, reconfiguring from 2 levels to 3 levels is a good alternative to gain similar or better performance than by adding buffers to the 2-level system.

Although buffering performs better than deflection routing, the simulator assumes no limit on the size of the buffers. Since the simulation results indicate the buffers are empty most of the time, perhaps the best loss-free flow control scheme is a combination of buffering and deflection routing. Thus, the buffer size could be relatively small; and whenever the buffers are full, a contending packet is deflected. Therefore, it is interesting to study the performance of networks using combined buffering and deflection routing.

References

[1] V. Carl Hamacher and Hong Jiang. Comparison of mesh and hierarchical networks for multiprocessors. *Proceedings of International Conference on Parallel Processing*, pages I:67–71, August 1994.

[2] M. Holliday and M. Stumm. Performance evaluation of hierarchical ring-based shared memory multiprocessors. *IEEE Transactions on Computers*, C-43(1):52–67, January 1989.

[3] Clement Lam, Hong Jiang, and V. Carl Hamacher. Design and analysis of hierarchical ring networks for shared-memory multiprocessors. *TR-95-08, Dept. of CSE, University of Nebraska-Lincoln*, April 1995.

[4] Z. G. Vranesic, M. Stumm, D. M. Lewis, and R. White. Hector: A hierarchically structured shared-memory multiprocessor. *IEEE Computer*, 24(1):72–79, January 1991.

Hierarchical Ring Topologies and the Effect of their Bisection Bandwidth Constraints

G. Ravindran and M. Stumm
Department of Electrical and Computer Engineering
University of Toronto
Toronto, Ontario, Canada M5S 1A4
Email: ravin@eecg.toronto.edu

Abstract -- *Ring-based hierarchical networks are interesting alternatives to popular direct networks such as 2D meshes or tori. They allow for simple router designs, wider communications paths, and faster networks than their direct network counterparts. However, they have a constant bisection bandwidth, regardless of system size. In this paper, we present the results of a simulation study to determine how large hierarchical ring networks can become before their performance deteriorates due to their bisection bandwidth constraint. We show that a system with a maximum of 128 processors can sustain most memory access behaviors, but that larger systems can be sustained, only if their bisection bandwidth is increased.*

1.0 Introduction

Shared memory multiprocessors based on hierarchical ring networks such as those for Hector [9], KSR [1], and the NUMAchine [10] are interesting alternatives to those based on popular direct networks such as 2D meshes or tori. Their simple node to ring interface allows them to be clocked at a much faster rate and the smaller number of connections at each node allow for wider data paths. An important parameter of an interconnection network is its bisection bandwidth, which is defined as the bandwidth provided by the minimum number of wires cut when the network is divided into two equal halves [2]. The bisection bandwidth of a hierarchical ring network is less when compared to a k-ary 2-cube network of equal size, and more importantly the bisection bandwidth is constant and does not scale with the size of the network. This raises the question of how large ring-based multiprocessors can become and what the best topologies are, given the constraint on the bisection bandwidth.

We use a bottom-up approach to address this question. First, we start with a single, *local* ring and determine how many nodes it can accommodate before the performance of the ring deteriorates. We then consider two level hierarchies and determine how many such local rings can be connected to a second level ring while still maintaining reasonable performance. We then proceed to determine how many such second level rings can be sustained by a third level and so on. We show that the bisection bandwidth constraint severely affects the performance, allowing at most three levels of hierarchy and

approximately 128 processors, unless there is a great deal of memory access locality. We also explore how sensitive the performance of the network is to its bisection bandwidth. For example, an interesting observation we make is that increasing bisection bandwidth can, at times, also hurt performance by increasing congestion at the local-ring level.

We use a detailed flit-level simulator to study these issues. We simulate hierarchical blocking networks with register insertion rings, where the processor-to-ring and ring-to-ring interfaces are similar to that of SCI [3] node interfaces[1]. The simulator is driven by a synthetic micro-benchmark that generates memory reference sequences with different access and sharing patterns. Thus we are able to emulate a wide spectrum of memory access behaviors, from high to low cache miss rates and from high degrees of memory locality to almost no locality.

There have been only few studies on performance of scalable hierarchical ring networks so far. Holliday and Stumm [5] studied the performance of large scale hierarchical slotted ring architectures. Throughout their study they assumed very large degrees of locality in their workloads which makes their results applicable only for well-behaved applications. Hamacher and Jiang [4] used analytical models and compared the performance of hierarchical ring interconnects with 2D meshes and concluded that a 3-level hierarchical ring performs somewhat better than a 2D mesh.

2.0 Simulated System

2.1 System Description

Figure 1 shows a two-level hierarchical ring interconnect. It consists of processing modules (PM) connected by a hierarchy of unidirectional rings. A processing module is connected to the lowest level *local* ring and contains a processor, a local cache, and a part of the main memory. A highest-level *global* ring connects several of these local rings. This is similar to the Hector architecture [9]. The system provides a flat, global address space and each PM is assigned a unique contiguous portion of that address space, determined by its location. All processors can transparently access all memory locations in

---

1. Our channel width and packet sizes are different from SCI standard. Also, we do not model SCI bandwidth allocation and queue allocation protocols.

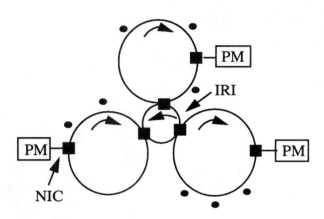

Figure 1: Two Levels of Ring Hierarchy

the system. Local memory accesses do not involve the network. Remote memory accesses require that a request packet be sent from the requesting processor to the target memory, followed by a response packet from the target memory to the requesting processor.

The packets are of variable size and are transferred in a bit-parallel format along a unique path through the ring hierarchy. If a packet is larger than the width of the ring, then it is sent as a contiguous sequence of flits, with the header flit containing routing and sequencing information. We assume *Wormhole* Routing, where a packet may become spread across contiguous links of the network as flits are forwarded, but the packet cannot be interleaved with the flits of other packets [2].

There are five main types of packets, namely read and write requests, read and write responses and negative acknowledgment packets. If a requesting module receives a negative acknowledgment in response to one of its requests, then it resends the request after a short delay. In our simulations, we assumed a cache line size of 16 bytes and that a processor can have at most 3 outstanding (prefetching) requests. The channel is assumed to be 128 bits wide and based on the NUMAchine multiprocessor [10]. Read responses and write requests require 256 bit packets (two flits), whereas read requests and write responses require 128 bit packets (single flit).

There are two main types of interfaces. The Network Inteface Controller (NIC) connects PMs to local rings and the Inter-Ring Interface (IRI) connects two rings. The NIC switches i) incoming packets to the input queue for the PM, ii) outgoing packets from the PM to the ring and iii) continuing packets from the input link to the output link. Each NIC has a bypass buffer capable of temporarily storing packets arriving from the previous ring interface while it is in the process of transmitting a packet from the local PM. The packets on the ring have priority over packets from the local PM.

The IRI controls the traffic between two rings. It is modelled as a 2x2 crossbar switch with input and output

FIFO queues which can hold 10 flits each. The routing delay at a NIC is assumed to be 1 network cycle while it is assumed to be 2 network cycles at an IRI.

2.2 Simulator

We constructed a simulator that reflects the behavior of a system on a cycle-by-cycle basis, using the *smpl* simulation library [7]. The batch means method [7] of output analysis was used with the first batch discarded to account for initialization bias. The batch termination criterion was that each processor had to complete at least some minimum number of requests (in our simulations it is 200 requests per processor). The base simulator was validated against measurements taken from the Hector prototype [5]. It was then extended to model features not present in Hector, such as the insertion ring interface, flit level simulation, wormhole routing and flow control.

2.3 Benchmark Description

In order to evaluate the performance of the interconnection network under controlled conditions, we used synthetic benchmarks to drive our simulator. It was adopted from the Multiprocessor Memory Reference Pattern (M-MRP) address generator of Saavedra, et. al. [8], originally developed to measure real system performance. A M-MRP is a set of P Uniprocessor MRPs, one for each processor, each accessing memory in its own region. (The access regions of each processor may overlap.) Each M-MRP in our simulation is characterized by three attributes: 1) the number of processors, P, generating memory accesses, 2) the size of the memory region, R, accessed by each processor, 3) the cache miss rate, C, of each processor. By varying each of these attributes we can exercise the network in a specific and predictable way and can measure how the network responds under controlled conditions. Throughout our study, P is set to the number of processors in the system. Parameter R allows us to model different memory access patterns by varying it to control the degree of locality and thus the sharing between different processors, and indirectly R is used to control the amount of bisection traffic, i.e. traffic through the global ring. We assume that the memory region of a processor starts with its local memory module and that the sequence of memory references in a given region is uniformly distributed and independent across the region.

The micro benchmark generates a series of memory references at each processor as a result of cache misses. In our simulations, C is varied from 1/100 (1 miss in 100 cycle) to 1/20 (1 miss in 20 cycles), and R is varied from 1/P (the access region is contained entirely in local memory) to 1 (the access region covers all memories of the system). This range gives us sufficient variation in the amount of interconnect network traffic. In our simulations we do not model (dirty) cache line write backs, nor do we model invalidation packets.

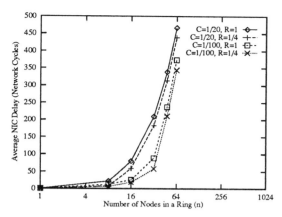

Figure 2: Single Ring Behavior: NIC Delay vs. Nodes

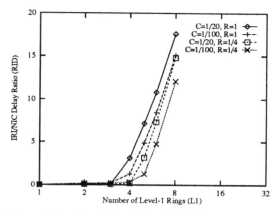

Figure 3: Two Level Ring Behavior: RID vs. L1 Rings

2.4 Measures of Network Performance

We use the following measures of performance in our study. The first two metrics gives us a direct measure of congestion in the network, the third metric measures the average latency in the network while the last one measures the severity of bisection bandwidth constraints:

Network Interface Controller (NIC) Delay is the total average time a packet spends in the Network Interface controllers. It is measured in terms of network cycles

Inter-Ring Interface (IRI) Delay is the total average time a packet spends in IRI controllers, also measured in network cycles.

Round-Trip Latency of a memory access is the elapsed time between when a request packet is generated and the corresponding response packet is received. This measure includes memory access time and is also measured in network cycles.

Ratio of Interface Delays (RID) is the ratio of IRI and NIC delays. This ratio measures the severity of bisection bandwidth constraints relative to local ring bandwidth constraints.

3.0 Hierarchical Ring Network Construction

In this section, we use a bottom-up approach to find hierarchial ring topologies that perform well. We start with a single local ring and determine the maximum number of nodes, n, it can accommodate while maintaining reasonable performance. We then add an extra level in the hierarchy and determine the maximum number of local rings L1, a second level ring can accommodate and so on.

Figure 2 shows the NIC delay plotted against the number of nodes in a single local ring for different values of R and C. This metric is chosen over average latency, since it is a direct measure of network congestion. The NIC delay for a single ring remains small and grows slowly up to 16 nodes for a high cache miss rate C

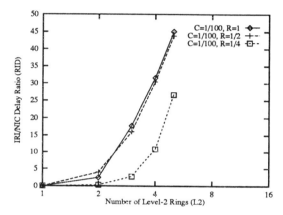

Figure 4: Three Level Ring Behavior: RID vs. L2 Rings

of 1/20 and up to 32 nodes for a low cache miss rate C of 1/100. After this point, the NIC delay starts to grow rapidly. We also observe in Figure 2 that the performance of a single ring is less sensitive to the values of R and C if there are not more than 16 nodes. We therefore conclude that a local ring with n set to 16 should perform well for almost all access patterns and hence assume n to be fixed at 16 for the following discussions.

As a next step, we add a second level ring to the hierarchy. We would like to determine how many 16-node local rings, L1, can be sustained in a two-level hierarchy without major performance degradation. The results (not shown) indicate that with n is fixed at 16, NIC delays are small compared to IRI delays and are almost constant over the entire range of R and L1. The IRI delays are reasonable only for systems with $L1 \leq 4$.

Figure 3 plots the RID (the ratio of IRI and NIC delay) against L1. We choose this metric because the performance of the total system is dominated by the performance of global ring and RID measures how much worse the congestion at the global ring is compared to the local rings. There are two sets of curves in Figure 3: one for C=1/20 and another for C=1/100. In each of these sets

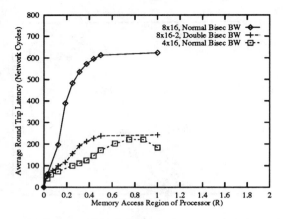

Figure 5: Two Level Ring Behavior: (Latency vs. R, C=1/100)

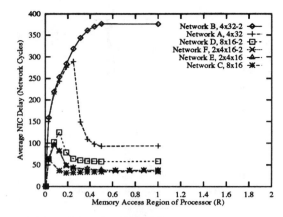

Figure 6: Hierarchical Ring Behavior(NIC Delay vs. R, C=1/20)

there are two curves: one for R = 1 (no memory locality) and one for R = 1/4 (high memory locality). Since it is apparent from Figure 3, that with $L1 \leq 4$, the RID values tend to be smaller, and therefore we fix L1=4 for our larger systems. Again the performance of the system is less sensitive to values of R and C when up to four local rings are connected to a global ring.

Now, we introduce a third level in the hierarchy and proceed to determine how many Level-2 rings, L2, can be sustained in that level. Each L2 ring now consists of a second-level ring connected to 4 L1 rings of 16 nodes each, for a total of 64 nodes. Thus a L2 ring can be represented as 4x16. Figure 4 shows RID curves plotted against the number of L2 rings where it is apparent that RID values are smaller for $L2 \leq 2$.

From the results obtained so far, we see a pattern that is easy to identify. With n=16, L1=4 and L2=2, we can no longer add a fourth level in the hierarchy. At this juncture, we observe that the bisection bandwidth constraint limits the size and scalability of the system for many access patterns. But a maximum size with 128 nodes three level of hierarchy could sustain most of the memory access patterns.

Now, instead of increasing the height of the network, we explore whether increasing the bisection bandwidth of a network would allow us to increase the system size without jeopardizing performance. The results are shown in Figure 5. It shows that a two-level, 64 processor system with a normal bisection bandwidth (4x16) has almost the same performance as a two-level, 128 processor system with double the bisection bandwidth (8x16-2). That is, we can increase the size of a hierarchical ring network, without increasing its height, by increasing its bisection bandwidth. In our model we double the bisection bandwidth by clocking the global ring at twice the speed. However, we will show in the next section that increasing the bisection bandwidth of a system, without increasing its size, may not always improve performance.

There are several options available for increasing the bisection bandwidth of a hierarchical ring system.

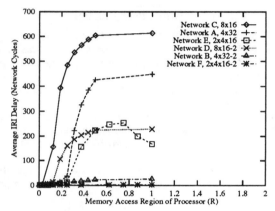

Figure 7: Hierarchical Ring Behavior (IRI Delay vs. R, C=1/20)

Above, we clocked the global ring twice as fast as the local ring. Alternatively, one could widen the channel width of the global ring, or have two global rings, that each connect to all next lower level rings. This is similar to the fat tree architecture [6], except that here we do not widen the channel width of the intermediate or lower-level rings.

4.0 Verification

In the previous section we attempted to compose as large a system as possible, using a bottom-up approach. To verify that the system we composed does indeed have the best topology for the given number of nodes, we compare its performance against a number of other topologies. We assume our goal is a system with 128 processors. The bottom-up approach resulted in a 3 level, 2x4x16 network. Alternatively, we could compose a 2-level, 8x16 network, if we double the bisection bandwidth. We compare these two networks with the four other topologies listed in Table 1.

Figures 6 and 7 show the NIC and IRI delay profiles for Networks A (4x32), B (4x32-2), C (8x16), D (8x16-2), E (2x4x16) and F (2x4x16-2) assuming a high cache miss rate of 1/20. Networks A, C and E are the same as

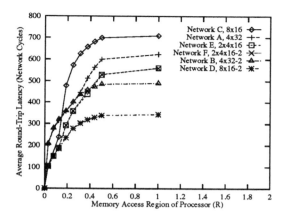

Figure 8: Hierarchical Ring Behavior (Latency vs. R, C=1/20)

Table 1: Hierarchical Ring Configurations

NETWORK	TOPOLOGY
A	4x32, Normal Bisection BW
B	4x32-2, Double Bisection BW
C	8x16, Normal Bisection BW
D	8x16-2, Double Bisection BW
E	2x4x16, Normal Bisection BW
F	2x4x16-2, Double Bisection BW

networks B, D and F except that the latter have twice the bisection bandwidth. We observe in Figure 6 that for networks other than A and B, NIC delays are small for all R, except for small peaks around R=1/16. This behavior is quite consistent with our earlier results, which predict a low NIC delay for values of n less than or equal to 16. Here, the NIC delays increase until R=1/16, at which point appreciable traffic starts flowing through IRI interfaces causing severe congestion at those points. The high IRI delay caused by IRI congestion relieves congestion at the network interface controllers, causing a drop in NIC delay after R=1/16. For network A, the peak is much higher and occurs at R=1/32 due to 32-node rings. For network B, on the other hand, there is no peak, and the NIC delay remains large for all values of R. This observation indicates that increasing bisection bandwidth does not necessarily increase the performance of the system, but merely shifts the bottleneck from the global ring to the local rings.

In Figure 7, networks C and A have high IRI delays, networks E and D have moderate IRI delays and networks B and F have small IRI delays. Network E, which is our network of choice has lower IRI delays than any other network running the global ring at normal speed. (networks A and C). In network C, the IRI delay is much higher due to the fact that L1=8 is twice our recommended value. This causes severe IRI congestion. Network D, which is the same as network C, but with twice bisection bandwidth, shows much improved performance, where IRI delay is 60% lower. In this case, increasing bisection bandwidth improves performance considerably and reduces the congestion at the inter-ring interfaces. It is interesting to note that for both networks C and D, the NIC delays are small.

Finally, Figure 8 depicts the average latency against R, for all networks listed in Table 1, assuming a cache miss rate of C = 1/100. These curves gives us an idea of the overall impact of the local and bisection bandwidth constraints. We observe that network C, 8x16, performs worse than any of the other networks, while network D, 8x16-2, with twice the bisection bandwidth performs best. These latency curves supports our earlier hypothesis that network D with double bisection bandwidth is one of the best configurations for a system with 128 nodes, while network E is one of the best configuration if the global rings run at normal speed.

5.0 Concluding Remarks

In this paper, we used a bottom-up approach to determine how large hierarchical, ring based networks can become before the constant bisection bandwidth property of these networks begin to severely degrade performance. We showed that a system with 128 processors can sustain many memory access behaviors, but that larger systems perform adequately only if the memory access patterns exhibit good locality. Without locality, larger systems can be sustained, only if their bisection bandwidth is increased.

REFERENCES

[1] H. Burkhardt et al., *Overview of KSR1 Computer System*, Technical Report KSR-TR 9202001, (Feb 1992), Kendall Square Research

[2] W.J. Dally, "Performance Analysis of k-ary n-cube interconnection networks," *IEEE Transaction on Computers*, (June,1990), pp. 775-785

[3] D.B. Gustavson, "The Scalable Coherent Interface and related standards projects," *IEEE Micro*, (Feb 1992), pp. 10-22

[4] V.C. Hamacher and H. Jiang, "Comparison of Mesh and Hierarchical Networks for Multiprocessors," *In Proceedings of the ICPP*, (August 1994), pp. 67-71

[5] M. Holliday, M. Stumm, "Performance Evaluation of Hierarchical Ring-Based Shared Memory Multiprocessors," *IEEE Transactions on Computers*, (Jan 1994), pp. 52-67

[6] C.E. Leiserson, "Fat-Trees: Universal Networks for Hardware-Efficient Supercomputing," *IEEE Transactions on Computers*, (October, 1985), pp. 892-901

[7] M.H. MacDougall, *Simulating Computer Systems: Techniques and Tools*, MIT Press, 1987

[8] R.H. Saavedra et. al., "Characterizing the performance space of shared memory computers using Micro-Benchmarks," *In Proceedings of Hot Interconnects*, (Aug 1993), pp. 3.3.1-3.3.5

[9] Z.G. Vranesic et. al., "Hector: A hierarchically structured shared-memory multiprocessor," *IEEE Computer*, (January 1991), pp. 72-78

[10] Z.G. Vranesic et. al., *The NUMAchine Multiprocessor*, Technical Report CSRI-TR-324, (March 1995), Computer Science Research Institute, University of Toronto, Toronto

Enhancing the Performance of HMINs using Express Links*

Yashovardhan R. Potlapalli and Dharma P. Agrawal

Department of Electrical and Computer Engineering

Box 7911, North Carolina State University

Raleigh, N.C. 27695-7911.

(ypotlap, dpa@eos.ncsu.edu)

Abstract

We present several techniques that enhance the performance of HMINs under low locality without adversely affecting the overall cost-effectiveness of HMINs. We show that "express links" are the most effective method to reduce the average distance of HMINs under low locality.

1 Introduction

With the developement of large multiprocessor systems, it has become necessary to have a cost-effective communication method between the processors. Interconnection networks are costly in terms of chip area as they are wiring intensive. However, the improvement in the performance makes MINs a cost-effective solution. Hierarchical networks have a lower cost and average delay due to locality[1].

In [2], multistage interconnection networks (MINs) had been used to build a hierarchical network, called HMIN, that are more cost-effective than MINs under any amount of locality However, the performance of HMINs is observed to be poorer than MINs when the locality of reference was low (Figure 3).

In this paper, we show that though multiple hierarchical links improve the performance minimally, the increase in cost makes HMINs with multiple hierarchical links less cost-effective as compared to HMINs with a single hierarchical link. We, therefore, introduce architectural modifications to enhance the performance under low locality, viz. express links and non-tree-like HMINs. The use of express links is similar to the concept of using express connections in a k-ary n-cube[3]. In non-tree-like MINs, a lower-level MIN may be connected to several MINs in the next higher level of the hierarchy. The optimal number of parents can be determined using analysis similar to [4]. We show that while both methods improve the performance of HMINs under low levels of locality without adversely affecting the cost-effectiveness of HMINs, substantial performance improvement is achieved only by using express links.

The paper is organized as follows. Section 2 presents the terminology used in the paper. Section 3 presents the structure of HMINs with and without "express links". We also discuss the structural differences between tree-like and non-tree-like HMINs. Section 4 compares the performance of HMINs with multiple hierarchical links, HMINs with express links and non-tree-like HMINs and discusses the results of such comparisons. Section 4 presents the concluding remarks.

2 Preliminaries

An HMIN with L levels can be represented bye an L-tuple $(d_1, d_2, \cdots, d_L)$ where d_i represents the number of stages of switches in a MIN at level i in the hierarchy. The number of hierarchical links between levels i and j, $i < j$, is given by $k_{i,j}$. Usually, $j = i+1$. If $j > i+1$, then $k_{i,j}$ represents the k "express links" between level i and level j.

The performance of HMINs is characterized using Average Distance Ratio (ADR) [2][1]. A lower average distance, and hence ADR, would imply better performance, i.e., the performance is inversely proportional to ADR.

The cost of a MIN is assumed to be proportional to the number of switches used. The cost-effectiveness of HMINs is measured by the product of the cost and the ADR which is called the Cost Effectiveness Ratio (CE). This essentially gives us the cost to performance ratio because ADR is inversely proportional to the performance. Therefore, a lower value of CE implies more cost-effectiveness.

We characterize the locality using a term called *degree of locality* and represent it using α. α_i is defined as the fraction of the number of messages at the input of a MIN at level i destined for the outputs of the same MIN and not using any hierarchical links to go higher in the hierarchy

3 Structure of HMINs

There are two basic strategies for interconnection between the different levels of hierarchy - a tree-like strategy[2] and a non-tree-like strategy. The tree-like strategy considers the lowest-level MINs as the

---

*This work has been partially supported by the National Science Foundation under grant MIP-9403191

[1]The number of stages of switches is used a measure of distance

leaves of a tree. Some outputs as well as inputs are connected to a single parent MIN. Several lowest-level MINs are connected to the same parent MIN. These parent MINs form the second layer of hierarchy (the lowest-level MINs form the first level of hierarchy). These parent MINs can be connected to MINs at the third level in the hierarchy in the same way.

In the non-tree-like HMIN, a child MIN is connected to several parent MINs. The advantage is that this strategy increases the fault-tolerance of the HMIN and reduces the average distance ratio. A non-tree-like HMIN is shown in Figure 1. The basic connection is similar to a tree-like HMIN. The connections that are different are shown in bold lines.

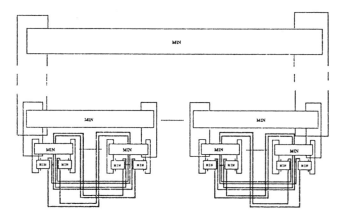

Figure 1: The basic connection for a non-tree-like HMIN

It should be noted that the non-tree-like connections are shown only for the leaves. Non-tree-like connections may exist at several levels of the hierarchy. Such connections exist uniformly in each level of the hierarchy.

Express links are hierarchical links which connect MINs which are more than one level apart in the hierarchy. The primary advantage of this scheme would be to reduce the distance when a message has to pass through several levels of the hierarchy, i.e., under low locality. Therefore, the average distance for all messages is reduced. Note that express links may be connected in a tree-like as well as a non-tree-like fashion. An HMIN with tree-like connected express links (bold lines) is shown in Figure 2.

4   Performance Analysis

Figure 3 shows the performance and cost-effectiveness of several configurations of a two-level HMIN. It is seen that the performance of HMINs is poorer than MINs when the degree of locality is low.

The first approach to improve the performance is to have multiple hierarchical links. The Average Distance (AD) for a two-level HMIN with k hierarchical

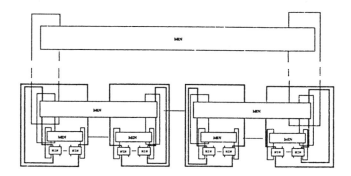

Figure 2: An HMIN with tree-like connected express links

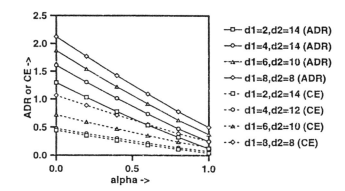

Figure 3: Performance and cost-effectiveness of two-level HMINs with 1 hierarchical link

links is given by:

$$AD = \begin{cases} \alpha d_1 + (1-\alpha)[2d_1 + \\ \quad log_2(k2^{d_2})] & \text{if } (1-\alpha)(2^{d_1}-k) \leq k \\ \alpha d_1 + (1-\alpha)[2log_2\{(1-\alpha)\times \\ \quad (2^{d_1}-k) + (2^{d_1}-k)\}+ \\ \quad log_2\{(1-\alpha)(2^{d_1}-k)2^{d_2}\}] & \text{otherwise} \end{cases}$$

Figure 4 compares the performance (ADR) and cost-effectiveness (CE) of HMINs with one hierarchical link and several hierarchical links. It is seen that the some performance increase is seen for low α's - an improvement which is noticeable when the number of hierarchical links is increased from 1 to 32 and then to 64. The overall performance gain is very small and the increase in cost (larger lower and upper level MIN) makes the approach less cost-effective. Therefore, increasing the number of hierarchical links it not a very practical solution to the problem of reducing the average distance under low locality.

An alternative would be to connect all the k hierarchical links to different parents in the next level. This would result in a non-tree-like HMIN described in the previous section. However, a two-level HMIN is always tree-like. Therefore, we investigate the performance of three-level HMINs. Before we present the expression for the AD of a three-level HMIN, we introduce a

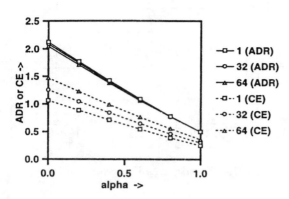

Figure 4: Performance and cost-effectiveness of 8,8 HMINs with multiple hierarchical links

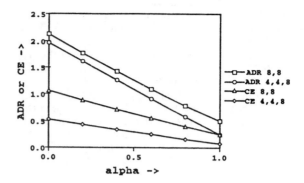

Figure 5: Comparison between a two level (8,8) and a three level (4,4,8) HMIN

term called *effective number of hierarchical links* represented by $\hat{k}$. It is a measure of the blocking at the hierarchical link. The average distance is increased if the number of messages that need to use the hierarchical link exceeds the number of hierarchical links because the extra messages need to be buffered till the next cycle. This effect is captured in the AD expression by using $\hat{k}$ instead of k in the expression. The expression for $\hat{k}_{i,j}$ is as follows:

$$\hat{k}_{i,j} = \begin{cases} k_{i,j} & (2^{d_1} - k_{i,j})(1 - \alpha_i) \le k_{i,j} \\ (2^{d_1} - k_{i,j})(1 - \alpha_i) & \text{otherwise} \end{cases}$$

The AD for a three-level tree-like HMIN is as follows:

$$\begin{aligned} AD &= \alpha_1 d_1 + (1 - \alpha_1)\alpha_2[2log_2(2^{d_1} - k_{1,2} + \hat{k}_{1,2}) + \\ &\quad log_2\{(2^{d_2} - k_{2,3})\frac{\hat{k}_{1,2}}{k_{1,2}} + k_{2,3}\}] + \\ &\quad (1 - \alpha_1)(1 - \alpha_2)[2log_2(2^{d_1} - k_{1,2} + \hat{k}_{1,2}) + \\ &\quad 2log_2(2^{d_2} - k_{2,3} + \hat{k}_{2,3}) + log_2(2^{d_3}\frac{\hat{k}_{2,3}}{k_{2,3}})] \end{aligned}$$

Consider a three level non-tree-like HMIN in which the level 1 MINs have 2 hierarchical links connected to two parent MINs, say MIN_1 and MIN_2, in level 2. Clearly, any processor can reach all processors connected to the clusters rooted at MIN_1 and MIN_2 as compared to the processor in a tree-like MIN which can reach only those clusters rooted at MIN_1. Thus, α_2 is higher for non-tree-like HMINs as compared to tree-like HMINs. For uniform communication at higher levels of hierarchy, the relation between α_i's of non-tree-like and tree-like HMINs is as follows:

$$\alpha_i^{non-tree} = k_{i-1,i} \times \alpha_i^{tree}$$

We first compare the performance and cost-effectiveness of a two level HMIN to a three level

HMIN in Figure 5. We see that the three level HMIN always has a better performance and is more cost-effective than a a 2 level HMIN. This is because the higher level MIN is effectively replaced by an HMIN.

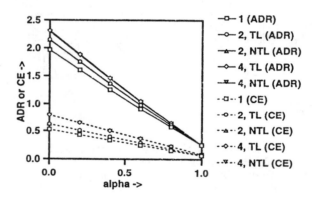

Figure 6: Performance and cost-effectiveness of tree-like and non-tree-like 4,4,8 HMINs

Figure 6 shows the performance and cost-effectiveness of tree-like(TL) and non-tree-like (NTL) HMINs. It is seen that non-tree-like connections provide a small increase (of the order of 0.1%) due to the increase in locality.

An important point to note is that connecting multiple links between levels 2 and 3 is better than connecting multiple links between levels 1 and 2 (Figure 7). The reason is as follows: Both approaches reduce the conflicts for access to hierarchical links. However, in the former method, the size of only the highest-level MIN is substantially increased. In the latter case, the size of the second-level MIN is increased and messages need to pass through this level more often (messages destined only to this level as well as messages that cross over to the third level).

The only drawback of the above two approaches (non-tree-like connections and multiple links at higher levels) is that they do not reduce the ADR to 1 at lower levels of locality. For this reason, we now consider

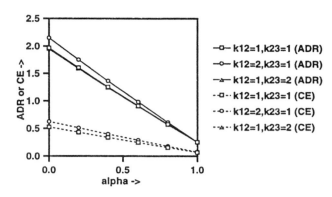

Figure 7: Performance and cost-effectiveness of 4,4,8 HMINs with multiple hierarchical links between levels 2 and 3

Express HMINs, i.e., HMINs with express links. The expression for AD of a three level express HMIN is as follows:

$$AD = \alpha_1 d_1 + (1 - \alpha_1)\alpha_2[2log_2(2^{d_1} - k_{1,2} + \hat{k}_{1,2}) +$$

$$log_2\{(2^{d_2} - k_{2,3})\frac{\hat{k}_{1,2}}{k_{1,2}} + k_{2,3}\}] +$$

$$(1 - \alpha_1)(1 - \alpha_2)[2log_2(2^{d_1} - k_{1,3} + \hat{k}_{1,3}) +$$

$$log_2\{(2^{d_2+d_3}\frac{\hat{k}_{1,3}}{k_{1,3}}) + 2^{d_3}\}]$$

Figure 8 shows the performance and cost-effectiveness of an HMIN with multiple express links as compared to an HMIN with 0 or no express links. Clearly, the performance is substantially increased when α is low. Furthermore, adding express links makes the HMIN even more cost-effective.

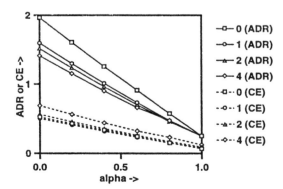

Figure 8: Performance and cost-effectiveness of 4,4,8 HMINs with no(0), 1, 2 and 4 express links

We present Figure 9 to show that Express HMINs with any structure always have a better performance than HMINs with the same structure but with no

express links. However, we also note that Express HMINs do not always increase the cost-effectiveness as seen in the case of the 2,4,10 HMIN.

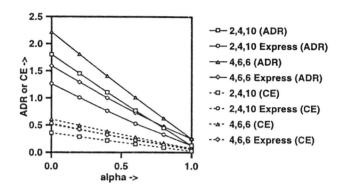

Figure 9: Performance of HMINs with and without express links

5   Conclusions

In this paper, we discuses several methods of reducing the ADR under low levels of locality. We showed that increasing the number of hierarchical links, changing the structure to a non-tree-like structure and adding more hierarchical links at higher levels of the hierarchy have only small, if any, positive impact on the performance and almost always have a negative impact on the cost-effectiveness. However, we demonstrated that increasing the number of levels of the hierarchy improved both performance as well as cost effectiveness. Furthermore, we showed that any more substantial increase in performance could be done only by adding express links. We also showed that adding multiple express links usually increases the cost-effectiveness.

References

[1] S.P. Dandamudi and D.L. Eager, "Hierarchical interconnection networks for multicomputer systems," IEEE Trans. Computers, Vol. C-39, No. 6, pp. 786-797, Jun. 1990.

[2] Y.R. Potlapalli and D.P. Agrawal, "HMIN: A new method for hierarchical interconnection of processors," Proc. 1993 Intl. Conf. Parallel Processing, Vol. 1, pp. 303-306, Aug. 1993.

[3] W.J. Dally, "Express Cubes: Improving the performance of k-ary n-cube interconnection networks," IEEE Trans. Computers, Vol. C-40, No. 9, pp. 1016-1023, Sep. 1991.

[4] J.R. Goodman and C.H. Sequin, "Hypertree: A multiprocessor interconnection topology," IEEE Trans. Computers, Vol. C-30, No. 12, pp. 923-933, Dec. 1981.

Drop-and-Reroute: A New Flow Control Policy for Adaptive Wormhole Routing*

Ji-Yun Kim, Hyunsoo Yoon, Seung Ryoul Maeng, Jung Wan Cho
Department of Computer Science, CAIR
Korea Advanced Institute of Science and Technology (KAIST)
373-1 Kusong-Dong Yusung-Gu, Taejon 305-701, Korea
E-mail: {jykim,hyoon,maeng,jwcho}@camars.kaist.ac.kr

Abstract

In this paper, we propose a new flow control policy, called *drop-and-reroute*, for adaptive wormhole routing and describe architectural supports and its operation. Drop-and-reroute flow control further exploits the adaptivity achieved by adaptive routing by allowing a blocked packet to be routed again in the preceding node. We also measure the effect of drop-and-reroute flow control on the network performance by extensive simulation. Our simulation results show that drop-and-reroute flow control considerably improves the network performance of the adaptive routing algorithm under nonuniform traffic patterns.

1 Introduction

Multicomputers [1] have been expected as the most promising way to construct large-scale parallel machines. A multicomputer is commonly organized as a set of nodes that communicate over an interconnection network, each node containing a processor and some memory. The interconnection network is often the critical component of multicomputers because their performance is very sensitive to the network latency and throughput.

The performance of an interconnection network is mainly affected by the topology, switching technique, flow control, and routing. The *topology* defines how the nodes are interconnected by channels and is usually modeled as a graph. A popular network topology in multicomputers is the k-ary n-cube. A k-ary n-cube network consists of k^n nodes and has n dimensions with k nodes in each dimension.

Switching technique is the actual information transmission mechanism that provides a facility for moving data through the network. In order to reduce the network latency and buffer requirements, *wormhole routing* [4] is being used in most current multicomputers [1, 7, 14]. In wormhole routing, a message is broken into one or more packets for transmission and each packet contains a few flow control digits or *flits*. The flit at the head of a packet, or the *header flit*, governs the route. As the header flit advances along the specified route, the remaining flits follow it in a pipeline fashion. If the header flit encounters a channel already in use, it is blocked until the channel is freed; the flow control within the network blocks the trailing flits.

Flow control of a network determines how resources such as buffers and channel bandwidth are allocated and how packet collisions are resolved. A resource collision occurs when a packet is unable to proceed because any resource it needs is held by another packet. Whether the packet is buffered, blocked in place, dropped, or misrouted to a channel other than the one it requires depends on the flow control policy. A good flow control policy should avoid channel congestion while reducing the packet latency.

Routing specifies the path of a packet through the network. For maximum system performance, a routing algorithm should have high throughput and low latency, avoid deadlocks, livelocks, and starvation, and be able to work well under various traffic patterns.

Routing can be classified as *deterministic* or *adaptive*. In deterministic routing, which is also referred to as *oblivious* routing, the path is completely determined by the source and destination addresses. Adaptive routing allows packet routing to take advantage of the network resources. Typically, in adaptive schemes, packets can take one of a number of paths. Packet routes are adapted based on dynamic network conditions or whether channels are busy or not when a header flit arrives. If the desired physical channel is busy, another channel leading towards the destination may be chosen. Several deadlock-free adaptive routing algorithms have been developed for wormhole routing [2, 6, 9, 10, 12].

Adaptive routing is one of the promising approaches to overcome the packet contention problem in the network using wormhole routing. In wormhole routing, a packet becomes blocked when it cannot acquire a desired channel because the channel is being used by another packet, and the blocked packets continue to hold previously acquired channels while waiting for others. With adaptive routing adopted, we observe that the packet is blocked even though

*This work was supported in part by the Ministry of Science and Technology under grant number NN12800 and in part by the Korea Science and Engineering Foundation through the Center for Artificial Intelligence Research in KAIST.

the preceding nodes it has visited may have available output channels. That is, in adaptive wormhole routing, the header flit discovers congestion when all forward paths are blocked and this is "too late" since the header flit must be blocked in place, adding to the congestion. This may result in low degree of adaptiveness and low utilization of channels, and hence low performance.

In this paper, we propose a new flow control policy, called the *drop-and-reroute*, for adaptive wormhole routing. The main idea behind the proposed flow control policy is that we may further exploit the adaptivity achieved by adaptive routing by allowing a blocked packet to be routed again in the preceding node. When a header flit fails to advance to the next node, it is dropped and routing is retried in the previous node from which it was routed. Drop-and-reroute flow control can be made possible by employing some buffering facility in the input channels. We present architectural supports and operation for drop-and-reroute flow control and show that the proposed flow control considerably improves the network performance of adaptive wormhole routing through extensive simulations.

A method similar to drop-and-reroute flow control, called *backtracking*, is described in [3]. Backtracking differs from drop-and-reroute flow control in that it is used in the *path setup* phase of circuit switching. The drop-and-reroute flow control is based on wormhole routing. Yet another similar approach, called the *B*-network, can be found in [13]. However, the *B*-network, which is a multistage interconnection network, employs backward links to provide the backward paths for the requests blocked.

The rest of this paper is organized as follows. Section 2 presents basic definitions and assumptions used in this paper. In Section 3, we introduce the main idea of drop-and-reroute flow control, and describe the architecture and the operation which support it. Extensive simulations studying the performance of drop-and-reroute flow control for various traffic patterns, packet lengths, and network sizes are described in Section 4. Finally, concluding remarks are given in Section 5.

2 Preliminaries

In this section, we represent basic definitions and assumptions used throughout this paper. These are very similar to the ones proposed by Dally [6] and Duato [9] with some exceptions. In the following, a channel may represent either physical or virtual channel [8, 5].

Definition 1 An *interconnection network* I is a strongly connected directed multigraph, $I = G(N, C)$. The vertices of the multigraph, N, represent the set of processing nodes, and the edges of the multigraph, C, represent the set of communication channels. Each channel c_i has an associated buffer denoted $buf(c_i)$ with capacity $cap(c_i)$.

In general, adaptive routing is decomposed into a *routing relation* and a *selection function*.

Definition 2 A packet is assigned a route through the network according to a *routing relation*, $R \subseteq C \times N \times C$. Given the channel occupied by the header flit of a packet and the destination node of the packet, the routing relation specifies a set of channels that may be used for the next step of the packet's route.

Definition 3 A selection function, $\rho(C, \alpha) \mapsto C$, is used to pick the next channel of the route from the elements of this set using some additional information α. This additional information may include the status of output channels.

Adaptive routing can be classified in various ways. A routing algorithm is said to be *minimal* if the path selected is one of the shortest paths between the source and destination nodes. Using a minimal routing algorithm, every channel visited will bring the packet closer to the destination. A *nonminimal* routing algorithm allows packets to follow a longer path, usually in response to current network conditions. If nonminimal routing is used, care must be taken to avoid *livelock*. An adaptive algorithm can be either *fully* or *partially* adaptive. Fully adaptive algorithms do not impose any restrictions on the choice of paths to be used in routing packets; in contrast, partially adaptive algorithms allow only a subset of available paths in routing packets.

Definition 4 The number of flits in the buffer for channel c_i will be denoted $size(c_i)$. Note that all of them belong to the same packet[see the following assumption 5].

Definition 5 The numbers of flits, which belong to the same packet, routed from and arrived at the channel c_i are denoted $outf(c_i)$ and $inf(c_i)$, respectively.

Definition 6 Consider a sequence of channels, $c_s \rightarrow \cdots \rightarrow c_i \rightarrow c_j$, where c_s represents the first channel routed by a packet p_i and c_j the current channel occupied by the header flit of p_i. Then, the preceding channel of c_j, which is c_i, is denoted $prec(c_j)$.

The following assumptions are made in this paper.

1. Wormhole routing is used.

2. Deadlock-free, minimal, and fully adaptive routing is adopted. Therefore, the proposed flow control policy does not cause a deadlocked configuration since packets are always routed through the permissible output channels by the deadlock-free routing algorithm.

3. The head of a packet occupies a single flit.

4. Each channel has an input buffer of fixed capacity and the capacity of each buffer in the network is equal.

5. A buffer cannot contain flits belonging to different packets. After accepting a tail flit, a buffer must be emptied before accepting another header flit. The reason for this assumption is explained in Section 3.

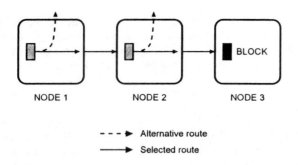

Figure 1: The packet is blocked while the preceding nodes which it has visited have available outgoing channels

3 Drop-and-Reroute Flow Control

In this section, we introduce drop-and-reroute flow control and describe architectural supports and operation for it

3.1 Observation

In wormhole routing, a packet is blocked in place where resource collision is occurred and waits until a channel is available. This situation is not a problem with deterministic routing since the blocked packet does not have any alternative route. However, when adaptive routing is used, one may observe that the packet is blocked while the preceding nodes which it has visited may have available output channels. Fig. 1 illustrates this situation. In Fig. 1, a fragment of a network is depicted with rounded boxes denoting nodes and small boxes denoting flit buffers. Black and gray boxes denote flit buffers that are filled with a header flit and the following data flits, respectively. The header flit is blocked in node 3 holding the input buffer, and hence the input channel, while the nodes 1 and 2 which it has passed through have available outgoing channels.

Therefore, if the header flit can release the holding channel (drop) and it can be routed again in one of the preceding nodes (reroute), it may bypass the congestion and thus reduce the packet latency. We call this flow control policy *drop-and-reroute*. With drop-and-reroute flow control, if the header flit is unable to proceed, it is dropped, instead of being blocked, and rerouting is conducted in the preceding node from which it was routed.

3.2 Architectural Supports

In most multicomputers using wormhole routing, each node contains a separate *router* to handle communication among nodes. Fig. 2 shows a generic wormhole router with five input channels and five output channels [15]. One of the input/output channel pairs are connected to the local processor. A crossbar switch is normally used to allow all

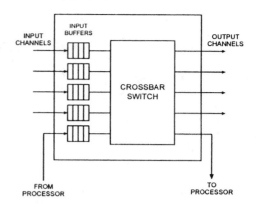

Figure 2: A generic wormhole router architecture

possible connections between the input and output channels within the router.

To support drop-and-reroute flow control, three architectural changes are made to the generic router. The first is the modification of the input buffer as shown in Fig. 3. Fig. 3(a) shows a conventional input buffer organized into a first-in, first-out (FIFO) queue. A network using drop-and-reroute flow control partitions the original input buffer into an *input buffer* and a *reroute buffer* as shown in Fig. 3(b) and (c). Fig. 3(b) and (c) show the cases that the capacities of reroute buffers are one and two, respectively. To discriminate between the original input buffer and the partitioned one, we call the former a *channel buffer* for the remainder of this paper. Note that, in spite of this change, the crossbar switch does not need to be modified.

For this modification, we assume that an input buffer can only contain flits belonging to the same packet. That is, after accepting a tail flit, the buffer must be emptied before accepting another header flit. This assumption is required because, otherwise, the capacity of a reroute buffer must be larger than the one of an input buffer. This assumption is not novel and has been used in [9].

The role of a reroute buffer is to preserve the routed flits from an input channel up to its capacity. When rerouting is necessary, the flits from this reroute buffer, not from the input buffer, is used. The reroute buffer also operates in the FIFO manner with the exception that the flits in it must be preserved even after they have been rerouted successfully since the rerouted packet may experience blocking and require drop-and-reroute flow control again. The reroute buffer is freed as soon as the number of routed flits from the channel exceeds the capacity of it. Note that all those flits belong to the same packet.

To prevent the previously selected routes from being chosen again by the rerouted packet, a set of registers is necessary for storing them. We call the set of registers *history vector* and this is the second change. Note that, in 2D meshes and with minimal adaptive routing, this storage requirement is not an additional overhead since a register for preserving the route selected by the header flit is needed so that it can be used by subsequent data flits which do

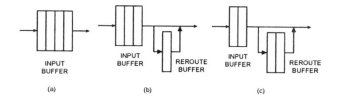

Figure 3: (a)A conventional input buffer organized into a single FIFO queue. (b)(c) A network using drop-and-reroute flow control organizes its buffers into input and reroute buffers.

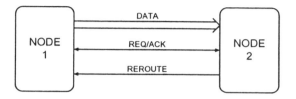

Figure 4: Signals between two routers.

not carry the routing information.

The third change is the addition of the "reroute" signal to each channel, which sends, in the reverse direction, the information indicating that the flit dropping is occurred and the packet must be rerouted. This additional signal is illustrated in Fig. 4. In this figure, we assume that each router communicates asynchronously with a neighbor one, and each router is clocked by its own local clock. The REQ/ACK signal associated with the channel represents this asynchronous operation between routers.

3.3 Operation

Before describing the operation of drop-and-reroute flow control, some definitions are needed.

Definition 7 For each channel c_i, the channel buffer $buf(c_i)$ is partitioned into an input buffer $buf_i(c_i)$ and a reroute buffer $buf_r(c_i)$ with capacities $cap_i(c_i)$ and $cap_r(c_i)$, respectively.

Definition 8 The numbers of flits in the input and reroute buffers will be denoted $size_i(c_i)$ and $size_r(c_i)$, respectively. Note that all the flits in the buffer for channel c_i belong to the same packet.

A router which supports drop-and-reroute flow control operates in three modes, *normal mode, reroute mode,* and *unified mode,* on a channel base. The major difference among the three modes is that which buffer is used. Normal mode uses the input buffer, reroute mode the reroute buffer, and the unified mode uses both input and reroute buffers as a single channel buffer. Detailed operation protocol of drop-and-reroute flow control is as follows.

Normal Mode Initially, each input channel c_j in node y operates in normal mode. When a router receives a header flit, either from the local processor or from an input channel connecting it to another router, a routing relation R and a selection function ρ are invoked to select the next channel on which to forward the packet. If an available outgoing channel is found, the header flit is copied into $buf_r(c_j)$ and routed through the channel, simultaneously. If the header flit fails to obtain an outgoing channel, it is dropped, i.e., $buf_i(c_j)$ associated with it is cleared, and a "reroute" signal is sent back to the preceding node x. Note that this sequence effectively releases the input channel c_j in node y which was once held by the header flit.

If the flit in the input channel c_j is not a header, it is transmitted along the path preserved by its header flit copying itself into $buf_r(c_j)$ while $size_r(c_j) < cap_r(c_j)$. If $outf(c_j) > cap_r(c_j)$, $buf_r(c_j)$ is overflowed and cannot be used for the purpose of rerouting any more for the packet. Therefore, $buf_r(c_j)$ is cleared and the mode of c_j is changed to unified.

Unified Mode During the unified mode, the input channel c_j operates in the same way as normal mode except that both $buf_i(c_j)$ and $buf_r(c_j)$ are used as a single channel buffer $buf(c_j)$. This effectively increase the capacity of the input buffer by the size of the reroute buffer. When a tail flit is routed out from this unified (channel) buffer $buf(c_j)$, the input channel c_j is returned to normal mode.

Reroute Mode As soon as the preceding node x receives the reroute signal, it releases the output channel c_j associated with the signal and sets the mode of the input channel c_i (which has a connection in the crossbar switch with the output channel c_j) to reroute mode. Note that $c_i = prec(c_j)$. This connection has been established in the previous cycle by the header flit. In reroute mode, flits from $buf_r(c_i)$ are considered. First, the header flit in $buf_r(c_i)$ is decoded and permissible output channels are selected again according to the routing relation R. Among several available output channels, the one which has not been chosen by the current packet is selected. Once a route is set up, flits in $buf_r(c_i)$ are routed. After successful routing of all the flits in $buf_r(c_i)$, the input channel c_i is reset to normal mode.

If the header flit fails to find such an output channel, because either the required output channels are used by other packets or there is no available flit buffer in the neighbor node, it is blocked in place or dropped and rerouted again according to $inf(c_i)$. If $inf(c_i) > cap_r(prec(c_i))$, the packet is blocked in place waiting for a channel to be available since the condition means that $buf_r(prec(c_i))$ is overflowed and used as a channel buffer. Note that the above condition can be determined locally since $cap_r(prec(c_i)) = cap_r(c_i)$. Otherwise, more rerouting is possible, that is, the channel c_i is released, the reroute signal is sent backward, and the mode of channel c_i is changed to normal. When

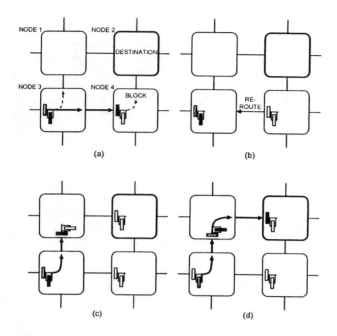

Figure 5: An example of drop-and-reroute flow control: (a) The packet is blocked at node 4. (b) The blocked header flit is dropped and the reroute signal is sent to node 3. (c) The header flit in the reroute buffer of node 3 is rerouted to node 1. (d) The packet arrives at its destination node, which is node 2.

the packet has tried all the available output channels, the history vector is reset so that it may be routed without restriction in selecting output channels.

Example Fig. 5 depicts a simplified example of the steps involved in drop-and-reroute flow control. In the figure, each input channel consists of input and reroute buffers of capacity one, respectively. Black and gray small boxes denote buffers that are filled with a header flit and the following data flits, respectively. Thick solid and thin dashed arrows denote a selected and another available routes, respectively. Thin solid arrow represents a reroute control signal. Fig. 5(a) shows the case that the packet is blocked at node 4 because the output channel toward its destination node (node 2) is not available. Note that, in the previous cycle, at node 3, the packet could be routed to two output channels, one is toward node 1 and the other toward node 4. The blocked header flit is dropped and a reroute signal is transmitted to node 3 in Fig. 5(b). In Fig. 5(c), the header flit in the reroute buffer of node 3 is rerouted to node 1, and finally, the packet arrives at its destination node in Fig. 5(d). In Fig. 5(d), node 3 shows the case that the input and reroute buffers are used as a unified channel buffer. Note that the channel outgoing from node 3 toward node 4 is free and hence can be allocated to other requesting packets. Therefore, no channel is wasted with drop-and-reroute flow control.

4 Experimental Results

In this section, we measure the effect of drop-and-reroute flow control on network performance by simulation.

4.1 Routing Algorithm Used

To show the effect of the proposed flow control, we simulate a minimal fully adaptive routing algorithm. The routing algorithm used is the one which Jesshope et al. [12] have proposed for 2D meshes. A 2D mesh network has at least four pairs of network input/output channels, one each along the +X(or East), -X(or West), +Y(or North), and -Y(or South) directions. First, packets are classified according to whether they need to be routed NE, SE, SW, or NW and a separate *virtual network* is created to route each class with full adaptiveness. For example, for the NE class, an N or E virtual channel is created for each N or E physical channel. With two virtual channels for each physical channel, the four virtual networks can be constructed. Then, the algorithm routes a NW packet along the first set of N and W channels, a NE packet along the second set of N and E channels, a SE packet along the the first set of S and E channels, and a SW packet along the second set of S and W channels. We call this algorithm the *double-xy routing algorithm* as is called in [11].

4.2 Simulation Model

We simulate 2D meshes at the flit-level. A flit transfer between two nodes is assumed to take place in one time unit. The network is simulated synchronously moving all flits that have been granted channels in one time step and then advancing time to the next step [5]. A packet consists of a header, some data, and a tail flits. For the sake of simplicity, we assume that the header occupies a single flit; thus a packet may be routed as soon as the first flit arrives at a node. The packet generation rate is constant, and the same for all the nodes. Packets that are blocked from immediately entering the network are queued at the source node. Packets that arrive at a destination node are immediately consumed. When multiple input channels contain header flits waiting for the same available output channel, we select one in a *round-robin* fashion. This policy is fair and hence prevents *starvation*. Since we use adaptive routing, a header flit in an input channel has multiple available output channels. When selecting an output channel, we decide in favor of the dimension closest to the current dimension [6].

Physical channels are split into two virtual channels according to the routing algorithm used. Each virtual channel has buffers of equal size. The total buffer size associated with each physical channel is held constant, which is eight flits of storage per physical channel, whether drop-and-reroute flow control is used or not. Virtual channels are assigned to the physical channel cyclically, only if they can transfer a flit. So, channel bandwidth is shared among the virtual channels requesting it.

4.3 Traffic Patterns

Network performance is greatly affected by the packet traffic pattern, which is application dependent. We consider one random, which is *uniform*, and two permutation, which are *bit-reversal* and *transpose*, patterns. In the uniform random pattern, each packet is sent to any of the other nodes with equal probability. Although uniform random traffic may not represent any actual workload, it is useful because it is the "natural" workload thought of, and it is the standard workload used in comparing router performance. In the permutation patterns, a node x sends packets with destination $\sigma(x)$, where σ is the permutation of the node's address. Two permutation patterns are simulated. In the bit-reversal pattern, the destination address is the bit-reversal of the source address. In the transpose pattern, the destination address is obtained by splitting the source address into halves and swapping these halves.

4.4 Performance Measures

The most important performance measures of the network are latency and throughput. The latency of a packet is the elapsed time from packet creation until the last flit of the packet is accepted at the destination. Network latency is the average packet latency under specified conditions. Source queueing time is included in the latency measurement. Network throughput is the number of packets the network can deliver per time unit and is given as a fraction of theoretical network capacity [5].

4.5 Simulation Results

We simulate 16×16 2D mesh networks with uni-directional channels. We also simulate 32×32 meshes to show that the proposed flow control policy works equally well in the large-scale networks. Fixed length packets are simulated and 20-flit packets are mainly used. To investigate the effect of the packet length on the network performance, we simulate additional two different lengths, 10 and 30, which represent short and long packets, respectively. Each simulation was run for a total of 40,000 flit times. Statistics gathering was inhibited for the first 10,000 flit times to avoid distortions due to the start up transient [5].

The first simulations investigate the performance improvements caused by drop-and-reroute flow control. The results from the uniform random, bit-reversal, and transpose traffic patterns are shown in Figs. 6 through 8, respectively. The figures show the latencies as a function of the throughputs when the network has 256 nodes and the packet length is 20 flits. Recall that the throughput is measured as a fraction of the network capacity. In the figures, "double-xy" and "drop-and-reroute" represent the results without and with drop-and-reroute flow control, respectively. Note that, not only at high throughputs but also at low throughputs, drop-and-reroute flow control achieves lower latencies than without it under all traffic patterns considered.

Fig. 6 shows the result under uniform traffic pattern. Under uniform random traffic, drop-and-reroute flow control shows only slight performance improvement. The reason for this slight improvement is that more adaptiveness is not much productive under uniform traffic pattern [11, 6]. However, under bit-reversal and transpose traffic patterns, the use of drop-and-reroute flow control greatly increases the throughput achieved by the double-xy routing algorithm, as is shown in Figs. 7 and 8. Drop-and-reroute flow control improves the throughput by around 25% and 20% under bit-reversal and transpose traffic patterns, respectively.

The next simulations explore how the performance of drop-and-reroute flow control varies with the reroute buffer capacity, packet length, and network size. In the following, we only show the results under bit-reversal traffic pattern. However, we had similar results under other traffic patterns.

Fig. 9 shows the latency as a function of throughput varying the capacity of the reroute buffer for 256 nodes and 20-flit packet length under bit-reversal traffic pattern. Although drop-and-reroute flow control improves the performance of the adaptive routing algorithm, increasing the capacity of the reroute buffer, i.e., the steps allowed to be dropped and rerouted, does not affect the performance significantly.

It is interesting to see the effect of the packet length. Fig. 10 shows the results varying the packet length for 256 nodes under bit-reversal traffic pattern. In the figure, the solid and dashed curves represent the results when packet lengths are 10 and 30, respectively. As can be seen, drop-and-reroute flow control achieves considerable reductions in latency for both packet lengths of 10 and 30.

The last simulation is to know whether this performance improvement continues for the large network size. Commonly considered network size for a large network is 1024 nodes. Fig. 11 shows the latency as a function of throughput for 256 and 1024 nodes when the packet length is 20 under bit-reversal traffic pattern. Note that the latency increases as the network size is increased. However, drop-and-reroute flow control still improves the performance of the double-xy routing algorithm by around 25%.

5 Conclusion

In this paper, we have proposed a new flow control policy, called drop-and-reroute, for adaptive wormhole routing and described architectural supports and its operation. When a packet experiences the resource collision, drop-and-reroute flow control allows it to be rerouted in one of the preceding nodes, and hence increases the channel utilization and further improves the network performance of adaptive routing.

To support drop-and-reroute flow control, 1) input channel buffers have been partitioned into *input* and *reroute* buffers, 2) *history vector*, which is a set of registers for storing previous routes that have been selected by the specific packet, has been added, and 3) a *reroute* signal line per channel has been added. However, even with these changes, the crossbar switch is not changed. In wormhole routing, at least a single register is necessary for preserving the route that the header flit has set up since the trailing data flits do not have the routing information. Therefore, in 2D meshes and with minimal adaptive routing, the hardware overheads for drop-and-reroute flow control are only extra control logic and a single signal line per channel.

To show the performance improvement by drop-and-reroute flow control, we have simulated a minimal fully adaptive routing algorithm under various parameters. Our simulation results say that under nonuniform traffic patterns, bit-reversal and transpose, drop-and-reroute flow control considerably improves the network performance of the adaptive routing algorithm without regard to the packet length and the network size. However, it was observed that the effect of increasing the capacity of the reroute buffer was insignificant.

There are several directions for future research. One of the most important is applying the proposed flow control policy to nonminimal adaptive routing algorithms and evaluating the performance under faulty networks. Also, identifying realistic workload distributions is very important, so that the results of future simulations can be more meaningful. Finally, measuring the effect of drop-and-reroute flow control for various topologies, such as high-dimensional meshes, tori, and hypercubes, would be an obvious extension of our work.

References

[1] W. C. Athas and C. L. Seitz, "Multicomputers: Message-Passing Concurrent Computers," *IEEE Computer*, vol. 21, no. 8, pp. 9–24, August 1988.

[2] A. A. Chien and J. H. Kim, "Planar-Adaptive Routing: Low-cost Adaptive Networks for Multiprocessors," in *Proceedings of the 19th Annual International Symposium on Computer Architecture*, pp. 268–277, IEEE Computer Society, April 1992.

[3] E. Chow, H. Madan, J. Peterson, D. Grunwald, and D. A. Reed, "Hyperswitch Network for the Hypercube Computer," in *Proceedings of the 15th International Annual Symposium on Computer Architecture*, pp. 90–99, May 1988.

[4] W. J. Dally, "Performance Analysis of *k*-ary *n*-cube Interconnection Networks," *IEEE Transactions on Computers*, vol. 39, no. 6, pp. 775–785, June 1990.

[5] W. J. Dally, "Virtual-Channel Flow Control," *IEEE Transactions on Parallel and Distributed Systems*, pp. 194–205, March 1992.

[6] W. J. Dally and H. Aoki, "Deadlock-Free Adaptive Routing in Multicomputer Networks Using Virtual Channels," *IEEE Transactions on Parallel and Distributed Systems*, vol. 4, no. 4, pp. 466–475, April 1993.

[7] W. J. Dally *et al.*, "The J-Machine: A Fine-Grain Concurrent Computer," *Information Processing 89*, pp. 1147–1153, 1989.

[8] W. J. Dally and C. L. Seitz, "Deadlock-Free Message Routing in Multiprocessor Interconnection Networks," *IEEE Transactions on Computers*, vol. C-36, no. 5, pp. 547–553, May 1987.

[9] J. Duato, "A New Theory of Deadlock-Free Adaptive Routing in Wormhole Networks," *IEEE Transactions on Parallel and Distributed Systems*, vol. 4, no. 12, pp. 1320–1331, December 1993.

[10] C. J. Glass and L. M. Ni, "Adaptive Routing in Mesh-Connected Networks," in *Proceedings of 12th International Conference on Distributed Computing Systems*, (Los Alamitos, California), pp. 12–19, IEEE Computer Society, June 1992.

[11] C. J. Glass and L. M. Ni, "Maximally Fully Adaptive Routing in 2D Meshes," in *Proceedings of the 1992 International Conference on Parallel Processing*, vol. I, pp. 101–104, August 1992.

[12] C. R. Jesshope, P. R. Miller, and J. T. Yantchev, "High Performance Communications in Processor Networks," in *Proceedings of the 16th International Annual Symposium on Computer Architecture*, pp. 150–157, IEEE Computer Society, May 1989.

[13] K. Y. Lee and H. Yoon, "The B-Network: A Multistage Interconnection Network with Backward Links," *IEEE Transactions on Computers*, vol. 39, no. 7, pp. 966–969, July 1990.

[14] S. L. Lillevik, "The Touchstone 30 Gigaflop DELTA Prototype," in *Proceedings of the 6th Distributed Memory Computing Conference*, pp. 671–677, IEEE Computer Society, April 1991.

[15] L. M. Ni and P. K. McKinley, "A Survey of Wormhole Routing Techniques in Directed Networks," *IEEE Computer*, vol. 26, no. 2, pp. 62–76, February 1993.

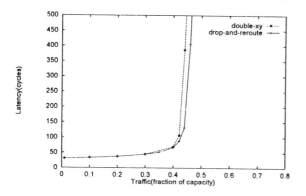

Figure 6: Latency versus offered traffic for 16 × 16 2D mesh networks under uniform traffic.

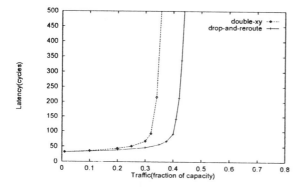

Figure 7: Latency versus offered traffic for 16 × 16 2D mesh networks under bit-reversal traffic pattern.

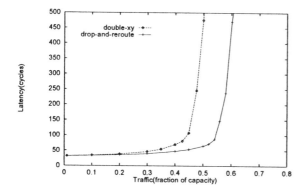

Figure 8: Latency versus offered traffic for 16 × 16 2D mesh networks under transpose traffic pattern.

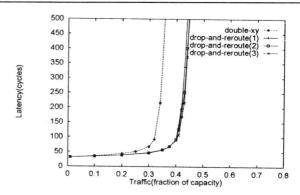

Figure 9: Effect of the reroute buffer capacity on drop-and-reroute flow control in 16 × 16 2D meshes under bit-reversal traffic pattern.

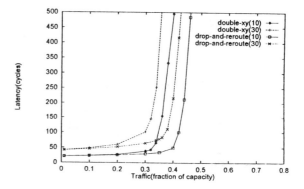

Figure 10: Effect of packet length on drop-and-reroute flow control in 16 × 16 2D meshes under bit-reversal traffic pattern.

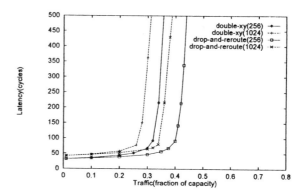

Figure 11: Effect of network size on drop-and-reroute flow control under bit-reversal traffic pattern.

Adaptive Virtual Cut-through as an Alternative to Wormhole Routing*

Hyun Suk Lee[†], Ho Won Kim[†], Jong Kim[‡], and Sunggu Lee[†]

[†]Dept. of Electrical Engineering

[‡]Dept. of Computer Science and Engineering

Pohang University of Science and Technology (POSTECH)

San 31 Hyoja Dong, Pohang 790-784, SOUTH KOREA

E-mail: {jkim,slee}@vision.postech.ac.kr

Abstract

An adaptive version of virtual cut-through is considered as an alternative to wormhole routing for fast and hardware-efficient inter-processor communications in multicomputers. Several possible versions of adaptive virtual cut-through are investigated using simulations, and it is shown that a fully adaptive minimal-path version is the most desirable. Our simulations also show that this method outperforms both deterministic and adaptive wormhole routing methods under both uniform random message distributions and clustered distributions such as matrix transpose. A hardware-efficient implementation of adaptive virtual cut-through has been implemented as a semi-custom-designed VLSI chip that requires only 2.3% more area than a comparable deterministic wormhole routing chip.

1 Introduction

The interconnection network is a critical component of a massively parallel computer system because information must be communicated quickly in order for the processing nodes of the parallel computer to cooperate in solving a given problem. Information may be sent as messages between the processing nodes of a distributed-memory multicomputer or as data between the processors and memories of a shared-memory multiprocessor.

The first-generation multicomputers of the mid-1980's (NCUBE/ten, iPSC, etc.) used "store-and-forward type" packet switching methods borrowed from the computer network field in order to perform the necessary message passing. However, since this communication was typically handled by software in the processor responsible for the computations, inter-processor communication was extremely slow (on the order of a few milliseconds per hop) and was a major limiting factor in the scalability of parallel programs.

Realizing this deficiency, the second-generation multicomputers (NCUBE2, iPSC/860, etc) turned to hardware-supported "cut-through" switching methods for fast inter-processor communications (on the order of microseconds). In this type of method, a dedicated hardware device, termed a *router*, is used to perform flow control without the intervention of the computation processor.

Two popular "cut-through" switching methods are *wormhole routing* [1] and *virtual cut-through switching* [2]. In wormhole routing (WR) and virtual cut-through (VCT), messages arriving at an intermediate node are immediately forwarded to the next node on the path without buffering, provided that a channel to the next node is available. This is referred to as a *cut-through* operation. A cut-through may be performed at an intermediate node as soon as the header of the message arrives with the destination information. Thus, with cut-through, only an on-line *flit*[1] buffer is required to examine the message header at each intermediate node. Thus, when the message has cut-through several nodes, the flits of the message become spread out in the flit buffers along the path used.

If cut-throughs are established through all intermediate nodes, a *circuit* is established to the destination, and the switching method resembles traditional *circuit switching*. If the message header is *blocked* at an intermediate node because the requested outgoing channel is unavailable, the message is kept in the network (i.e., in the on-line flit buffers at each node along the path up to the current node) in WR and completely buffered at the current node in VCT. Since sufficient buffer storage must be available for any message that reaches a node, VCT typically requires packetization of messages and buffer space partitioning. Wormhole routing, on the other hand, contributes to network congestion when busy channels are encountered. In the rest of this paper, it will be assumed that messages are sent in fixed-size packets.

Wormhole routing is the most popular communication method used in current commercial multicomputers due to its low hardware overhead (requiring only a

---

* This paper was supported in part by the Inter-University Semiconductor Research Center (Grant ISRC-94-E-2033), funded by the Education Ministry of Korea.

---

[1]A *flit* as defined in [1] is the smallest unit of information that a node may refuse to accept. Thus, in a simple implementation, the number of bits in a flit may correspond to the number of data lines in a single physical channel.

single flit buffer per incoming channel) and high performance. However, VCT can typically achieve higher performance than WR for the following reason: when a packet cannot cut-through an intermediate node, the packet is buffered at that node instead of being kept in the network as in WR. This effect becomes particularly evident with heavy network traffic. The main reason that VCT has not been as popular as WR is that the required buffering capability of the former method has been seen as a large hardware overhead.

Due to its relative unpopularity, VCT has not been as heavily researched as WR. In this paper, we study several adaptive forms of virtual cut-through as possible alternatives to WR. It is shown that adaptive VCT has several attractive features when compared to WR: negligible overhead for avoiding deadlock (a situation in which no packet can proceed towards its destination), use of 100% of the free channels, good performance with heavy network traffic, and good performance with random or arbitrary locations of destination nodes. The hardware complexity issue is also addressed by showing that VCT can be implemented with little hardware overhead (in some cases, less than WR), by exploiting the source buffer in the network interface to the router.

The rest of the paper is organized as follows. Section 2 presents background concepts. Section 3 describes the shortcomings of WR and presents the proposed adaptive VCT methods. Section 4 describes a hardware-efficient implementation of adaptive VCT and the design and testing of an adaptive VCT router chip. Section 5 presents the results of simulations on a 2-dimensional mesh network. The paper concludes with Section 6.

2 Background

The type of parallel computer assumed will be a multicomputer with a *direct interconnection network*, in which a communication link can only be used by the two end-nodes that it is connected to. Examples of direct interconnection networks include 2-dimensional meshes, tori, hypercubes, cube-connected cycles, etc. A *k-ary n-cube* is a generalization of 2-dimensional meshes, tori, and hypercubes in which there are n dimensions and k nodes in a single dimension. K-ary n-cubes can be constructed as torus or mesh networks, where the nodes along one dimension are connected in a cycle in torus networks and in a linear array in mesh networks.

Adjacent processing nodes are assumed to be connected by two directed communication links in opposite directions (although undirected links are also possible, directed links are assumed for simplicity). A communication link consists of one or more data lines and several control lines. One or more *physical channels* may occupy a single communication link by using time multiplexing or by partitioning the set of data lines. One or more *virtual channels* may occupy a single physical channel by using separate flit buffers and control lines for each virtual channel. When two or more virtual channels are active over a single physical channel, they must be time-multiplexed; however, when only one of the virtual channels is active, it can utilize the full bandwidth of the physical channel. For simplicity of discussion, unless otherwise stated, it will henceforth be assumed that there is one physical channel per communication link and one virtual channel per physical channel.

All communication methods must deal with the problems of *deadlock*, *livelock*, and *starvation*. *Deadlock* occurs when no packet can progress toward its destination because it is blocked by some other packet, *livelock* occurs when a packet moves continuously without ever reaching its destination, and *starvation* occurs when a node is never able to inject its packet into the network because other packets occupy all available outgoing channels. Wormhole routing is particularly susceptible to deadlock because blocked packets remain in the network, thereby reducing the number of free channels. Although much less likely, deadlock is also possible with VCT as the packet buffers at intermediate nodes can become full. Livelock can occur with non-minimal algorithms, in which a packet is permitted to move further away from its destination. Starvation can be prevented by using a fair scheme (such as round robin) for assigning packets to output channels.

The use of virtual channels was introduced by Dally and Seitz [1] as a means of guaranteeing *deadlock-free* WR. The WR algorithms introduced in [1] are *deterministic* algorithms, in which the routing path used is completely determined by the source and destination node addresses. In [3], Dally showed that virtual channels could also be used with a deterministic WR algorithm to significantly reduce the probability of blocking, and thus increase the throughput of the network.

Realizing the need for adaptability in routing around congested or faulty nodes and links, several researchers [4, 5] used virtual channels to produce *adaptive* deadlock-free WR algorithms, in which a packet is permitted to change its routing path if a blocked channel or faulty node/link is encountered. Adaptive algorithms can be classified into *minimal* [4, 6] and *non-minimal* [5, 7] algorithms, where a minimal algorithm always uses shortest-length paths to each destination, and *fully adaptive* [4, 5] and *partially adaptive* [6, 7] algorithms, where a partially adaptive algorithm differs from a fully adaptive algorithm in that only a subset of the possible paths may be used. Some adaptive algorithms use one *virtual network* (a set of virtual channels) to route adaptively and another *virtual network* to route deterministically in order to escape possible deadlock situations [5, 8]. In all of these WR methods, even though several physical or virtual channels may be unoccupied on minimal paths to the destination, only a subset of the unoccupied channels may be used to route any given packet (the remaining physical or virtual channels are necessary to guarantee freedom from deadlock).

3 Adaptive Virtual Cut-through

Wormhole routing methods have several deficiencies when compared to VCT. First, WR in general has a lower saturation throughput than a comparable VCT method as blocked packets contribute to network congestion in WR. This problem is significantly alleviated by the use of multiple virtual channels. Second, even with the use of virtual channels, 100% of the free channels are not available for use by a given packet as some of the channels have to be reserved in order to prevent deadlock. Third, fully adaptive routing in torus networks is difficult and requires a large proportion of the virtual channels to be used for deadlock avoidance. Although virtual channels do not waste channel bandwidth, virtual channels do occupy valuable hardware resources because each virtual channel requires a separate on-line flit buffer and control line and contributes to the complexity of the arbitration logic within the router (refer to Section 4).

For the above reasons, adaptive VCT is proposed as an alternative to WR. When originally proposed in [2], VCT was described as a deterministic routing algorithm. In deterministic VCT, when the head of a packet is received at an incoming channel, the header is decoded to determine the "one" outgoing channel to connect to. If that outgoing channel is busy, then the packet is buffered at the current node. This method is analogous to deterministic WR as originally described by Dally and Seitz [1]. Thus, it is natural to consider an adaptive version of virtual cut-through in order to be able to navigate around congested or faulty nodes/links.

In order to facilitate discussion of the many possible forms of adaptive VCT, a general switching model (shown in Fig. 1) is proposed. When the head of a packet is received at an incoming channel, the destination node information is extracted from the header. A connection is then attempted to the first-choice outgoing channel. If successful, then a cut-through is established through the current node. However, if the requested outgoing channel is occupied by another packet as in Fig. 1(a), then a connection is attempted to the second-choice outgoing channel. This process is repeated until a cut-through is achieved or all permissible outgoing channels are exhausted. In the latter case, which corresponds to the situation in Fig. 1(a), the packet blocks in the network (as in WR) until one of the permissible outgoing channels becomes available or the "permitted waiting time" expires. In the latter case, which corresponds to the situation in Fig. 1(b), the packet is received at the current node and inserted into the source buffer of the current node just as if the packet originated there (the circuit implementation is discussed in Section 4). The packet at the head of the source buffer waits until one of the outgoing channels permitted to it becomes available, and then exits through that outgoing channel, as shown in Fig. 1(e–f).

The main philosophy behind the development of adaptive VCT is a twist on the common philosophy behind computer design: "make the fast case common." Thus, since the fastest case is when a cut-through is

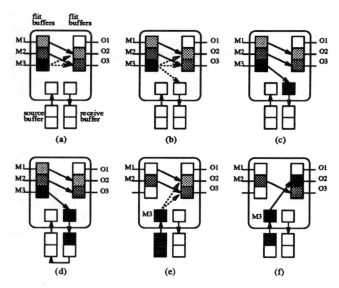

Figure 1: A general switching model for adaptive VCT.

established through all intermediate nodes, we would like to increase the probability of cut-through as much as possible. This naturally leads to the concept of *deflection VCT*, an extension of the *deflection routing* method originally proposed for computer networks with packet switching [9]. In this method, a packet arriving at an incoming channel is permitted to use *any* outgoing channel available, even the one directed "backwards," with preference given to channels directed toward the destination node. If all outgoing channels are occupied, then the entire packet must be buffered at the current node. Although conceptually desirable, deflection VCT was found to perform poorly in computer simulations (shown in Section 5) due to the many deflected packets present in the network. This led us to consider alternate forms of adaptive virtual cut-through.

Adaptive VCT variations can be developed on the basis of several parameters. For the sake of convenience, let us restrict the discussion to 2-dimensional mesh and torus topologies. Then, the number of permitted outgoing channels can be varied from one to four, with the corresponding algorithms named as A_{nk} ($k = 1$ to $k = 4$). Channels that take the packet closer to its destination will be considered before channels that move the packet away from its destination, with ties handled randomly. The A_{nk} algorithms are shown in Fig. 2 — A_{n4} corresponds to deflection routing. If the number of permitted outgoing channels is ≥ 2, then adaptive non-minimal routing is being used. For fully-adaptive minimal routing, the number of permitted channels is one or two, depending on the relative location of the destination node. If two outgoing channels are available on minimal paths to the destination node, two policies can be adopted. In order to maximize the probability of routing around congested or faulty nodes/links, a "zig-zag" pattern can be followed as shown in Fig. 3(a) (Algorithm A_{mc}). How-

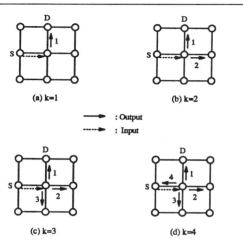

Figure 2: Algorithms A_{nk} with (a) $k = 1$, (b) $k = 2$, (c) $k = 3$, and (d) $k = 4$.

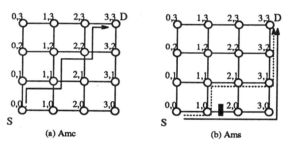

Figure 3: Fully-adaptive minimal algorithms (a) A_{mc} and (b) A_{ms}.

ever, since this involves a difficult control algorithm at the routers, a simpler "priority-X-direction" pattern can also be used as shown in Fig. 3(b) (Algorithm A_{ms}). Finally, all of the above variations on adaptive routing can be used with different values for the permitted waiting time before buffering blocked packets (to permit cut-through if a permitted outgoing channel becomes available quickly).

The above adaptive VCT algorithms deal with deadlock, livelock, and starvation in the following manner. Since VCT is being used, the deadlock conditions are identical to packet switching – thus the methods used to avoid deadlock in packet switching can be used with these algorithms also [10, 11]. The architectural design in Section 4 and the simulation results in Section 5 will show that we very rarely have to resort to these deadlock avoidance techniques. The fully adaptive minimal routing algorithms A_{mc} and A_{ms} are inherently livelock-free. For the non-minimal routing algorithms, a policy of giving priority to channels which reduce the distance to the destination results in probabilistically livelock-free routing [9, 12] — however, in general, non-minimal algorithms will suffer from the possibility of livelock. Finally, starvation will be avoided by using a round-robin polling policy to connect incoming channels and the source buffer to their

respective outgoing channels.

4 Communication Architecture and Router Design

The proposed interconnection network communication architecture for supporting adaptive VCT is based on the communication architecture of the Intel Paragon, shown in Fig. 4. As can be seen from this

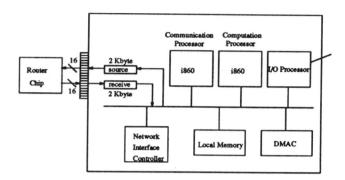

Figure 4: The network interface for the Intel Paragon.

figure, there is a network interface controller (NIC), source and receive buffers, and a dedicated communication processor (a type of I/O processor) responsible for preparing and sending the packets out onto the network. Packets are initially stored in the source buffer before being sent out to the router to be routed to the destination node. Packets to be received by the current node are initially buffered in the receive buffer before being transferred to local memory. In the Intel Paragon, the source and receive buffers are 2 Kbyte buffers and the links to the router are 16-bits wide.

To support adaptive VCT, the communication processor must handle some extra work and the router must be redesigned. Upon receiving a new packet in the receive buffer, the NIC must examine the header to determine if the packet is destined for the current node or "in transit." In the latter case, the packet must be rerouted to the source buffer to be transmitted when one of its permitted outgoing channels becomes available. (Note that bandwith matching requirements imply that the data transfer rates in the source and receive buffers must match the data transfer rates within the router network.) In order to perform this rerouting efficiently without having to contend for the local bus, a dedicated path must be established between the receive and source buffers. If the source buffer is full, the packet must overflow into a reserved portion of local memory, which is used as an "extended buffer." If the source buffer can hold several packets, buffer overflow is rare as our simulations show that the network saturates when an average of one packet is buffered at each node. In order to guarantee freedom from deadlock, the extended buffer must be partitioned as in [10] or [11].

To demonstrate the feasibility of adaptive VCT, the A_{ms} algorithm was implemented in a semi-custom-designed VLSI router chip. For a complete implementation, the NIC and source and receive buffers of the processing node must also be redesigned; this is the target of our next VLSI design project. A block diagram of our router chip, designed for a 2-dimensional mesh or torus network, is shown in Fig. 5. As can

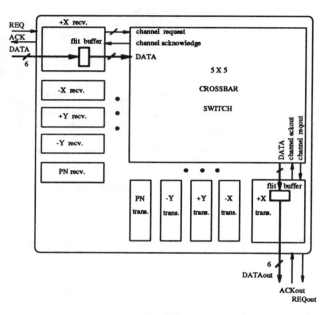

Figure 5: Block diagram of the A_{ms} adaptive VCT router chip.

be seen, there are flit buffers at the receivers and the transmitters and a crossbar switch connecting the receivers to the transmitters. The crosspoint logic includes arbitration logic to handle contention for common transmitters (outgoing channels). Finite state machine control logic (not shown for lack of space) in the receiver implements algorithm A_{ms} by first attempting to connect to the first-choice transmitter, then the second-choice transmitter (if one exists), then the PN (processing node) transmitter (for reception) if none of the first two choices are available. Note that in our prototype chip, 6-bit-wide data lines (instead of 16-bits as in the Intel Paragon) were used to connect to other routers and the local processing node due to a shortage of I/O pins. Note also that the "permitted waiting time" used is 0.

Our router chip is based on the torus routing chip (TRC) [13], which uses deterministic WR on a 2-dimensional torus network. The main differences are in the receiver control logic and in the use of one flit buffer for each incoming or outgoing channel. Because the TRC uses two virtual channels for each physical channel to guarantee freedom from deadlock, it requires two flit buffers and the associated control lines for each incoming or outgoing channel. In addition, with multiple virtual channels, either the crossbar switch has to become larger or multiplexers and more complex arbitration logic has to be used [3]. Thus, when compared to a deterministic WR chip, our router chip requires a

small overhead for the receiver control logic but does not require multiple virtual channels to avoid deadlock. If multiple virtual channels are used to enhance performance as in [3], *all* of the virtual channels may be used by any packet (unlike WR).

The router chip was designed using Mentor Graphics and GDT design tools, manufactured using Samsung's 1.0 μm CMOS process, and successfully tested in December of 1994. The arbitration capabilities of the crossbar switch were successfully tested and the A_{ms} adaptive routing algorithm worked as designed. There was one minor flaw in that the Y- transmitter was inadvertently connected to an input pad and thus failed to function. The chip required 33,000 transistors, consumed 800 mW of power, and was tested successfully up to a clock frequency of 12.5 MHz. The layout for this chip is shown in Fig. 6. The placement of the various blocks for the router design are approximately the same as in Fig. 5.

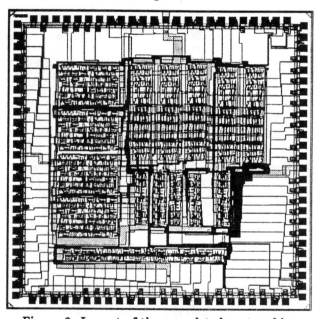

Figure 6: Layout of the completed router chip.

The hardware overhead required for adaptive VCT was estimated by measuring the layout area required by our chip (excluding the pads, which lie on the periphery of the die) as compared to the layout area required by a comparable deterministic WR chip. This comparison is shown in Table 1. When our chip is compared to a deterministic WR chip without multiple virtual channels, the overhead is 2.3%. When our chip is compared to a deterministic WR chip with two virtual channels per physical channel (to avoid deadlock), our chip is smaller by 85.0%, even though the size of the crossbar is kept constant [3]. Since this latter chip has higher performance, however, it should perhaps be compared to an adaptive VCT chip with two virtual channels per physical channel – in this case, the overhead for our chip is 9.5%. It should be noted that this comparison is only a rough estimate, as the layout is not optimized and the use of semi-custom-design tends to waste silicon area.

Table 1. Router Chip Layout Area Comparison

Routing algorithm	Total Area (mm^2)
(1) Deterministic WR	18.699 (4.278 × 4.371)
(2) Adaptive VCT	19.136 (4.367 × 4.382)
(3) Deterministic WR with 2 virtual chan.	35.405 (6.098 × 5.806)
(4) Adaptive VCT with 2 virtual chan.	38.763 (6.147 × 6.306)

5 Simulations

Computer simulations are used to compare the performance of the various adaptive VCT and WR methods. A mathematical analysis of adaptive VCT is extremely involved and is the subject of another paper which we are currently working on. Simulations were conducted for various topologies, network sizes, and multiple virtual channels. However, as all of our simulation results showed similar trends, the results for an 8 × 8 2-dimensional mesh topology with one virtual channel per communication link are shown in this section. A 2-dimensional mesh topology was chosen to allow meaningful comparison with deterministic and adaptive WR algorithms as most WR algorithms do not work on a 2-dimensional torus with only one virtual channel per communication link. Note that the adaptive VCT algorithms, by contrast, work efficiently (without deadlock) on any network topology, using only one virtual channel per communication link.

The following parameters and assumptions are used in our simulations. A fixed packet size of 20 flits is used. Packets are transmitted one flit at a time through the network, with one flit being transferred in one clock cycle time. Each node generates new packets according to a Poisson distribution. Destination nodes are selected using two methods: based on a random uniform distribution and based on a matrix transpose operation. Matrix transpose is an operation which is representative of a "clustered distribution" that is not efficiently routed by a deterministic routing algorithm. Finally, in order to avoid the issue of deadlock avoidance, it is assumed that a source buffer of infinite capacity is used to buffer blocked packets. In a practical implementation, a method such as [10, 11] must be used to guarantee freedom from deadlock. It is noted that the source buffer does not have to be too large in practice as our simulation results show that the network becomes saturated when an average of one packet is buffered at each node.

The simulation results are presented with the average packet inter-generation time (at each node) along the X-axis and the average latency on the Y-axis. The latency is defined as the time from when a packet is first generated and stored in the source buffer to when the tail of the packet leaves the network and enters the receive buffer of the destination node. All times are measured in terms of clock cycles. Each simulation is executed for 50,000 cycles. However, measurements

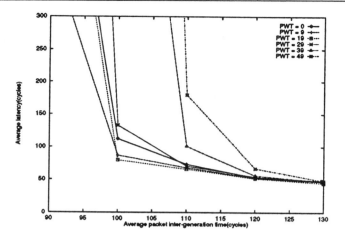

Figure 7: Average latency for the A_{mc} algorithm with different PWT values given an 8 x 8 mesh and a uniform destination node distribution.

are only taken after the first 10,000 cycles to avoid transient startup effects.

Fig. 7 shows the results of experiments with the A_{mc} algorithm, a uniform destination node distribution, and various values for the permitted waiting time (PWT) before buffering blocked packets. It can be seen that the best performance is obtained with PWT = 19 cycles, which corresponds to the packet length minus 1. However, the curves for PWT = 0 through PWT = 29 are close together until the average packet inter-generation time (at each node) is 100 cycles or less. Thus, since PWT = 0 is the simplest to implement and shows a minimal loss of performance at unsaturated conditions, it is the policy adopted in all of the following simulations.

Fig. 8 shows the results of experimenting with the A_{mc} and A_{ms} algorithms given uniform and matrix transpose destination node distributions. The A_{ms} algorithm is slightly better than the A_{mc} algorithm given a uniform distribution, while the A_{mc} algorithm is generally better than the A_{ms} algorithm given a matrix transpose distribution. Thus, since the results are mixed and the differences are small, in the general case, A_{ms} is recommended over A_{mc} due to its simpler circuit implementation.

Fig. 9 shows a comparison of the A_{mc} algorithm with the A_{nk} ($k = 1$ to 4) algorithms under a uniform destination node distribution. It can be seen that the A_{mc} algorithm saturates at the smallest value of the average packet inter-generation time. The A_{mc} algorithm is slightly worse than the A_{n2}, A_{n3}, and A_{n4} algorithms at light traffic conditions. The A_{n1} algorithm (deterministic VCT) is shown to perform significantly worse than any of the other algorithms.

Fig. 10 shows a comparison of the A_{mc} algorithm with deterministic and adaptive WR algorithms based on a uniform destination node distribution. The deterministic WR algorithm used is the X-Y routing algorithm in which the packet is first sent in the X di-

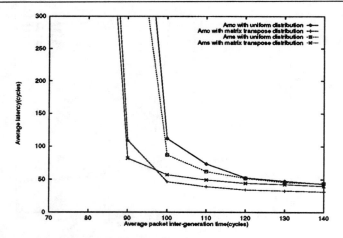

Figure 8: Average latency for the A_{mc} and A_{ms} algorithms given an 8 x 8 mesh with uniform and matrix transpose destination node distributions.

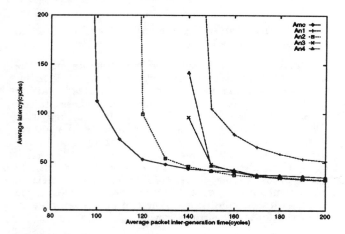

Figure 9: Average latency for the A_{mc} and A_{nk} algorithms given an 8 x 8 mesh with a uniform destination node distribution.

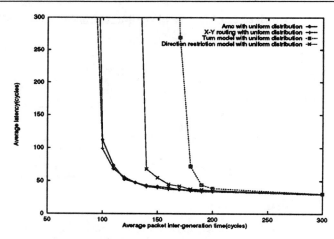

Figure 10: Average latency comparison with deterministic and adaptive WR algorithms given an 8 x 8 mesh with a uniform destination node distribution.

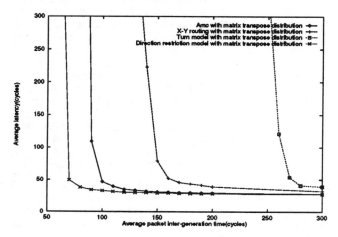

Figure 11: Average latency comparison with deterministic and adaptive WR algorithms given an 8 x 8 mesh with a matrix transpose destination node distribution.

rection and then in the Y direction. The adaptive WR algorithms simulated are Glass and Ni's Negative-first algorithm based on the turn model [7] and Boura and Das's $R_{7,0}$ algorithm (referred to as "zenith-routing" in [11]) based on the direction restriction model [6]. Both methods are partially-adaptive minimal algorithms which can be executed on a 2-dimensional mesh with single virtual channels. There do not exist any fully-adaptive minimal WR algorithms for the 2-dimensional mesh that work with single virtual channels. With a uniform destination node distribution, the A_{mc} algorithm performs as well or better than all of the alternative algorithms. It is noted that the deterministic X-Y algorithm performs surprisingly well. This is because the uniform distribution tends to randomize the paths followed by packets, thereby reducing the probability of blocking.

Fig. 11 shows the results for a matrix transpose destination node distribution. With a matrix trans-

pose destination node distribution, the A_{mc} algorithm again performs the best with the exception of Boura and Das's $R_{7,0}$ algorithm. This is because the $R_{7,0}$ algorithm tends to use paths which are optimal for the matrix transpose operation used. If, however, the matrix used is placed in the processing nodes such that the matrix diagonal is rotated 90 degrees, the $R_{7,0}$ algorithm will have much poorer performance. Similarly, Glass and Ni's negative-first algorithm performs exceptionally poorly because it tends to use paths which are extremely bad for the matrix transpose operation used. Thus, partially adaptive algorithms tend to perform well with only certain types of clustered distributions. It can also be seen that the deterministic X-Y algorithm performs poorly with this type of clustered distribution.

If multiple virtual channels are used, all of the algorithms considered perform significantly better. This has already been verified by Dally [3]. However, the relative performance of all of the algorithms show similar trends to the simulation results shown here. Thus, given the same number of virtual channels per communication link, the A_{mc} and A_{ms} algorithms perform as well or better than all of the other algorithms considered under uniform and clustered destination node distributions.

6 Discussion

This paper has investigated adaptive VCT as a possible alternative to WR. With WR, efficient prevention of deadlock is of utmost importance because the presence of blocked packets occupying network channels can very easily result in deadlock-inducing traffic patterns. With VCT, deadlock-inducing traffic patterns are much less likely because blocked packets are absorbed into buffers at intermediate nodes. With *adaptive* VCT, the possibility of a deadlock-inducing pattern is even rarer.

Many adaptive WR algorithms have been proposed in the literature. However, none of these algorithms is able to route adaptively in torus networks without the aid of multiple virtual channels. Even deterministic WR requires two virtual channels per communication link for torus networks. Thus, in torus networks, all WR algorithms must reserve some of the virtual channels in order to prevent the possibility of deadlock, which would otherwise occur frequently. In general, because blocked packets continue to occupy channels, WR requires some of the available channels to be reserved to prevent deadlock regardless of the type of network used.

Several possible adaptive VCT algorithms have been proposed and analyzed in this paper. It is shown that adaptive VCT has several attractive features when compared to WR: negligible overhead for avoiding deadlock (in terms of both performance and hardware), use of 100% of the free channels, good performance with heavy network traffic, and good performance with uniform or clustered locations of destination nodes. The hardware complexity issue is addressed by showing that VCT can be implemented with little hardware overhead (in some cases, less than WR), by exploiting the source buffer in the network interface to the router. A VLSI router chip for implementing adaptive virtual cut-through has been designed, manufactured, and tested. A quantitative comparison of the overhead for adaptive VCT has been attempted by showing that our router chip requires only 2.3% more layout area than a comparable deterministic WR chip. As our router chip has yet to be incorporated into a complete parallel system, simulations have been used to investigate the performance of several versions of adaptive VCT. The results show that a fully-adaptive minimal version has the best performance, outperforming all other deterministic and adaptive WR and VCT algorithms considered.

References

[1] W. Dally and C. L. Seitz, "Deadlock-free message routing in multiprocessor interconnection networks," *IEEE Trans. Comput.*, vol. C-36, pp. 547–553, May 1987.

[2] P. Kermani and L. Kleinrock, "Virtual cut-through: A new computer communication switching technique," *Computer Networks*, vol. 3, pp. 267–286, 1979.

[3] W. J. Dally, "Virtual-channel flow control," *IEEE Trans. Parallel and Distributed Systems*, vol. 3, pp. 194–205, Mar. 1992.

[4] D. H. Linder and J. C. Harden, "An adaptive fault-tolerant wormhole routing strategy for k-ary n-cube," *IEEE Trans. Comput.*, vol. C-40, pp. 2–12, Jan. 1991.

[5] W. J. Dally and H. Aoki, "Deadlock-free adaptive routing in multicomputer networks using virtual channels," *IEEE Trans. Parallel and Distributed Systems*, vol. 4, pp. 466–475, Apr. 1993.

[6] Y. M. Boura and C. R. Das, "A class of partially adaptive routing algorithms for n-dimensional meshes," in *Proc. of Int'l Conf. on Parallel Processing*, vol. III, pp. 175–182, Aug. 1993.

[7] C. J. Glass and L. M. Ni, "The turn model for adaptive routing," in *Proc. 19th Int'l Symp. Computer Architecture*, pp. 278–287, 1992.

[8] J. Duato, "A new theory of deadlock-free adaptive routing in wormhole networks," *IEEE Trans. on Parallel and Distributed Systems*, vol. 4, pp. 1320–1331, Dec. 1993.

[9] N. F. Maxemchuk, "Problems arising from deflection routing: live-lock, lockout, congestion and message reassembly," in *Proc. of NATO Workshop on Architectures and Performance Issues of High Capacity Local and Metropolitan Area Networks*, May 1990.

[10] D. Gelerter, "A dag–based algorithm for prevention of store–and–forward deadlock in packet networks," *IEEE Trans. Comput.*, vol. C-30, pp. 259–270, Oct. 1981.

[11] G. D. Pifarre, L. Gravano, S. A. Felperin, and J. L. C. Sanz, "Fully adaptive minimal deadlock-free packet routing in hypercubes, meshes, and other networks: algorithms and simulations," *IEEE Trans. Comput.*, vol. 5, pp. 225–246, Mar. 1994.

[12] S. Konstantinidou and L. Snyder, "The chaos router: a practical application of randomization in network routing," in *2nd Annual ACM SPAA*, pp. 21–30, 1990.

[13] W. Dally and C. L. Seitz, "The torus routing chip," *J. Distributed Computing*, vol. 1, no. 3, pp. 187–196, 1986.

All-to-All Personalized Communication in a Wormhole-Routed Torus

Yu-Chee Tseng

Department of Computer Science
Chung-Hua Polytechnic Institute
Hsin-Chu, 30067, Taiwan
yctseng@chpi.edu.tw

Sandeep K. S. Gupta

Department of Computer Science
Duke University
Durham, NC 27708-0129
sandeep@cs.duke.edu

Abstract — All-to-all personalized communication, or complete exchange, is at the heart of numerous applications in parallel computing. In this paper, we consider this problem in a torus of any dimension with the wormhole-routing or circuit-switching capability under the direct model. We propose complete exchange algorithms that use optimal numbers of phases (if each side of the tori is a multiple of eight) or asymptotically optimal numbers of phases (otherwise).

1 Introduction

The *all-to-all personalized communication*, or simply *complete exchange*, is an important communication pattern in a multicomputer network. It is at the heart of numerous applications, such as matrix transpose, fast Fourier transform (FFT), distributed table-lookup, etc. In this pattern, each of the N processors in the network has a distinct, but equal-size, message to be sent to each of the remaining $N - 1$ processors. It is thus a highly dense communication pattern.

Wormhole routing technology has been adopted by many new-generation parallel computers, such as Intel Touchstone DELTA, Intel Paragon, MIT J-machine, Caltech MOSAIC, and Cray T3D. Such networks are known to be insensitive to routing distance if the communication is contention-free. In this paper, we study the scheduling of contention-free complete exchange in a wormhole-routed multicomputer network. There are two models used in the literature: *indirect* and *direct*. The indirect model, which has been adopted by [1, 2, 5] for meshes and by [7] for 2D tori, needs the intermediate nodes to store the messages, which are then rearranged and further forwarded to subsequent nodes. On the contrary, the direct model, which has been adopted by [3, 4], tries to find a path from each source node to each destination, along which the message is sent directly. As the size of the memory required at intermediate nodes in the indirect model is proportional to the size of the network, the direct model is more scalable to network size.

In this paper, we study the complete exchange problem under the direct model in torus networks. A torus is a mesh with wrap-around connections. Examples of mesh-like networks include the MP-1 from MasPar, the Paragon from Intel, and T3D from Cray.

In the direct model, the communication is scheduled as a sequence of *phases*, where a phase is a set of source-destination pairs accompanied with a routing function for each pair. Within each phase, the routing must be path-disjoint. As wormhole routing is insensitive to routing distance and the transmitted data blocks are of the same size, all communications within a phase are assumed to be finished at the same time, so the sequence of phases can be executed in a lockstep manner. Note that the same assumption is used in [1, 2, 3, 4, 5, 7].

We propose novel complete exchange algorithms for a d-dimensional torus of size $n_1 \times n_2 \times \cdots \times n_d$, where without lost of generality $n_1 \geq n_2 \geq \ldots \geq n_d$. We show that complete exchange can be done in $2\lceil \frac{n}{4} \rceil^2$ phases if $d = 1$, and in $8^d \lceil \frac{n_1}{8} \rceil^2 \Pi_{i=2}^d \lceil \frac{n_i}{8} \rceil$ phases if $d \geq 2$. As any complete exchange algorithm will require at least $\frac{1}{2} \lceil \frac{n_1}{2} \rceil \lfloor \frac{n_1}{2} \rfloor \Pi_{i=2}^d n_i$ phases, our algorithm is asymptotically optimal. In particular, when each n_i is a multiple of four (if $d = 1$) or a multiple of eight (if $d \geq 2$), the tight bound is reached. Interestingly, in order to achieve this, we only make very weak assumptions — that a node is capable of sending and receiving at most one message at a time, and the network is capable of supporting the dimension-ordered (or e-cube) minimum routing. Thus, the result can be used immediately by current machines.

It should be mentioned that the ratio of the number of links in a torus to that in an equal-size mesh is close to 1, as the size of the torus/mesh increases. However, a torus can provide 2 times more links than a mesh on any hyperplane that cut the torus/mesh into two parts. Thus, we should expect a speedup of 2 in a torus as opposed to a mesh for a communication problem such as complete exchange. Optimal complete exchange algorithms, under the direct model, for 1-D and 2-D square meshes have been given in [3]. Indeed, we have made this speedup, and the result is for non-square, any-dimensional tori.

2 Preliminaries

A d-dimensional torus $T_{n_1 \times n_2 \times \cdots \times n_d}$ is an undirected graph of $n_1 n_2 \ldots n_d$ nodes with each node denoted as $p_{i_1, i_2, \ldots, i_d}$, where for each $j = 1..d$, $0 \leq i_j < n_j$ and n_j is called the *size* of the torus along the j-th dimension. Two nodes $p_{i_1, \ldots, i_j, \ldots, i_d}$ and $p_{i_1, \ldots, (i_j + 1) \bmod n_j, \ldots, i_d}$, which differ by 1 in exactly one index (in this example, the j-th index), are connected by an edge. Each edge has two directed links going in opposite directions. Without lost of generality, it is assumed $n_1 \geq n_2 \geq \cdots \geq n_d$ throughout this paper.

In this paper, complete exchange will be done by scheduling a sequence of *phases*. A phase is a set of (di-

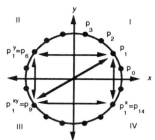

Figure 1: Arranging a T_n in an xy-plane.

rected) paths each following the dimension-ordered (or e-cube) minimum routing starting from lower to higher dimensions. Intuitively, through a path, a message is brought from a source node to a destination node. A routing is *minimum* if the shortest paths are always taken. A phase must be *legal* in the following sense: i) all its paths use disjoint links, and ii) all its paths have disjoint source and destination nodes. A path will be written as "$x_1 \rightarrow x_2 \rightarrow \cdots \rightarrow x_m$," where each x_i is a node in the torus and each transition $x_i \rightarrow x_{i+1}$ only routes the message in a *unique* dimension. As the routing is minimum, in each $x_i \rightarrow x_{i+1}$, there is normally one shortest path from x_i to x_{i+1}. However, if there are two shortest paths, we may write $x_i \overset{+}{\rightarrow} x_{i+1}$ or $x_i \overset{-}{\rightarrow} x_{i+1}$ to indicate that the routing goes in the positive or negative direction, respectively.

Lemma 1 *In* $T_{n_1 \times n_2 \times \cdots \times n_d}$*, where* $n_1 \geq n_2 \geq \cdots \geq n_d$*, any complete exchange algorithm will require at least* $\frac{1}{2}\lfloor \frac{n_1}{2} \rfloor \lceil \frac{n_1}{2} \rceil \Pi_{i=2}^{d} n_i$ *phases.*

3   1-D Tori

In this section, we consider a torus T_n. For ease of presentation, we assume that $n = 4t$ is a multiple of four, unless otherwise stated. We arrange the torus in an xy-plane as illustrated in Fig. 1. The x-axis and y-axis evenly partition the torus into 4 parts, with nodes $p_0, \ldots, p_{t-1}$ falling in quadrant I, nodes $p_t, \ldots, p_{2t-1}$ in quadrant II, etc. For any node p_i in the torus, define node p_i^x to be the node symmetric to p_i with respect to the x-axis. Similarly, define p_i^y to be the node symmetric to p_i with respect to the y-axis, and p_i^{xy} the one symmetric to p_i with respect to the origin.

Let p_i and p_j be any two distinct nodes in quadrant I. We use p_i and p_j to define two phases as follows:

$$R_1(p_i, p_j) = \{p_i \rightarrow p_j, p_j \rightarrow p_j^{xy}, p_j^{xy} \rightarrow p_j^{xy}, p_j^{xy} \rightarrow p_i,$$
$$p_i^x \rightarrow p_j^x, p_j^x \rightarrow p_i^y, p_i^y \rightarrow p_j^y, p_j^y \rightarrow p_i^x\} \quad (1)$$

$$R_2(p_i, p_j) = \{p_i \rightarrow p_j^x, p_j^x \rightarrow p_i^{xy}, p_i^{xy} \rightarrow p_j^y, p_j^y \rightarrow p_i,$$
$$p_i^x \rightarrow p_j, p_j \rightarrow p_i^y, p_i^y \rightarrow p_j^{xy}, p_j^{xy} \rightarrow p_i^x\}. \quad (2)$$

Fig. 2(a) and (b) illustrate the phases defined above. Note that in each transmission, the shortest path is unique. Clearly, both $R_1(p_i, p_j)$ and $R_2(p_i, p_j)$ are legal phases.

As there are $t(t-1)$ possibilities to select p_i and p_j, there are $2t(t-1)$ phases defined in Eq. (1) and

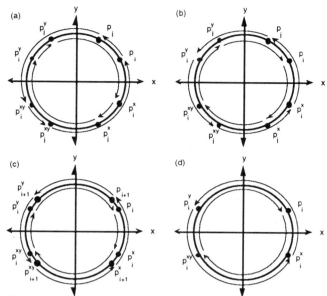

Figure 2: Definitions of (a) $R_1(p_i, p_j)$, (b) $R_2(p_i, p_j)$, (c) $R_3(p_i)$, and (d) $R_4(p_i)$.

Eq. (2). It is not hard to prove that if we perform all these $2t(t-1)$ phases for all possible p_i and p_j, then any node p in T_n will have communicated with every other node exactly once except nodes $\{p^x, p^y, p^{xy}\}$.

So it remains to schedule the communications between p and $\{p^x, p^y, p^{xy}\}$ for each p in T_n. This can be done by using another $2t$ phases. Specifically, assuming p_i to be any node in quadrant I and $p_j = p_{(i+1) \bmod t}$, we define

$$R_3(p_i) = \{p_i \rightarrow p_i^x, p_i^x \rightarrow p_i^{xy}, p_i^{xy} \rightarrow p_i^y, p_i^y \rightarrow p_i,$$
$$p_j \rightarrow p_j^y, p_j^y \rightarrow p_j^{xy}, p_j^{xy} \rightarrow p_j^x, p_j^x \rightarrow p_j\} \quad (3)$$

$$R_4(p_i) = \{p_i \overset{-}{\rightarrow} p_i^{xy}, p_i^{xy} \overset{-}{\rightarrow} p_i, p_i^x \overset{+}{\rightarrow} p_i^y, p_i^y \overset{+}{\rightarrow} p_i^x\} \quad (4)$$

Fig. 2(c) and (d) illustrate examples. In Eq. (3), node p communicates with nodes $\{p^x, p^y\}$, and in Eq. (4) it communicates with p^{xy}. Clearly, these are legal phases. So Eq. (1)–(4) together perform complete exchange in T_n using $2t^2$ phases when n is a multiple of four. If n is not a multiple of four, we can insert some imaginary nodes into the torus. Whenever the source or destination node of a path is an imaginary node, we simply remove this path from the phase. So we have the following theorem.

Theorem 1 *In* T_n*, where* n *is any positive integer, complete exchange can be done in* $2\lceil \frac{n}{4} \rceil^2$ *phases.*

As a result, the number of phases used by our algorithm is optimal when n is a multiple of four, and asymptotically optimal otherwise.

4   2-D Tori

A 2-D torus is the *graph product* of two 1-D tori. To schedule a complete exchange in a 2-D torus, we will also use the *product* (defined below) of two phases of 1-D tori.

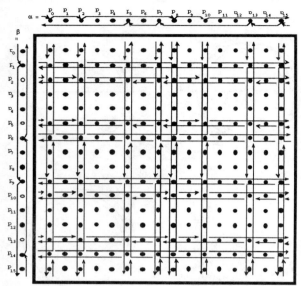

Figure 3: The product of two phases $\alpha = R_1(p_0, p_2)$ and $\beta = R_4'(p_1)$ in a $T_{16 \times 16}$, where the node located in the ith column and jth row is $p_{i,j}$. In β, each circled node has an implicit self-looped path.

4.1 Product of Two Phases

To start with the discussion, we need to generalize the definition of complete exchange. Specifically, for each node in the network, in addition to sending messages to other nodes, it also sends a dummy message to itself. In a T_n, we realize this by adding some self-looped paths into $R_4(p_i)$. Specifically, assuming $p_j = p_{(i+1) \bmod t}$, we define

$$R_4'(p_i) =$$
$$R_4(p_i) \cup \{p_j \to p_j, p_j^x \to p_j^x, p_j^y \to p_j^y, p_j^{xy} \to p_j^{xy}\} \quad (5)$$

Note that each node in the torus has exactly one self-looped path defined in $R_4'(p_i), i = 0..t - 1$. From now on, "a phase of T_n" will be referred to one belonging to $R_1(p_i, p_j)$, $R_2(p_i, p_j)$, $R_3(p_i)$, or $R_4'(p_i)$ with appropriate p_i and p_j.

Definition 1 Let α and β be any phases of T_{n_1} and T_{n_2}, respectively. Define the *product of α and β*, denoted as $\alpha \times \beta$, to be a set of paths in $T_{n_1 \times n_2}$ such that

$$\alpha \times \beta = \{p_{i,r} \overset{\sigma_1}{\to} p_{j,r} \overset{\sigma_2}{\to} p_{j,s} | p_i \overset{\sigma_1}{\to} p_j \in \alpha \wedge p_r \overset{\sigma_2}{\to} p_s \in \beta\},$$

where the symbols σ_1 and σ_2, if any, indicate the routing directions ($+$ or $-$).

Fig. 3 shows an example. Note that in $\alpha \times \beta$ if $i = j = r = s$, then the path is self-looped. The self-looped paths will be used later for generalizing our algorithm to higher dimensional tori.

Clearly, $\alpha \times \beta$ forms a legal phase. To perform a complete exchange in $T_{n_1 \times n_2}$, we can execute all $\alpha \times \beta$ phase by phase for all possible combinations of α and

β. As there are $2 \lceil \frac{n_1}{4} \rceil^2$ possible α's and $2 \lceil \frac{n_2}{4} \rceil^2$ possible β's, this will need $4 \lceil \frac{n_1}{4} \rceil^2 \lceil \frac{n_2}{4} \rceil^2$ phases. Unfortunately, this is an order higher than the lower bound given in Lemma 1. The main technique to reduce the gap is to combine multiple $\alpha \times \beta$'s into one phase. Observe that in Fig. 3, only the links along the i-th column (resp., row) will be utilized if p_i appears in α (resp., β). Furthermore, in $\alpha \times \beta$, only the node $p_{i,j}$ such that p_i appears in α and p_j appears in β may send or receive messages. This gives rise to the following definition.

Definition 2 Let α_1 and α_2 be two phases of T_n. Then α_1 and α_2 are said to be *compatible* if they share no source node (or equivalently, destination node) in common. (Note that a self-looped node is regarded as both source and destination.)

Lemma 2 *Let α_1 and α_2 be two compatible phases of T_{n_1}, and β_1 and β_2 two compatible phases of T_{n_2}. Then $\{\alpha_1 \times \beta_1, \alpha_2 \times \beta_2\}$ forms a legal phase in $T_{n_1 \times n_2}$.*

Corollary 1 *Let $\alpha_1, \ldots, \alpha_r$ be pairwise-compatible phases of T_{n_1}, and $\beta_1, \ldots, \beta_r$ pairwise-compatible phases of T_{n_2}. Then $\{\alpha_i \times \beta_i | i = 1..r\}$ forms a legal phase in $T_{n_1 \times n_2}$.*

4.2 Compatible Phases

To utilize Corollary 1, given a T_n, we need to find as many phases of T_n that are pairwise compatible as possible. As there are 8 sources in a phase, it is impossible to find a set having more than $\lfloor n/8 \rfloor$ pairwise-compatible phases. In this section, we show that if n is a multiple of eight, then the $n^2/8$ phases of T_n defined in Eqs. (1), (2), (3), (5) can be perfectly partitioned into n sets, each containing $n/8$ compatible phases. In the following presentation, we assume that $n = 8k$ is a multiple of eight, unless otherwise stated.

Definition 3 The *0-th skeleton set* of T_n is

$$K_0(T_n) = \{(p_i, p_{2k-1-i}) | i = 0..k - 1\}.$$

For $m = 1..2k - 2$, the *m-th skeleton set* of T_n is

$$K_m(T_n) = \{(p_m, p_{2k-1})\} \cup$$
$$\{(p_{(m+i) \bmod (2k-1)}, p_{(m-i) \bmod (2k-1)}) | i = 1..k - 1\}.$$

Each of the above sets has k pairs of nodes. For instance, if $k = 4$, then $K_0(T_{32}) = \{(p_0, p_7), (p_1, p_6), (p_2, p_5), (p_3, p_4)\}$, and $K_1(T_{32}) = \{(p_1, p_7), (p_2, p_0), (p_3, p_6), (p_4, p_5)\}$. Intuitively, $K_m(T_{32}), m \neq 0$, can be obtained by shifting the elements of $K_0(T_{32})$ for m positions along the arrows as shown in Fig. 4.

Recall that in Section 3, in order to perform a complete exchange in T_n, we need to exhaust all possible combinations of p_i and p_j in quadrant I. The skeleton sets indeed help us to enumerate all such combinations, as indicated by the following lemma.

Lemma 3 *For any two distinct elements p_i and p_j in T_n such that $0 \leq i, j \leq 2k - 1$, there exists exactly one skeleton set $K_m(T_n), 0 \leq m \leq 2k - 2$, such that either $(p_i, p_j) \in K_m(T_n)$ or $(p_j, p_i) \in K_m(T_n)$.*

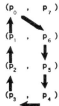

Figure 4: The construction of skeleton sets of a T_{32}.

Now we use the $2k - 1$ skeleton sets to generate $4(2k - 1)$ sets of phases (of T_n). Specifically, for each $K_m(T_n), m = 0..2k - 2$, we define 4 sets:

$$C_{4m}^{(1)}(T_n) = \{R_1(p_i, p_j)|(p_i, p_j) \in K_m(T_n)\}$$
$$C_{4m+1}^{(1)}(T_n) = \{R_1(p_j, p_i)|(p_i, p_j) \in K_m(T_n)\}$$
$$C_{4m+2}^{(1)}(T_n) = \{R_2(p_i, p_j)|(p_i, p_j) \in K_m(T_n)\}$$
$$C_{4m+3}^{(1)}(T_n) = \{R_2(p_j, p_i)|(p_i, p_j) \in K_m(T_n)\}$$

The superscript, 1, refers to 1-D torus. Each set contains k phases. It is easy to see that each set contains only compatible phases. So we have partitioned the $8k^2 - 4k$ phases defined in Eq. (1) and Eq. (2) into $8k - 4$ sets. It remains to partition the $4k$ phases defined in Eq. (3) and Eq. (5). The following 4 equations partition them into 4 sets of compatible phases:

$$C_{n-4}^{(1)}(T_n) = \{R_3(p_i)|\ 0 \leq i \leq 2k - 1 \text{ and } i \text{ is odd}\}$$
$$C_{n-3}^{(1)}(T_n) = \{R_3(p_i)|\ 0 \leq i \leq 2k - 1 \text{ and } i \text{ is even}\}$$
$$C_{n-2}^{(1)}(T_n) = \{R_4'(p_i)|\ 0 \leq i \leq 2k - 1 \text{ and } i \text{ is odd}\}$$
$$C_{n-1}^{(1)}(T_n) = \{R_4'(p_i)|\ 0 \leq i \leq 2k - 1 \text{ and } i \text{ is even}\}$$

Therefore, we have partitioned the $n^2/8$ phases of T_n into n sets, namely $C_i^{(1)}(T_n), i = 0..n - 1$.

4.3 Complete Exchange Algorithm

In this section, we consider a 2-D torus $T_{n_1 \times n_2}$, where $n_1 \geq n_2$. We assume that both $n_1 = 8k_1$ and $n_2 = 8k_2$ are multiples of eight, unless otherwise stated.

Definition 4 Let $C_{i_1}^{(1)}(T_{n_1}) = \{\alpha_0, \ldots, \alpha_{k_1-1}\}$ and $C_{i_2}^{(1)}(T_{n_2}) = \{\beta_0, \ldots, \beta_{k_2-1}\}$ be any two sets of phases, where $0 \leq i_1 < n_1$ and $0 \leq i_2 < n_2$. The product of $C_{i_1}^{(1)}(T_{n_1})$ and $C_{i_2}^{(1)}(T_{n_2})$ is defined to be

$$C_{i_1,i_2}^{(2)}(T_{n_1 \times n_2}) = \{S_0, S_1, \ldots, S_{k_1-1}\},$$

where for $i = 0..k_1 - 1$,

$$S_i = \{\alpha_{(i+j) \bmod k_1} \times \beta_j | j = 0..k_2 - 1\}.$$

Example 1 If $C_{i_1}^{(1)}(T_{32}) = \{\alpha_0, \alpha_1, \alpha_2, \alpha_3\}$ and $C_{i_2}^{(1)}(T_{24}) = \{\beta_0, \beta_1, \beta_2\}$, then $C_{i_1,i_2}^{(2)}(T_{32 \times 24}) = \{S_0, S_1, S_2, S_3\}$, where $S_0 = \{\alpha_0 \times \beta_0, \alpha_1 \times \beta_1, \alpha_2 \times \beta_2\}$, $S_1 = \{\alpha_1 \times \beta_0, \alpha_2 \times \beta_1, \alpha_3 \times \beta_2\}$, $S_2 = \{\alpha_2 \times \beta_0, \alpha_3 \times \beta_1, \alpha_0 \times \beta_2\}$, and $S_3 = \{\alpha_3 \times \beta_0, \alpha_0 \times \beta_1, \alpha_1 \times \beta_2\}$. $\diamond$

Note that as $C_{i_1}^{(1)}(T_{n_1})$ and $C_{i_2}^{(1)}(T_{n_2})$ contain only compatible phases, by Corollary 1, each S_i in Definition 4 can form a legal phase in $T_{n_1 \times n_2}$. Furthermore, every phase in $C_{i_1}^{(1)}(T_{n_1})$ will be coupled with every phase in $C_{i_2}^{(1)}(T_{n_2})$ exactly once in one of the S_i's. This gives rise to a simple algorithm as follows:

Algorithm: *2-D Complete-Exchange*
> **for** each $C_{i_1,i_2}^{(2)}(T_{n_1 \times n_2})$ **do**
>> Let $C_{i_1,i_2}^{(2)}(T_{n_1 \times n_2}) = \{S_0, S_1, \ldots, S_{k_1-1}\}$;
>> **for** $j = 0$ to $k_1 - 1$ **do**
>>> Perform S_j as a phase;
>>
>> **end for**;
>
> **end for**;

Theorem 2 *In $T_{n_1 \times n_2}$, where n_1 and n_2 are arbitrary positive integers, complete exchange can be done in $8^2 \lceil \frac{n_1}{8} \rceil^2 \lceil \frac{n_2}{8} \rceil$ phases.*

5 Higher Dimensional Tori

The result can be extended inductively to higher dimensional tori. Due to space limit, we only summarize the result as follows. More details can be found in [6].

Theorem 3 *In $T_{n_1 \times \cdots \times n_d}$, where $n_i, i = 1..d$, are arbitrary positive integers and $n_1 \geq n_2 \geq \cdots \geq n_d$, complete exchange can be done in*

$$8^d \cdot \left\lceil \frac{n_1}{8} \right\rceil^2 \cdot \Pi_{i=2}^d \left\lceil \frac{n_i}{8} \right\rceil$$

phases.

References

[1] S. H. Bokhari and H. Berryman. Complete exchange on a circuit switched mesh. In *Scalable High Performance Computing Conf.*, pages 300–306, 1992.

[2] S. Gupta, S. Hawkinson, and B. Baxter. A binary interleaved algorithm for complete exchange on a mesh architecture. Technical report, Intel Corporation, 1994.

[3] D. S. Scott. Efficient all-to-all communication patterns in hypercube and mesh topologies. In *IEEE Distributed Memory Conference*, pages 398–403, 1991.

[4] S. R. Seidel. Circuit switched vs. store-and-forward solutions to symmetric communication problems. In *4th Conf. Hypercube Concurrent Computers and Applications*, pages 253–255, 1989.

[5] N. S. Sundar, D. N. Jayasimha, D. K. Panda, and P. Sadayappan. Complete exchange in 2D meshes. In *Scalable High Perf. Comput. Conf.*, pages 406–413, 1994.

[6] Y.-C. Tseng and S. K. S. Gupta. All-to-all personalized communication in a wormhole-routed torus. Technical report, Dept. of Comp. Sci., Chung-Hua Polytechnic Inst., 1994.

[7] Y.-C. Tseng, S. K. S. Gupta, and D. K. Panda. An efficient scheme for complete exchange in 2D Tori. In *Int'l Parallel Processing Symp.*, 1995. to appear.

MINIMUM DEPTH ARCS-DISJOINT SPANNING TREES
FOR BROADCASTING ON WRAP-AROUND MESHES

P. Michallon
ETCA/CREA/SP
16 bis, Av. du Prieur de la Cote d'Or
F-94114 Arcueil Cedex (France)
Philippe.Michallon@etca.fr

D. Trystram
LMC-IMAG
46, Av. Felix Viallet
F-38031 Grenoble Cedex (France)
Denis.Trystram@imag.fr

Abstract – *This paper presents a new family of arc-disjoint spanning trees in wrap-around meshes which have the property of minimum depth. These theoretical results are applied to design broadcasting algorithms in a parallel machine based on Transputers.*

INTRODUCTION

Some collective communication routines have been identified as being useful in many scientific applications [1, 6]. Spanning trees are graph-theoretical concepts often used for designing broadcasting algorithms. This paper presents a new family of spanning trees of minimal depth in wrap-around meshes of processing nodes for implementing efficient broadcasting algorithms.

Problem definition

The interconnection network of a parallel distributed-memory machine is often represented by a graph whose vertices are the processors (or special hardware component dedicated to the communications). We consider a directed graph $G = (V, A(G))$. A spanning tree rooted at node $x \in V$ is a tree $T = (V, A(T))$ where $A(T) \subset A(G)$. Two spanning trees T and T' are said arc-disjoint if $A(T) \cap A(T') = \emptyset$. We will denote in short ADST for arc-disjoint spanning tree.

Edmonds gives in [4] a bound of the maximum number of ADST available in a graph G. Applied to the wrap-around mesh, this guaranties the existence of 4 ADST rooted at any node. Broadcasting is influenced by both number and depth of ADST. The problem is to obtain 4 such trees of minimal depth rooted at the same node. A lower bound on this depth is given by the following proposition:

Proposition 1 *Four ADST of common root in a wrap-around mesh of diameter D are of depth at least $D + 1$.*

Proof. Let r be the common root of the trees, and P_i $(1 \le i \le 4)$ its neighbor on each tree i. Under the arc-disjoint assumption, each tree uses exactly one outgoing arc from r that imposes one starting direction for each tree. The 4 nodes at distance D of the P_i's, in the wrap-around mesh, are at distance $D + 1$ from the root in respect to the spanning trees.   ◇

The main result of this paper is to construct a new family of ADST of minimal depth in wrap-around meshes. We will apply this result to design an optimal broadcasting algorithm. Some experiments are run on a parallel machine based on Transputers.

Properties of Wrap-around Meshes

Meshes are very popular for building interconnection networks of parallel computers. They are symmetric topologies simple to construct with VLSI. The use of a wrap-around meshes has received lot of interest recently inview of multiprocessor based parallel computers with bounded degree, like CRAY T3D, Transputer systems, iWARP, Fujitsu AP1000, etc.. In the following we will denote by $WM(n, n)$ the square wrap-around mesh whose number of nodes in each dimension is n.

Let us recall the definition and some basic properties of this (well-known) topology: A node in a $WM(n, n)$ is represented by an ordered pair of integers (i, j) where i and j are respectively the column and row indices ($0 \le i, j \le n - 1$). $WM(n, n)$ is the topology where edges connect the nodes which differ with 1 modulo n in exactly one coordinate (i or j). Each node has exactly 4 neighbors and the four arcs adjacent to each node are characterized as usually by the N, E, W, S pattern. According to most available machines, the links are supposed to be bidirectional. In reference to this pattern the four Spanning Trees will be denoted by NST, EST, WST and SST, according to the direction of the arc outcoming from the root. We will present the trees rooted at node $(0, 0)$ but the construction is obviously available at any node (i, j) (by simply add i and j modulo n respectively on each coordinate). The diameter of $WM(n, n)$ (defined as the maximum distance in the graph) is equal to $2\lfloor \frac{n}{2} \rfloor$. It will be denoted by D in the rest of the paper.

Related works

Hypercube was a very popular topology to build old generation parallel machines. Johnsson and Ho have presented a family of n ADST (the maximum number as possible according to Edmonds's proposition) of depth $n + 1$ (equal to the diameter plus one) on n-cubes [6].

A family of 2 edge-disjoint spanning trees has been proposed for undirected $WM(n, n)$ by Seidel et al. [10]. Their depth is equal to $\frac{3}{2}D$. This solution has been improved by Bermond et al. [2] under the same hypotheses leading to minimum depth trees (equal to $D + 1$).

When the arcs are bidirectional, Fraigniaud [5] has proposed a simple solution with 4 ADST of depth $2(D - 1)$. This is the best result known at this time.

Basic principle

We present below a basic principle on infinite meshes which guaranties that the arcs are disjoint by rotation. The basic principle will be described on EST (fig. 1). The 3 other ADST are deduced by successive rotations of $\frac{\pi}{2}$ over the directions (it means that the N,E,W,S incoming or outcoming arcs of node (i, j) become respectively W,N,S,E for node (j, i) if they exist). The idea is to fill one column over two on the right side and one row over two on the left side. As depicted in fig. 1 the fill-columns are even and fill-rows are odd in order to perfectly fit together by rotation without overlap. This result can be easily proved by considering the different status of the arcs in each region.

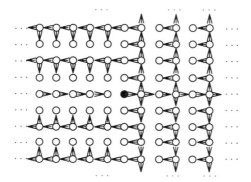

Figure 1: Basic principle for EST.

The "borders" of finite wrap-around meshes (defined as the farest rows and columns when the root is placed at the center) need a special treatment. The choice of the corresponding links have been defined according to same rules as those used in the basic principle. Remark that for square wrap-around meshes of even dimensions, the center does not exist, this imposes to distinguish 4 cases: $4k + 1$, $4k + 3$ for odd-sized WM and $4k$, $4k + 2$ for even-sized WM. In the following, we will restrict the presentation and analysis to the case $4k + 1$ and derive briefly the other cases.

A NEW FAMILY OF SPANNING TREES

A solution of depth D+2

We obtain easily a first construction of 4 ADST in $WM(4k + 1, 4k + 1)$ by simply applying the basic principle for all rows j as depicted in fig. 2. In this case, the number of rows in the upper part (j=1 to 2k) and lower part of the WM is even. The number of columns in the right and left is also even, but the basic principle refers to odd columns in the right of EST. It remains to explain how to reach the nodes of columns $-2k$ and $2k$. They can be attained by simple local patterns from column $2k - 1$ each reaching nodes of column $-2k$ using wraparound arcs.

As for other existing solutions, the 3 other ADST are obtained by rotations over the directions. We verify that this construction is arc-disjoint by considering that the different patterns are arc-disjoint in the rotation process. The depth of the previous ADST is $D + 2$. There are 4 nodes at distance $D + 2$

from the root, represented in square boxes in fig. 2. Assuming the center at $(0, 0)$ these nodes are $((-2k, 2k-1), (-2k, -2k+1), (-2k+1, 2k-1), (-2k+1, -2k+1))$.

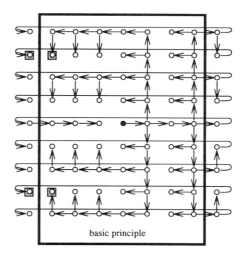

Figure 2: EST of depth $D + 2$ in $WM(9, 9)$.

Optimal depth ADST in $WM(4k + 1, 4k + 1)$

We show now how to remove some arcs on the longest paths and introduce other arcs to attain the cut-off nodes without conflicts between the arcs of the 3 other trees obtained by rotation. The local transformations on the arcs of the border have been allowed only if they generate no arc-conflicts [8]. However, it is easy to verify that the family built from EST by successive rotations of $\frac{\pi}{2}$ are arc-disjoint. An example is developed for EST on $WM(9, 9)$ on fig. 3. The generalization leads to the main result of the paper.

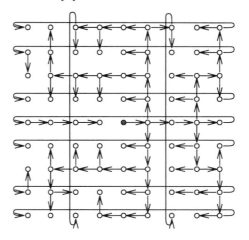

Figure 3: EST of depth $D + 1$ in $WM(9, 9)$.

Proposition 2 *There exist 4 ADST of minimal depth $4k + 1$ in any $WM(4k + 1, 4k + 1)$.*

The proof comes directly from the previous construction [8]. It can be generalized similarly to any $WM(4k + 1, 4k + 1)$ for any $k \geq 2$. To conclude this section, let us remark that $WM(5, 5)$ needs a particular construction because the previous one is not valid. It is easy to verify that the EST of $WM(5,5)$ built with our construction is of depth only 6. Now we derive the extensions to other square sizes and rectangular WM.

EXTENSIONS

Construction for WM(4k+3,4k+3)

This case is easy to derive from the previous analysis. The idea is to use as long as possible the basic principle, and as before to adapt the border with local conflict-free patterns. We give in figure the minimum EST for $WM(11, 11)$. Of course, the construction can be extended to any k.

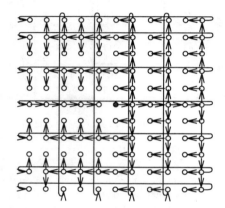

Figure 4: Minimum depth EST for $WM(11, 11)$.

Construction for WM(4k,4k) and WM(4k+2,4k+2)

The case of square WM of even dimensions is more difficult to obtain since the center does not exist. This imposes to distinguish two generators (EST and NST). For WM(4k,4k), SST is obtained from EST by a composition of a rotation of $-\frac{\pi}{2}$ and a symmetry (for instance, the arc in direction N of node (i, j) becomes W for node $(-j, -i)$) WST is obtained similarly from NST by the composition of a rotation of $\frac{\pi}{2}$ and a symmetry. It is depicted in fig. 5 for $k = 2$. The case of WM(4k+2,4k+2) is similar. We give in fig. 6 the two generators for $WM(10, 10)$.

Extension to rectangular torus

The generalization of the previous ADST of depth $D + 2$ to rectangular $WM(n, m)$ is easy to obtain [7]. It is also based on the use of the basic principle on interior nodes and some local adaptations on the border. More precisely, the sizes $n = 4k + s$ and $m = 4k' + s$ lead to simple adaptations since the borders fit together exactly. The other sizes are more difficult to derive. For $WM(4k + s, 4k' + s')$ with $s \neq s'$, we have 2 or 3 generators [8].

APPLICATION TO THE DESIGN OF EFFICIENT BROADCASTING ALGORITHMS

This section is devoted to the application of ADST for designing efficient broadcasting algorithms. It consists in sending one message from a given processor to all the others [3, 6, 11]. p is the number of processors.

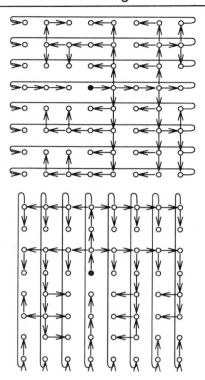

Figure 5: EST and NST of depth $D + 1$ in $WM(8, 8)$.

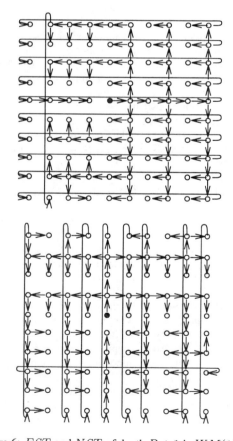

Figure 6: EST and NST of depth $D + 1$ in $WM(10, 10)$.

The time required for sending a message of length L between two neighbors can be expressed as: $\beta + \tau L$ where β is the start-up and τ the transmission rate factor. We assume that a processor can receive and send simultaneously on the same link and it can use its 4 neighbors in parallel. We also assume a store-and-forward routing. This model corresponds to the transputer machines.

The principle of the broadcasting algorithms is to split the message and pipeline the sending of the packets simultaneously along the ADST [6, 9]). Applied to $WM(\sqrt{p}, \sqrt{p})$ using the ADST described in this paper, we obtain the following broadcasting time:

$$\left(\sqrt{2 \lfloor \frac{\sqrt{p}}{2} \rfloor \beta} + \sqrt{\frac{L}{4} \tau} \right)^2$$

We report below some experiments on a *MegaNode* (a 128 reconfigurable parallel machine based on Transputers). We compare in figures 7 and 8 the performances of various strategies of broadcasting in square WM for $n = 11$. We will consider successively the influence of the degree (number of ADST) and diameter (depth of ADST), namely, using only one spanning tree of depth D+1, using 2 EDST of depth D+1 [2], using 4 ADST of depth 2(D-1) [5] and using 4 ADST of minimal depth (D+1) presented in this paper.

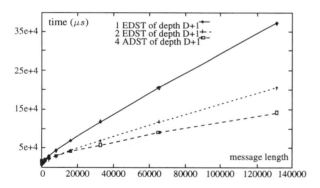

Figure 7: Comparison between broadcasting strategies.

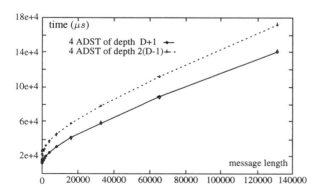

Figure 8: Comparison between broadcasting strategies.

These experiments show that the use of the maximum number of ADST of minimal depth is mostly the best strategy. However, for small messages, it is more interesting to use only

2 EDST (fig. 9) because the start-up factor decreases due to less management. The transmission rate remains the same.

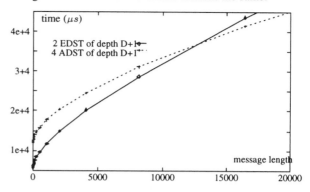

Figure 9: Comparison for small messages.

CONCLUSION

We have presented in this paper a new family of 4 ADST of minimum depth for square wrap-around meshes for designing efficient broadcasting algorithms. Some experiments are run on a Transputer-based parallel machine. They show the good behavior of the broadcasting algorithms based on pipelining simultaneously on ADST.

REFERENCES

[1] M. Barnett, S. Gupta, D. Payne, L. Shuler, R. Van de Geijn, and J. Watts. Interprocessor collective commmunication library. In IEEE, editor, *SHPCC, Knoxville*, 1994.

[2] J. C. Bermond, P. Michallon, and D. Trystram. Broadcasting in wraparound meshes with parallel monodirectionnal links. *Parallel computing*, (18):639–648, 1992.

[3] Jean de Rumeur. *Communications dans les réseaux de processeurs*. Masson, 1994.

[4] J. Edmonds. *Edges-disjoint branching, combinatorial algorithms*. Algorithms press, New-York, 1972.

[5] P. Fraigniaud. *Communications intensives dans les architectures à mémoire distribuée*. PhD thesis, Lyon (France), 1990.

[6] S.L. Johnsson and C.T. Ho. Optimum broadcasting and personalized communications in hypercubes. *IEEE Transaction on Computers*, 38(9):1249–1268, 1989.

[7] P. Michallon, D. Trystram, and G. Villard. Optimal broadcasting algorithms on torus. Technical Report 872-I-, IMAG, 1992.

[8] Ph. Michallon. *Schémas de communications globales dans les réseaux de processeurs : application à la grille torique*. PhD thesis, INP Grenoble (France), 1994.

[9] Y. Saad and M. Schultz. Data communication in parallel architectures. *Parallel Computing*, 11:131–150, 1989.

[10] S. Seidel and R. Bajwa. Communication algorithms for tori and grids. Technical Report CS-TR 89-04, Michigan Technological University (USA), 1989.

[11] Q. Stout and B.Wagar. Intensive hypercube communication; prearranged communication in link-bound machines. *JPDC*, 10(2):167–181, 1990.

DTN : A New Partitionable Torus Topology[†]

SangHo Chae*, Jong Kim*, DongSeung Kim*, SungJe Hong*, and Sunggu Lee**

*Department of Computer Science and Engineering
**Department of Electrical Engineering
Pohang University of Science and Technology (POSTECH)
San 31 Hyoja Dong, Pohang 790-784, SOUTH KOREA
E-mail : jkim@vision.postech.ac.kr

Abstract

In this paper, we propose a new topology named the Dual Torus Network (DTN), which is constructed by adding interleaved edges to a torus. The l-complete torus is the generalization of the DTN. The DTN has many advantages over the mesh and torus such as better extendibility, smaller diameter, higher bisection width, and robust link connectivity. An important property of the DTN is that it can be partitioned into sub-tori of different sizes (which is not possible for mesh and torus-based systems). Moreover, we show that if a certain topology can be embedded into a mesh or torus with dilation δ, then that topology is also embeddable into a DTN with smaller dilation. Specifically, we study the problem of embedding any rectangular grid into an optimal square DTN. Finally, we propose a method for embedding any rectangular grid into an optimal square DTN when there exist faulty nodes given that the number of faulty nodes is less than the number of unused nodes (in the host graph).

1   Introduction

Recently, scalable structures such as meshes and hypercubes have received much attention from academia and industry as a viable interconnection network. One advantage of scalable structures such as meshes and hypercubes is that these structures can be partitioned into several sub-structures which are identical to the original but smaller. Hence, in multitasking and multiprogramming environments, many tasks which require the same structure of different sizes can be allocated concurrently in such systems. This has led to the proliferation of many experimental and commercial systems based on mesh and hypercube structures such as the Intel Paragon, iPSC2, NCUBE2, and many transputer-based systems. Due to this increased interest, many aspects of hypercube and mesh structures have been extensively studied [1-6]. Important subjects are allocation, scheduling, embedding, and fault-tolerance. Allocation is the problem of partitioning the system and assigning the partitions to the requested tasks. Scheduling is the problem of assigning priorities in the sequence of allocating partitions

to tasks. Embedding is the problem of mapping one topology into another. Fault-tolerance treats all of these aspects in the presence of a faulty node(s).

Subcube allocation and scheduling strategies in a hypercube have been reported by many researchers [7]. Examples include the Gray code, Buddy, Free List, and Tree Collapsing methods [7]. Similarly, the same subject has been studied for the mesh structure [1-5]. The PMCS (Partitionable Mesh Connected System) has been proposed for better partitioning of a mesh structure [6].

A parallel architecture and its computational model can be represented by various topologies. Many parallel algorithms are designed with one specific topology in mind since this can reduce computation time by reducing communication overhead. When the available parallel system has a different structure, then the algorithm has to be redesigned to run efficiently in the available architecture. Rather than redesigning an algorithm, it would be better if we could map the original topology to the currently available structure and run the previous algorithm on the available structure.

Embedding is the process of mapping one topology to another topology. The topology to be used for mapping is called the host graph and the topology to be mapped is called the guest graph. The efficiency of embedding is measured by dilation and expansion. Dilation is the maximum distance in the host graph between two nodes which are adjacent in the guest graph before embedding, and expansion is the ratio of the number of nodes needed in the host graph to the number of nodes in the guest graph. Expansion is related to the cost efficiency of embedding and dilation is related to the performance of embedded algorithms in the host graph. Hence, much research is focused on minimizing dilation and expansion.

The torus is a mesh with wraparound edges which has better performance characteristics than a mesh because of its smaller diameter and regular structure. However, one drawback of a torus structure is that one cannot find a sub-torus from a torus structure. Many parallel and distributed algorithms based on a torus structure cannot be used efficiently because of this reason.

In this paper, we propose a new torus-based struc-

---

† This work was supported in part by NONDIRECTED RESEARCH FUND, Korea Research Foundation, 1993 and ETRI under Contract 94231.

ture, which is named the *Dual Torus Network (DTN)*. The DTN is constructed by adding extra edges to a torus structure. The DTN has superior performance characteristics when compared to the mesh or torus. Moreover, it supports the allocation of sub-tori of arbitrary size. The idea of the DTN is generalized to the *l-complete torus*. Then we study the performance characteristics of the proposed DTN topology. Next, we show the relationship between embedding in the mesh or torus and embedding in the DTN. The embedding method is extended so that faulty nodes can be tolerated (fault-tolerant embedding). We also show that the torus is a subgraph of the DTN, which is important for allocation of sub-tori. Finally, we show that allocation and scheduling strategies developed for the mesh or torus can also be used in the DTN.

This paper is organized as follows. Section 2 overviews the work done on embedding and the partitionable mesh. In Section 3, the DTN and *l*-complete torus are defined. In Section 4, we first analyze the performance characteristic of the DTN and *l*-complete torus. Then, we present the problem of embedding other topologies to the DTN. The problem of embedding any rectangular grid into an optimal square DTN when faults exist is discussed in Section 5. In Section 6, the partitionability of the DTN is discussed. We conclude the paper with Section 7.

2 Previous Work

In this section, we review the previously reported work on the mesh in the areas of allocation, embedding, and node index ordering.

Li and Cheng have proposed the PMCS (Partitionable Mesh Connected System) and have introduced the scheduling problem for the PMCS [6]. Since Li and Cheng, many researchers have proposed various strategies for searching, allocating, and scheduling sub-meshes in PMCS. These include TDBS (Two dimensional Buddy System) [1], FS (Frame Sliding) [2], FF (FirstFit), BF (BestFit) [3], and AS (Adaptive Scanning) [4]. TDBS has much internal and external fragmentation and has imperfect sub-mesh recognition ability. FS has removed the internal fragmentation problem but still has severe external fragmentation and has imperfect sub-mesh recognition ability. Both FF and BF have imperfect sub-mesh recognition ability. AS was proposed to eliminate imperfect sub-mesh recognition problem in FS. However, AS also have external fragmentation problem since it uses a first fit method for allocation of free sub-meshes and a first fit method generates external fragmentation.

Recently, Sharma and Pradhan [5] proposed a strategy which searches candidate sub-meshes by pruning the list of allocated sub-meshes. It use BestFit to prevent fragmentation and has perfect sub-mesh recognition ability. The proposed strategy has higher time complexity for allocation than the other strategies since it searches all sub-meshes exhaustively in order to achieve perfect recognition ability.

Embedding is very important for the exchange of parallel algorithms and architectures in parallel and distributed computing. The efficiency of embedding is measured by expansion and dilation [8]. Embedding of rectangular grids into square grids has been studied by many researchers [9, 8, 10-12]. Embedding techniques used in these works are based on primitive embedding methods such as break-and-fold [13], folding, and compression [9]. The bound on the dilation of embedding is analyzed by Shen [14].

Recently, embedding of any rectangular grid into an optimal square grid with dilation 6 was proposed by Huang [11]. An optimal square represents the smallest square that can accommodate all nodes of a rectangular grid. A brief description of Huang's method is as follows : Let $G = h \times w$ be the guest graph and $G' = h' \times w'$ be an ideal rectangular grid of G. Let $H = s \times s (s = \lceil \sqrt{hw} \rceil)$, the host graph, be the optimal square grid of G. Ideal rectangular grids represent the most suitable rectangles for embedding into square grids. The entire embedding procedure consists of two steps when $\rho = \frac{w}{h} \geq 9$. First, G is embedded to G' by using modified compression. This compression can be represented by $h \times w'$ embedding matrix C. An element in C denotes the distribution of a node of G when mapped to G'. All elements in C are $\lceil \eta \rceil$ or $\lfloor \eta \rfloor$, where η is the compression ratio $\frac{w'}{w}$. All elements in C satisfy $c_{i,j} = c_{i+1,j+1}$ for all $1 \leq i < h, 1 \leq j < w'$. The most attractive characteristic of C is that any two adjacent equal-length partial row or column sums differ by at most 1. Because the compression ratio is less than 2, the dilation of the resulting compression is no more than $3 (= \lceil \eta \rceil + 1)$. The second step is to embed G' into H by using modified folding. In this modified folding method, the unmapped nodes in the last row after applying modified compression are truncated. This folding process produces dilation 2. Hence, the total dilation is $2 \times 3 = 6$. If $\rho = \frac{w}{h} < 9$, pure compression can embed G into H with dilation 3.

Interleaved ordering is a method of connecting the nodes in one dimension of a torus such that the wire length between any two adjacent nodes is constant [15]. Fig. 1 shows a torus connected using interleaved ordering. Interleaved ordering is useful in building a torus which would otherwise have a long wraparound edges or for mesh variants such as the fault-tolerant mesh proposed by Bruck et al. [16]. Because interleaved ordering of nodes between all adjacent nodes in a torus results in short wire length, it is very useful for applications which are sensitive to communication delay. One such application is VLSI implemented processor arrays.

3 Dual Torus Network(DTN)

In this section, we present the *DTN*, which uses the interleaved ordering scheme, and its generalized form. As discussed in the previous section, node placement using interleaved ordering on a torus topology can guarantee short and bounded distance between any two neighboring nodes. However, interleaved ordering incurs the overhead of converting the given torus address to the actual location address, and vice versa.

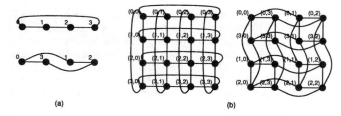

Figure 1: Ordering : normal ordering and interleaved ordering.

For example, in a 1-dimensional torus of node size N, the node located at the index i on torus does not have the same index in many cases. The index $I(N, i)$ on interleaved ordering (i.e., the actual location of node) can be found using the following conversion function.

$$I(N, i) = \begin{cases} \frac{i}{2} & \text{if } i \text{ is even,} \\ N - \lceil \frac{i}{2} \rceil & \text{if } i \text{ is odd} \end{cases} \quad (3.1)$$

Fig. 1(a) shows the difference between normal ordering and interleaved ordering for the 1-dimensional torus. Fig. 1(b) shows an example of extending 1-dimensional interleaved ordering to construct a 2-dimensional torus with interleaved ordering. It should be noted that the node index of an r-dimensional torus is represented as $(i_{r-1}, i_{r-2}, \ldots, i_0)$.

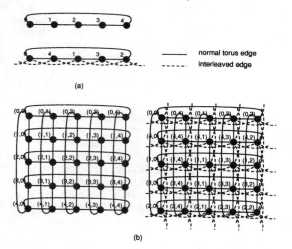

Figure 2: DTN and Torus.

Now, let us formally define the *Dual Torus Network(DTN)* as follows.

Definition 1: **(DTN)** Let a graph $G = (V, E)$ represent an r-dimensional torus. Then, the DTN, denoted as $G' = (V', E')$, is defined as follows: The set of vertices of DTN G', i.e., V', is exactly the same as V. The set of edges of DTN G', i.e., E', consists of E and E_I where E_I is a set of interleaved edges which are defined as follows. For any two nodes

$x = (x_{r-1}, x_{r-2}, \ldots, x_0)$, $y = (y_{r-1}, y_{r-2}, \ldots, y_0) \in V$, an edge $e \in E_I$ exists between x and y if $|(x_i - y_i) \bmod N| \leq 2$ and $x_j = y_j$ for all $0 \leq j \neq i < r$.

It is obvious that the DTN graph $G' = (V', E')$ has a torus with a normal ordering as a subgraph since the DTN is an extension of the torus graph $G = (V, E)$. One interesting point of the DTN graph is that we can also find a torus with an interleaved ordering by choosing a different set of edges. Hence, the name DTN. Fig. 2 shows construction of DTNs from the corresponding tori. Fig. 2(a) shows a 1-dimensional torus and DTN of size 5 and Fig. 2(b) shows a 2-dimensional torus and DTN of size 5×5. As we can see from the figure, the DTN is constructed by adding interleaved edges to the corresponding tori.

The *l-complete torus*, a generalized form of the DTN, is defined as follows.

Definition 2: **(l-complete Torus)** The *l*-complete torus is a graph constructed by adding extra edges between any two nodes in a torus when the distance between the two nodes is less than or equal to l in one dimension and is zero in the other dimensions.

Every node in the *l*-complete torus has an out-degree of $4l$ if the number of nodes in each dimension is greater than $2l$. The DTN is a special case of the *l*-complete mesh when l is 2. Hence, the DTN is a regular structure of out-degree 8 when the number of nodes in each dimension is greater than 4. When the number of nodes is less than or equal to 4, some nodes have a smaller out-degree since they have overlapped interleaved edges.

The concept of adding extra edges to a torus to form an *l*-complete torus is applicable to a torus of any dimension. Also, it is applicable to the mesh topology which does not have wraparound edges. In such a case, the resulting topology is called an *l*-complete mesh.

4 Characteristics of DTN and *l*-complete Torus

4.1 General Properties

Static interconnection networks are commonly evaluated by parameters such as diameter, connectivity, bisection width, and link cost [17]. The *Diameter* of a network is the maximum distance between any two processors in the network. The *Connectivity* is a measure of the multiplicity of paths between any two processors. More specifically, *arc-connectivity* represents the minimum number of edges that must be removed to break the network into two disconnected networks. The *Bisection width* is the minimum number of communication links that have to be removed to partition the network into two equal halves. The *Link cost* is the number of communication links in the network. The diameter is related to routing, the connectivity and bisection width are related to the reliability of the

Table 1: Comparison of various static interconnection topologies connecting p processors.

Network	Diameter	Bisection Width	Arc Connectivity	Cost (No. of links)
Complete graph	1	$p^2/4$	$p-1$	$p(p-1)/2$
Star	2	1	1	$p-1$
Complete binary tree	$2\log((p+1)/2)$	1	1	$p-1$
Linear array	$p-1$	1	1	$p-1$
Ring	$\lfloor p/2 \rfloor$	2	2	p
2-D Mesh	$2(\sqrt{p}-1)$	$\sqrt{p}$	2	$2(p-\sqrt{p})$
2-D Torus	$2\lfloor \sqrt{p}/2 \rfloor$	$2\sqrt{p}$	4	$2p$
Hypercube	$\log p$	$p/2$	$\log p$	$(p\log p)/2$
2-D DTN	$2\lfloor \sqrt{p}/4 \rfloor$	$6\sqrt{p}$	8	$4p$
r-D l-complete Torus	$r\lfloor p^{\frac{1}{r}}/2l \rfloor$	$l(l+1)\lfloor p^{\frac{(r-1)}{r}} \rfloor$	$2rl$	rlp
r-D l-complete Mesh	$r\lceil (p^{\frac{1}{r}}-1)/l \rceil$	$\frac{l(l+1)}{2}\lfloor p^{\frac{(r-1)}{r}} \rfloor$	rl	$rlp - rl\frac{(l+1)}{2}p^{\frac{r-1}{r}}$

network, and the link cost is one of the representative cost measures for networks.

Table 1 shows the characteristics of various topologies[1] and the proposed DTN and l-complete torus and mesh. As we can see from the table, when compared to the 2-D torus, the DTN requires twice as many links, half of the diameter, twice the connectivity, and three times the bisection width. This implies that the DTN is a more robust structure.

4.2 Embedding Properties

In this subsection, the embedding property of the DTN is discussed.

Theorem 1: Let there be a dilation-δ embedding of a graph G into an r-dimensional mesh or torus. Then, the graph G can be embedded into an r-dimensional DTN with dilation at most $\frac{\delta}{2} + r$ and at least $\frac{\delta}{2}$.

In this paper, the proofs for most theorems and lemmas have been omitted due to lack of space. The interested reader can obtain the proofs from [18].

Corollary 1: Let there be a dilation-δ embedding of a graph G to an r-dimensional mesh or torus. Then we can embed the graph G to an r-dimensional l-complete torus with dilation at most $\frac{\delta}{l} + r$ and at least $\frac{\delta}{l}$.

Now, as an example, we will show the process of embedding a rectangular grid into an optimal square DTN. First, we explain how to embed a rectangular grid into an optimal square torus with dilation 4 and apply Theorem 1 to embed rectangular grids into an optimal square DTN with dilation 3. Our embedding scheme is based on the method proposed by Huang [11], which embeds any rectangular grid into an optimal square mesh with dilation 6. Note that the problem studied by Huang is embedding any rectangular grid into an optimal square mesh whereas our

problem is embedding any rectangular grid into an optimal square torus.

Definition 3: (Torus Folding) Let (a, b) be the index of a node in a rectangular grid of size $h \times w$ and (a', b') be the index of the node in a torus of size $h' \times w'$ into which the node (a, b) will be mapped. Then, the mapping function $\phi(a, b) = (a', b')$ is defined as follows.

$$b' = b \bmod w',$$
$$a' = a'' \bmod h', \quad \text{where}$$
$$a'' = a + h\lfloor b/w' \rfloor + \begin{cases} 0 & \text{if } b' \leq w' - h, \\ b' - w' + h & \text{if } b' > w' - h \end{cases}$$

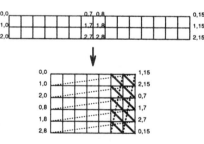

Figure 3: Embedding a 3×16 mesh system into a 6×8 mesh system using torus folding.

Fig. 3 shows an example of embedding a rectangular grid to a torus using torus folding. Torus folding has the following property.

Theorem 2: Let the guest graph be $G = h \times w$ and the host graph be $G = h' \times w'$. If $\rho = \frac{w}{w'}$ is an integer, then torus folding embeds G into G' with dilation 2.

Corollary 2: If two nodes are vertically adjacent in the guest graph G, then the corresponding two nodes in the host graph G' (torus) after torus folding are vertically adjacent.

---

[1]The characteristics of various topologies listed in Table 1 are quoted from the reference [17].

The following theorem shows that the proposed torus folding embeds a rectangular grid into a torus with dilation 4.

Theorem 3: When the folding scheme used in Huang's embedding method is replaced with torus folding, any rectangular grid can be embedded into an optimal square torus with dilation 4.

Proof : As discussed by Huang in [11], pure compression embeds a rectangular grid $G = h \times w$ into an optimal square mesh $G' = h' \times w'$ with dilation at most 3 if $\rho = \frac{w}{w'} < 9$. When $\rho \geq 9$, the two-step embedding procedure is required to embed G into G' with dilation 6. It was also mentioned that the first step of *Huang's* scheme is applying modified compression, which embeds a rectangular grid into an ideal rectangular grid with dilation 3. The second step of Huang's scheme is applying folding to embed an ideal rectangular grid into an optimal square mesh. We replace this part with torus folding. When torus folding is applied, we need to consider only three cases of adjacent node embeddings which have dilation 3 as shown in Fig. 5. As stated in Corollary 2, torus folding does not increase the dilation between vertically adjacent nodes. Hence, the dilation increase is only due to the dilation increment in the horizontal edge, which is at most two as claimed in Theorem 2. Therefore, the total dilation is not greater than 4. □

An example of embedding a 3×37 mesh into an 11×11 torus by using compression and torus folding is given in Fig. 4.

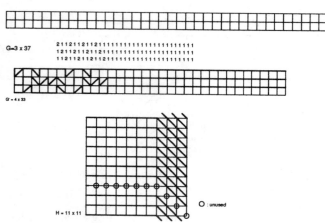

G = 3 × 37

2112112112111111111111111111111111
1211211211211111111111111111111111
1121121121121111111111111111111111

G' = 4 × 33

H = 11 × 11

O : unused

Figure 4: Embedding 3×37 into 11×11 by compression and torus folding.

Theorem 4: Any rectangular grid can be embedded into an optimal square DTN with dilation 3.

Proof : We found that any rectangular grid can be embedded into an optimal square torus with dilation 4. As a next step, we apply Theorem 1 to the embedding

of a rectangular grid into an optimal square DTN. In this embedding, a rectangular grid is embedded into an optimal square DTN by using only the regular edges of the torus (which is a subgraph of the DTN). Then if $\rho \leq 9$, dilation is not greater than 3, and if $\rho \geq 9$, dilation is at most 4. As defined in Section 3, DTN has interleaved edges that connect two nodes of distance 2 in the same dimension. Since the vertical dilation in the embedded regular torus is either 2 or 3, vertical dilation reduces to 1 or 2 by using interleaved edges. Hence, the dilation in embedding a rectangular grid into an optimal square DTN is at most 3. □

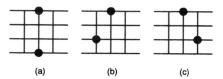

(a)   (b)   (c)

Figure 5: Possible cases of dilation 3 by modified compression.

5 Fault Tolerant Embedding in DTN

One commonly used method to tolerate a fault in a mesh- or torus-based system is reconfiguration with spare nodes [19]. In this scheme, the system excludes a faulty node by setting switches to replace the faulty node with the spare. Reconfiguration can also be done with no spare nodes (this kind of reconfiguration is called graceful degradation). For example, in the *l*-complete torus, there are links between any two nodes which are separated by a distance less than or equal to *l* in the corresponding torus. Hence, we can tolerate a maximum of $l - 1$ consecutive faulty nodes by simply changing the indices of the nodes. In a DTN, the system cannot tolerate two faults located consecutively in the same row or column. However, if faults are scattered, then the DTN system can tolerate a certain number of faulty nodes. In this section, we consider the problem of embedding a rectangular grid into an optimal square DTN or *l*-complete torus with a few faulty nodes, assuming that the system has enough available working nodes to accommodate a rectangular grid.

In Section 4.2, we considered the problem of embedding a rectangular grid into an optimal square DTN. Let us assume that faults have occurred in an optimal square DTN, but the system still has enough working nodes to accommodate a rectangular grid. We need a systematic method of embedding a rectangular grid into a faulty optimal square DTN. One possible method is to solve the problem of embedding a rectangular grid into an optimal square faulty torus since the torus is a subgraph of the DTN. Hence, we propose a fault tolerant embedding into a torus. We then extend the embedding scheme proposed in Section 4.

The fault-tolerant embedding problem into an optimal square torus can be viewed as the problem of finding a method to map a faulty node to an unused node in the embedding when the number of faulty nodes is

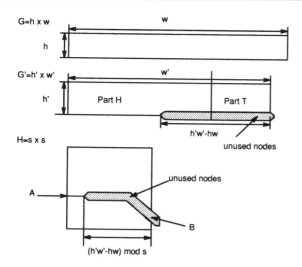

Figure 6: Unused nodes in embedding.

$1\ U_H = U\phi_2 - 1;$
$2\ \text{for } i = 0 \text{ to } w' - h'w' + hw + U_{\phi_2} - 1 \text{ do}$
$3\quad \text{if } U_H \leq 0 \text{ then}$
$4\qquad \text{return FALSE};$
$5\quad \text{if } i\text{th column contains } f_i \text{ faulty nodes}$
$6\qquad \text{if the number of faulty nodes in } c_i \text{ is less than or}$
$\qquad\quad \text{equal to } (h - h')$
$7\qquad\quad \text{Shift right } w' - i \text{ columns from } c_i \text{ to } c_{w'-1}$
$\qquad\qquad \text{and delete } c_{w'};$
$8\qquad\quad \text{Insert } c_i \text{ with } [111...1]^T;$
$9\qquad\quad \text{Generate } (h - h') \text{ more 2's in } C;$
$10\qquad\quad U_H = U_H - f_i;$
$11\qquad \text{else if the number of faulty nodes in } c_i \text{ is greater}$
$\qquad\quad \text{than } h' \text{ or } c_i \text{ has adjacent faulty nodes}$
$12\qquad\quad \text{Unmap the column } i \text{ from } G' \text{ when applying}$
$\qquad\qquad \text{modified compression};$
$13\qquad\quad \text{Generate } h' \text{ more 2's in } C;$
$14\qquad\quad U_H = U_H - h';$
$15\ \text{Apply modified compression } \phi_1 \text{ by this modified } C;$
$16\ \text{Apply modified torus folding } \phi_2 \text{ to this modified } G';$

Figure 7: Column insertion and skipping procedure in Part H.

not more than the number of unused nodes. There exist unused nodes due to the fact that the number of mapped nodes(optimal expansion) by embedding is not always the same as the number of node in the entire torus.

The problem is defined formally as follows. Let $\phi_1 : G = h \times w \rightarrow G' = h' \times w'$ and $\phi_2 : G' = h' \times w' \rightarrow H = s \times s$, where G is a rectangular grid, G' is an ideal rectangular grid of G, and H is an optimal square torus. ϕ_1 is a modified compression and ϕ_2 is a modified torus folding. Then the number of unused nodes in ϕ_1 is $U_{\phi_1} = \lceil hw/w' \rceil \times \lfloor w/s \rfloor \times s - hw$ and the number of unused nodes in ϕ_2 is $U_{\phi_2} = U_{\phi_1} \bmod s = \{\lceil hw/w' \rceil \times \lfloor w/s \rfloor \times s - hw\} \bmod s = s^2 - hw$.

The fault-tolerating embedding scheme consists of three steps. They are rotation, column insertion, and column skip. The first step is rotation and this step is applied only when U_{ϕ_2} is greater than or equal to 1. This is because we can map a faulty node to the last one of the unused nodes (node B in Fig. 6) by rotating the torus. Fig. 6 shows the range of unused nodes. The index of row A is $A = \lceil (w' - h'w' + hw)/s \rceil h'$ and the index of node B is $(i, j) = (A + h' - 1, s - 1)$. The rotation of the torus to match the faulty node to the location B is performed by following circular index shifting $\phi_3 : (X, Y) \rightarrow (X', Y')$, where $X' = (X - i) \bmod s$ and $Y' = (Y - j) \bmod s$. Thus, 'torus rotation' results in an index change.

After the rotation, column insertion is applied if the number of unused nodes left is greater than $(h - h')$, the number of faulty nodes in each column is not greater than $(h - h')$, and faulty nodes in the same column are not adjacent to each other. When column insertion cannot be applied and the number of unused nodes left is greater than h', column skip is applied. Fig. 7 shows the column insertion and column skip procedures.

Before applying the second and third steps, which correspond to the column insertion and skip steps, the indices of the faulty nodes in an optimal square torus H are converted to the indices in an ideal rectangular grid G'. Let Part H represent the nodes of G' from column 0 to column $w' - h'w' + hw + U_{\phi_2} - 1$. Note that the last unused node B is mapped to the last row of $(w' - h'w' + hw + U_{\phi_2} - 1)^{th}$ column in G'. When a faulty node is mapped to Part H in the grid G' shown in Fig. 6, the column which contains a faulty node is shifted and a null mapping column (denoted as $[111...1]^T$ in Fig. 7) is inserted. The variable C in Fig. 7 represents the node distribution in G' after the mapping ϕ_1 and c_i represent the ith column of C.

It should be noted that column insertion (lines 7-10 of the procedure shown in Fig. 7) and column skip (lines 12-14) do not change the property of the embedding matrix C. Therefore, dilation does not increase by column insertion and skip. Column insertion needs $(h - h')$ unused nodes and column skip needs h' unused nodes. By using rotation followed by this mapping, we can tolerate at most $\lfloor \frac{(U_{\phi_2} - 1)}{(h - h')} \rfloor$ faulty nodes.

If the number of remaining unused nodes is more than h' after rotation and insertion, we can tolerate more faulty columns in Part T in Fig. 6 by using column skip. Let the number of unused nodes be U_T after applying rotation and the above procedure. Then we can tolerate at most $\lfloor \frac{U_T}{h'} \rfloor$ faulty columns. Fig. 8 shows the column skip procedure in Part T in G'.

Fig. 9 shows an example embedding of a 3×37 rectangular grid into its optimal square torus when there are 6 faulty locations. The node which is in the 21st column of the last row is mapped to the last unused

```
 1 for i = w' − h'w' + hw + U_{φ_2} to w' − 1 do
 2   if ith column contains f_i faulty nodes then
 3     if U_T ≤ 0 then
 4       return FALSE;
 5     else
 6       the column i from G' when applying
            modified compression;
 7       Generate h' − 1 more 2's in C;
 8       U_T = U_T − h' + 1;
 9 Apply modified compression φ_1 by this modified C;
10 Apply modified torus folding φ_2 to this modified G';
```

Figure 8: Column skip procedure in Part T.

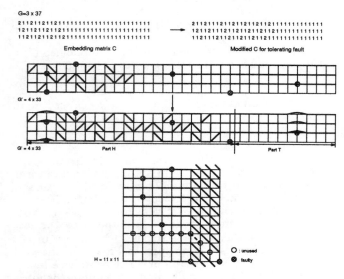

Figure 9: Example of 6-fault tolerant embedding 3×37 rectangular grid to its optimal square torus.

node in H by index rotation. For column 2, column skip is used for two faulty nodes, and for columns 5 and 15, column insertion is used. Column skip is also used for column 28.

6 Partition and Allocation

An important feature of the PMCS proposed by Li and Cheng [6] is that it can be dynamically partitioned into many rectangular sub-meshes of different sizes; the resulting sub-meshes can be assigned to independent jobs. These partitioned sub-meshes may execute any parallel algorithms either in SIMD/MIMD manner or in systolic fashion. However, the PMCS does not support the allocation of sub-tori even though it is known that a torus has many advantages when compared to a mesh. In this Section, we show that the DTN supports the allocation of sub-torus and sub-mesh structures.

Theorem 5: DTN supports the allocation of sub-torus.

Sub-mesh allocation strategies have been studied by many researchers. The sub-torus allocation problem in the DTN can be solved by solving the sub-mesh allocation problem in the PMCS. If we are able to find a sub-mesh of the required size in the given PMCS, then we can construct the same size sub-torus in the corresponding DTN by using interleaved edges. Hence, the previously proposed sub-mesh allocation algorithms for the PMCS can be used as sub-torus allocation algorithms for the DTN.

One advantage of the DTN over the PMCS is that sub-torus or sub-mesh availability of the DTN is higher than that of the PMCS. Sub-tori and sub-meshes can be represented by the index of the base node, width, and height.

The base node is the one that has the lowest row index number and the lowest column index number in the sub-mesh or sub-torus. Assume that a task requires an $n \times m$ sub-mesh or sub-torus. Then, in the PMCS, the node in the reject set cannot serve as a base node of the sub-mesh. The reject set denotes the set of nodes which cannot be a base node of sub-mesh of the required size because of the boundary conditions of the PMCS. For example, if the index of a node is represented by (x, y), where $0 \leq x \leq (N − 1)$ and $0 \leq y \leq (M − 1)$, the index range of the nodes (i, j) in the reject set is $(N − n + 1) \leq i \leq (N − 1)$ or $(M − m + 1) \leq j \leq (M − 1)$. However, a node in the DTN which is in the reject set can also serve as a base node of the sub-mesh or sub-torus since the sub-mesh or sub-torus represented by nodes in the reject set can be constructed by connecting opposite parts of torus using wraparound edges. Hence, if the size of the DTN is $N \times M$, the searching plane for finding candidates for the base node of the sub-mesh is extended from $(N − n + 1) \times (M − m + 1)$ to $N \times M$. This results in increased availability for sub-meshes of the required size.

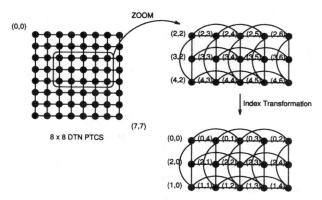

Figure 10: Node index reassignment when allocating a sub-torus in a DTN.

To execute a task in a sub-torus, the node indices must be converted. If a task requires an $n \times m$

torus and the index of the base node is (p, q), then the node index (i, j) in the base graph (mesh) is transformed to $(I(n, i - p), I(m, j - q))$ in the sub-torus. The index conversion function is $\psi((i, j) = (I(n, i - p), I(m, j - q))$. Fig. 10 shows the conversion of indices of allocated nodes to the indices in a sub-torus when a 3×5 sub-torus is allocated in an 8×8 DTN. Given this index conversion, one cycle in the top row of the 3×5 sub-torus, for example, is $(0, 0) \to (0, 1) \to (0, 2) \to (0, 3) \to (0, 4) \to (0, 0)$.

7 Conclusion

In this paper, we have proposed the Dual Torus Network (DTN) as a viable multiprocessor interconnection network and generalized the concept to the l-complete torus. Also, we analyzed many aspects of the proposed DTN topology and showed that the DTN has many advantages over the mesh and torus. In addition, we showed that if a certain topology can be embedded into a mesh or torus, then that topology is also embeddable into a DTN with smaller dilation. As a special case, we showed the embedding of a rectangular grid into an optimal square DTN with dilation 3. Note that the currently reported smallest dilation when embedding a rectangular grid into an optimal square mesh is 6. For fault-tolerant embedding, we modified the method used to embed a rectangular grid into an optimal square DTN.

Another salient feature of the DTN is it supports partitioning of sub-tori, which is not possible in the mesh or torus. Since the DTN has the mesh and torus as subgraphs, all of the research work developed for the mesh and torus can be used directly for a DTN. One such area is a partioning, scheduling, and allocation of sub-mesh and sub-torus structures in a DTN. Scheduling and allocation strategies in a mesh can be used with almost no modification.

References

[1] K. Li and K. H. Cheng, "A two dimensional buddy system for dynamic resource allocation in a partitionable mesh connected system," in *Proc. of ACM Computer Science Conference*, pp. 22–28, 1990.

[2] P.-J. Chuang and N.-F. Tzeng, "An efficiet sub-mesh allocation strategy for mesh computer systems," in *Proc. 11th International Conf. on Distributed Computing Systems*, pp. 256–262, 1991.

[3] Y. zhu, "Efficient processor allocation strategies for mesh-connected parallel computers," *Journal of Parallel and Distributed Computing*, pp. 328–337, 1992.

[4] J. Ding and L. N. Bhuyan, "An adaptive submesh allocation strategy for two-dimensional mesh connectd systems," in *Proc. 1993 International Conf. on Parallel Processing*, volume II, pp. 22–28, 1993.

[5] D. D. Sharma and D. K. Pradhan, "A fast and efficient strategy for submesh allocation in mesh connected parallel computers," in *Proceedings of 1993 The 5th IEEE Symposium on Parallel and Distributed Processing*, pp. 682–689, 1993.

[6] K. Li and K. H. Cheng, "Complexity of resource allocation and job scheduling problem in partitionable mesh connected systems," in *First Ann. IEEE Symp. Parallel and Distributed Processing*, pp. 358–365, 1989.

[7] C. Yu, P. Mohapatra, C. R. Das, and J. Kim, "A lazy scheduling scheme for improving hypercube performance," in *Proc. 1993 International Conf. on Parallel Processing*, volume I, pp. 110–117, 1993.

[8] J. A. Ellis, "Embedding rectangular grids into square grids," *IEEE Trans. on Computers*, no. 1, pp. 46–52, 1991.

[9] R. Aleliunas and A. L. Rosenberg, "On embedding rectangular grids in square grids," *IEEE Trans. on Computers*, no. 9, pp. 907–913, 1982.

[10] R. Methern and G. Hwang, "Embedding rectangular grids into square grids with dilation two," *IEEE Trans. on Computers*, no. 12, pp. 1446–1445, 1990.

[11] S.-H. S. Huang, H. Liu, and R. M. Verma, "On embeddings of rectangles into optimal squares," in *International Conf. on Parallel Processing*, volume III, pp. 73–76, 1993.

[12] S.-H. S. Huang, H. Liu, and R. M. Verma, "A new combinatorial approach to optimal embedding of rectangles," in *8th International Parallel Processing Symposium*, pp. 715–722, 1994.

[13] C.E.Leiserson, "Area-efficient graph layouts(for vlsi)," in *Proc. 21st IEEE Symp. on Foundations of Computer Science*, pp. 270–281, 1980.

[14] X. Shen, W. Liang, and Q. Hu, "Embedding between 2-d meshes of the same size," in *Proc. of the Fifth IEEE Symposium on Parallel and Distributed Processing*, pp. 712–719, 1993.

[15] W. Dally and C. L. Seitz, "Deadlock-free message routing in multiprocessor interconnection networks," *IEEE Trans. on Computers*, vol. C-36, pp. 547–553, May 1987.

[16] J. Bruck et al., "Fault-tolerant meshes and hypercubes with minimal numbers of spares," *IEEE Trans. on Computers*, no. 9, pp. 1089–1104, 1993.

[17] V. Kumar et al., *Introduction to Parallel Computing : Design and Analysis of Algorithms*, The Bejamin/Cumming Publishing Company, inc, 1994.

[18] S. Chae, J. Kim, D. Kim, S. Hong, and S. Lee, "DTN : A new partitionable torus topology," Technical Report CS-95-002, Department of Computer Science and Engineering, Pohang University of Science and Technology, 1995.

[19] B. W. Johnson, *Design and Analysis of Fault Tolerant Digital Systems*, Addison Wesley, 1989.

Sparse Hypernetworks Based on Steiner Triple Systems

S.Q. Zheng

Dept. of Computer Science and Dept. of Electrical & Computer Engineering
Louisiana State University
Baton Rouge, LA 70803
Email: zheng@bit.csc.lsu.edu

Abstract

We propose several linear, uniform and regular hypernetworks based on Steiner triple systems. The most attractive features of these hypernetworks include small vertex degree, small diameter, low rank, and increased wire-sharing. These hypernetworks have excellent potential to be used as interconnection networks for massively parallel computer systems.

1 Introduction

Designing high bandwidth, low latency and scalable interconnection networks is a great challenge faced by parallel computer designers. It is expected that in the near future hundreds of thousands of processors in a massively parallel computer system execute fine-grained concurrent programs in parallel to solve a large-scale problem. The interprocessor communication performance is the most critical aspect for the success of such a computing system. Recently, a new class of interconnection networks, the hypernetworks, was proposed [9]. The class of hypernetworks is a generalization of point-to-point networks. It includes point-to-point interconnection networks as a subclass. The underlying graph theoretic model of hypernetworks is hypergraph theory [1]. In a hypernetwork, a communication link (called a hyperlink) connects two or more processors. The major advantages of hypernetworks over point-to-point networks include increased wire sharing, design flexibilities and network scalability. With increased wire sharing, the cost effectiveness of interconnection network systems can be improved. The relaxation on the number of vertices that can be connected by a link provides more design alternatives so that greater flexibilities in trade-offs of contradicting design goals are possible. The hypernetworks are particularly suitable for optical interconnects. Interested readers may refer to [9] for more justifications,

design issues, implementation aspects, and some hypernetwork design examples.

In this paper, we propose several linear, uniform and regular hypernetworks based on Steiner triple systems. The most attractive features of these hypernetworks include small vertex degree, small diameter, low rank, and increased wire-sharing. These sparse hypernetworks are good alternatives for high performance interprocessor communications.

2 Preliminaries

Hypergraphs are used as underlying graph models of hypernetworks. A *hypergraph* $H = (V, E)$ consists of a set $V = \{v_1, v_2, \cdots, v_n\}$ of vertices, and a set $E = \{e_1, e_2, \cdots, e_m\}$ of *hyperedges* such that each e_i is a non-empty subset of V and $\cup_{i=1}^{m} e_m = V$. If $e_i \subset e_j$ implies that $i = j$, then H is a *simple hypergraph*. We assume that the cardinality of each edge is no less than 2. By doing so we rule out the possibility of a selfloop edge. If each of all edges has cardinality 2, then H is a graph that corresponds to a point-to-point network.

For a subset J of $\{1, 2, \cdots, m\}$, we call the hypergraph $H'(V', E')$ such that $E' = \{e_i | i \in J\}$ and $V' = \cup_{e_i \in E'} e_i$ the *partial hypergraph generated by the set J*. For a subset U of V, we call the hypergraph $H''(V'', E'')$ such that $E'' = \{e_i \cap U | 1 \leq i \leq m, e_i \cap U \neq \emptyset\}$ and $V'' = \cup_{e \in E''} e$ the *sub-hypergraph induced by the set U*. For $v_i \in V$, we define the *star* $H(v_i)$ with center v_i to be the partial hypergraph formed by edges containing v_i. The degree $d_H(v_i)$ of v_i in H is the number of edges of $H(v_i)$. A hypergraph in which all vertices have the same degree is said to be *regular*. The *degree of hypergraph H*, denoted by $\Delta(H)$, is defined as $\Delta(H) = \max_{v_i \in V} d_H(v_i)$. A regular hypergraph of degree k is called *k-regular hypergraph*. The *rank* $r(H)$ and *antirank* $s(H)$ of a hypergraph H is defined as $r(H) = \max_{1 \leq j \leq m} |e_j|$ and $s(H) = \min_{1 \leq j \leq m} |e_j|$, respectively. We say that

H is a *uniform hypergraph* if $r(H) = s(H)$. A uniform hypergraph of rank k is called *k-uniform hypergraph*. In a hypergraph H, a chain of length q is defined as a sequence $(v_{i_1}, e_{j_1}, v_{i_2}, e_{j_2}, \cdots, e_{j_q}, v_{i_{q+1}})$ such that (1) $v_{i_1}, v_{i_2}, \cdots, v_{i_{q+1}}$ are all distinct vertices of H; (2) $e_{j_1}, e_{j_2}, \cdots, e_{j_q}$ are all distinct edges of H; and (3) $v_{i_k}, v_{i_{k+1}} \in e_{j_k}$ for $k = 1, 2, \cdots, q$. A path from v_i to v_j, $i \neq j$, is a chain in H with its end vertices being v_i and v_j. A hypergraph is *connected* if the intersection graph of its edges is connected. We only consider connected hypergraphs. A hypergraph is *linear* if $|e_i \cap e_j| \leq 1$ for $i \neq j$. For any two distinct vertices v_i and v_j in a hypergraph H, the distance between them, denoted by $dis(v_i, v_j)$, is the length of the shortest path connecting them in H. The *diameter* of a hypergraph H, denoted by $\delta(H)$, is defined by $\delta(H) = \max_{v_i, v_j \in H, i \neq j} dis(v_i, v_j)$. More concepts in hypergraph theory can be found in [1]

A *hypernetwork* N is a network whose underlying structure is a hypergraph H, in which each vertex v_i corresponds to a unique processor P_i of N, and each hyperedge e_j corresponds to a *connector* that connects the processors represented by the vertices in e_j. A connector is loosely defined as an electronic or a photonic component though which messages are transmitted between pairs of connected processors, not necessarily simultaneously, in constant time. We may call a connector a *hyperlink*. The simplest implementation of a hyperlink is by a bus [9]. In this paper, the following pairs of terms are used interchangeably: (hyper)edges and (hyper)links, vertices and processors, point-to-point networks and graphs, and hypernetworks and hypergraphs.

3   Steiner Hypernetworks

In this section, we first introduce the notion of Steiner triple systems. Then, we show how to construct hypernetworks using these systems.

3.1   Steiner Systems

A *Steiner system* (V, E), denoted $S(t, k, v)$, is a set V of cardinality v, and a collection E of k-subsets of V such that every t-subset of V is contained in exactly one member of E, where t is a constant less than k. The question that for which integers t, k, and v is it possible to construct such a system was originally asked by W.S.B. Woolhouse in 1844. Steiner systems $S(2, 3, v)$ are called *Steiner triple systems*. The investigations on Steiner triple systems were stimulated by

the famous 15-schoolgirl problem posted by T.P. Kirkman [2] in 1847, which corresponds to the Steiner triple system $STS(2, 3, 15)$. The subject has drawn interest from researchers in several mathematics areas such as algebraic numbers theory, combinatorial systems, hypergraph theory, etc. Interested readers may refer to [6] for a collection of papers, and a bibliography that consists of 730 titles of the articles on the subject, from mind-1840's to 1980. To our knowledge, no one has considered using the concept of Steiner systems to design interconnection networks.

In terms of hypergraph, $S(t, k, v)$ is a uniform hypergraph of order v and rank k, with each element in V being a vertex, and each k-subset in E being a hyperedge. Clearly, any $S(2, k, v)$ is a linear hypergraph, and each pair of vertices is contained in exactly one hyperedge. We are interested in Steiner systems with $t = 2$; in particular, the Steiner triple systems. We use $STS(n)$ to represent the Steiner triple system of order n, and its corresponding hypergraph. The necessary and sufficient condition for the existence of $STS(n)$ is $n \equiv 1$ or $3 \pmod 6$ [7]. $STS(n)$ is a uniform, regular hypergraph. The number of hyperedges, the degree, the rank, and the diameter of $STS(n)$ are $\frac{n(n-1)}{6}$, $\frac{n-1}{2}$, 3, and 1, respectively. The following is an example of $STS(7)$.

$$STS(7) = \{\{1,2,4\},\{1,3,7\},\{1,5,6\},\{2,3,5\},$$
$$\{2,6,7\},\{3,4,6\},\{4,5,7\}\}.$$

We say that an integer n is *admissible* if it is positive and $n \equiv 1$ or $3 \pmod 6$. We may use $STS(n)$, where n is admissible, as a hypernetwork. We call the hypernetwork whose underlying hypergraph is $STS(n)$ the *Steiner hypernetwork* of order n, and denote it by $SH(n)$. Note that for $n > 3$ $STS(n)$ is not unique. For a multiprocessor system with a small number of processors, a Steiner hypernetwork can be a better alternative as an interconnection network than point-to-point network of the same size.

3.2   Multidimensional Steiner Networks

Based on STS's, we propose a class of hypernetworks, called the *multidimensional Steiner hypernetworks*. Let $STS^*(b)$ be a selected Steiner triple system of order b (b must be admissible). The d-multidimensional Steiner hypernetwork of base b, denoted by $MSH(b, d)$, consists of b^d vertices, each is labeled by a unique d-tuple $(v_1, v_2, \cdots, v_d)$ such that $1 \leq v_i \leq b$. We call such a d-tuple the *vertex label* of a vertex in $MSH(b, d)$, and each element in it a *digit*. Each hyperlink of $MSH(b, d)$ consists of exactly 3 vertices. Three vertices x, y and z in $MSH(b, d)$ are connected by a hyperlink if and only if their vertex labels differ in exactly one digit, v_i, and $\{v_i(x), v_i(y), v_i(z)\} \in STS^*(b)$; such a hyperlink is called an *i-dimensional hyperlink* of $MSH(b, d)$. It is easy to verify the following claim:

Theorem 1 $MSH(b, d)$, *b is admissible and $d \geq 1$, is a 3-uniform, $(d\frac{b-1}{2})$-regular linear hypernetwork. Its diameter is d and it has b^d vertices and $\frac{d}{6}b^d(b-1)$ hyperlinks.*

Given a point-to-point network G and a hypernetwork H. An *embedding* of G into H is a mapping of the vertices of G into the vertices of H in conjunction with a specification of a path in H that connects the two images of two adjacent vertices of G in H for all pairs of adjacent vertices in G. The maximum length of the paths connecting the images of adjacent vertices of G in H is called the *dilation* of the embedding, and the maximum of mapped paths that include an hyperedge of H is called the *congestion* of the embedding. Similarly, we define the embedding of a hypergraph H_1 into another hypergraph H_2. Graph embedding is used as a tool for simulating one network (the embedded one) by another.

Consider the $SH(7)$ based on the Steiner triple system $STS(7)$ given in the previous section. We can identify the following cycle:

$$1\{1, 5, 6\}6\{2, 6, 7\}7\{4, 5, 7\}5\{2, 3, 5\}2$$
$$\{1, 2, 4\}4\{3, 4, 6\}3\{1, 3, 7\}1$$

Each hyperlink of $SH(7)$ is used exactly once. Therefore, a point-to-point ring network of 7 vertices can be embedded into $SH(7)$ with dilation 1 and congestion 1. Now let us compare $MSH(7, d)$ with 7-ary d-cube. For the definition of k-ary d-cube, refer to [3]. By the above ring embedding scheme, it is easy to show that the point-to-point 7-ary d-cube, which has 7^d vertices, can be embedded into $MSH(7, d)$, which also has 7^d vertices, with dilation 1 and congestion 1. This implies that the bus-implemented $MSH(7, d)$ can simulate each parallel step, either computation step or communication step, of the 7-ary d-cube in one step. This simulation is optimal, since both have the same number of vertices and the same number of (hyper)links. Since k-ary d-cube contains a Hamiltonian cycle, a point-to-point ring network of 7^d vertices can be embedded into $MSH(7, d)$ with dilation 1 and congestion 1.

3.3 Generalized Steiner Hypernetworks

In this section, we propose a class of hypernetworks, called the *generalized multidimensional Steiner hypernetworks*, which are shown having advantages over the multidimensional Steiner hypernetworks introduced in the previous section.

The d-dimensional generalized multidimensional hypernetwork of base b, denoted by $GMSH(b, d)$, is defined recursively. We use $n(b, d)$ and $m(b, d)$ to denote the number of vertices and the number of hyperlinks in $GMSH(b, d)$, respectively. Each vertex of $GMSH(b, d)$ is labeled by a d-tuple $(v_1, v_2, \cdots, v_d)$. $GMSH(b, 1)$, where b is admissible, is a b-vertex hypergraph with each vertex w having a unique v_1 label $v_1(w) = i$, $1 \leq i \leq b = n(b, 1)$, such that three vertices x, y and z are connected by a hyperlink if and only if $\{v_1(x), v_1(y), v_1(z)\} \in STS^*(b)$, where $STS^*(b)$ is a selected Steiner triple system of order b. That is, $GMSH(b, 1)$ is $SH^*(b)$. The hyperlinks in $GMSH(b, 1)$ are called the *1-dimensional hyperlinks*. $GMSH(b, d)$ is a hypernetwork that is constructed using $2n(b, d-1) + 1$ copies, $GMSH^1(b, d-1), GMSH^2(b, d-1), \cdots, GMSH^{2n(b,d-1)+1}(b, d-1)$, of $GMSH(b, d-1)$. Each of these component $GMSH(b, d-1)$'s is called a $(d-1)$-*dimensional sub-GMSH*. For each vertex w in $GMSH^i(b, d-1)$, $v_d(w) = i$, and $v_l(w)$, $1 \leq l \leq d-1$, remain the same as in $GMSH(b, d-1)$. All hyperlinks connecting vertices in the same $GMSH^i(b, d-1)$ follow the connectivity conditions specified by their vertex labels formed by digits $v_1, v_2, \cdots, v_{d-1}$, i.e. the l-dimensional hyperlinks in each $GMSH^i(b, d-1)$ are the l-dimensional hyperlinks of $GMSH^i(b, d)$. The only additional hyperlinks, which are called *d-dimensional hyperlinks*, are those connecting vertices from different copies of $GMSH(b, d-1)$. More specifically, a hyperlink $\{x, y, z\}$ is an d-dimensional hyperlink of $GMSH(b, d)$ if and only if (1) x, y and z are in different copies $GMSH^i(b, d-1)$, $GMSH^j(b, d-1)$, and $GMSH^k(b, d-1)$, respectively, of $GMSH(b, d-1)$; (2) $v_l(x) = v_l(y) = v_l(z)$ for $1 \leq l \leq d-1$; (3) $\{v_d(x), v_d(y), v_d(z)\} \in STS^*(2n(b, d-1) + 1)$; and (4) each vertex w of $GMSH(b, d)$ is included in exactly one d-dimensional hyperlink. Here, $STS^*(2n(b, d-1) + 1)$ is a selected Steiner triple system of order $2n(b, d-1) + 1$. An algorithm for constructing $STS((k-1)v + 1)$ from $STS(v)$ described in [5] can be used to find all STS's that are needed in the construction of a $GMSH(b, d)$. The assignment of each element of a d-tuple vertex label (subject to recursive conditions (1) through (4)) is rather arbitrary. In an actual design of $GMSH$, one may need to choose a particular vertex labeling scheme for the ease of data routing among vertices. It is not difficult to prove the following claim:

Theorem 2 *The generalized d-dimensional Steiner hypernetwork of base b, $GMSH(b,d)$, is a 3-uniform, $(\frac{b-1}{2} + d - 1)$-regular, linear hypernetwork that has $n(b,d) = O((2b)^{2^{d-1}})$ vertices, and $m(b,d) = O(2^{d-1}b^2(2b)^{2^{d-1}})$ hyperlinks. The degree $\Delta(GMSH(b,d)) = d = O(\log_2 \log_{2b} n(b,d)) = O(\log_2 \log_{2b} m(b,d))$, and the diameter $\delta(GMSH(b,d)) = 2^d - 1 = O(\log_{2b} n(b,d)) = O(\log_{2b} m(b,d))$.*

The $GMSH$'s are uniform and regular, and they have a constant rank 3, sublogarithmic degrees and logarithmic diameters with respect to the number of vertices and hyperlinks. For example, $GMSH(3,4)$ has 1,631,721 processors, 2,175,628 hyperlinks of rank 3, and its degree and diameter are 4 and 15, respectively. The hyperlink/processor ratio is $m(3,4)/n(3,4) = 1.3$. For comparisons, consider the hypercube. The point-to-point hypercube network whose size is closest to that of $GMSH(3,4)$ is Q_{21}, the 21-dimensional hypercube. It has $2^{21} = 2,097,152$ processors, $22,020,096$ links. Both of its degree and diameter are 21. Its link/processor ratio is 15.5.

The gap between d-dimensional $GMSH$ and $(d+1)$-dimensional $GMSH$ is too big so that it becomes unpractical when d is large. In fact, for any given N such that $n(b,d) < N < n(b,d+1)$, we can construct a subhypergraph H of $GMSH(b,d+1)$ such that $n(H) = N$. Such a subhypergraph can be called an *incomplete* $GMSH(b,d+1)$, and it may no longer to be uniform and regular.

Theorem 3 *For any given N such that $n(b,d) < N < n(b,d+1)$, one can construct an incomplete $GMSH(b,d+1)$, H, which is a subhypernetwork of $GMSH(b,d+1)$ such that H has exactly N vertices, and the diameter of H is no more than the diameter of $GMSH(b,d+2)$.*

4 Concluding Remarks

We proposed several linear, uniform and regular hypernetworks based on Steiner triple systems. The most attractive features of these hypernetworks include small vertex degree, small diameter, low rank, and increased wire-sharing. These sparse hypernetworks have excellent potential to be used as interconnection networks for massively parallel computer systems.

All the proposed hypernetworks have rank 3. The small rank of our hypernetworks results in simple hyperlink implementations. Using a 3×3 crossbar switch to implement a hyperlink may be considered practical. The 3-way switchs (where is called concentrators) used in the binary fat tree of [4] can also be used to implement a hyperlink of rank 3.

There are many unexploited aspects of hypernetworks based on Steiner systems. For example, can we construct a $GMSH(3,d)$ that is vertex and/or edge symmetric? Such symmetries are very important for a hypergraph to be used as a hypernetwork, because they allow for all processors (hyperlinks) to be treated as identical so that simple and efficient message routing schemes can be derived. A good understanding of the topological structures of our hypernetworks is also important in analyzing the the fault-tolerance features of these hypernetworks.

References

[1] Berge, C. *Hypergraphs*. North-Holland, 1989.

[2] Kirkman, T. P. On a Problem in Combinations. *Cambridge and Dublin Math J..*, 2, 1847, pp. 191-204.

[3] Leighton, F. T *Introduction to Parallel Algorithms and Architectures: Arrays · Trees · Hypercubes*. Morgan Kaufmann Publishers, San Mateo, CA, 1992.

[4] Leiserson, C.L. Fat Trees: Universal Networks for Hardware-Efficient Supercomputing. *IEEE Trans. Computers*, 34, (1985), pp. 892–901.

[5] Lindner, C.C. A Survey of Embedding Theorems for Steiner Systems. Topics on Steiner Systems., edited by C. C. Lindner and A. Rosa, North-Holland, 1980

[6] Lindner and A. Rosa, C. C *Topics on Steiner Systems*. North-Holland, 1980.

[7] Ray-Chaudhui, D.K., and Wilson, R.M. Solution of Kirkman's School-girl Problem. *Proc. Symp. Pure Math. 19.*, 1971.

[8] Woolhouse, W.S.B. Prize question 1733. *Lady's and Gentleman's Diary.*, 1844.

[9] Zheng, S. Q Hypernetworks - A Class of Interconnection Networks with Increased Wire Sharing: Part I - Part IV. Technical Reports, Dept. of Comp. Sci., LSU, Baton Rouge, Dec., 1994.

Permutation Routing and Sorting on Directed de Bruijn Networks

D. Frank Hsu
Dept. of Computer and Info. Science
Fordham University
Bronx, New York 10458
e-mail: hsu@murray.fordham.edu

David S. L. Wei
School of Computer Science and Engrg.
The University of Aizu
Fukushima, 965-80 Japan
e-mail: d-wei@u-aizu.ac.jp

Abstract

We show that any deterministic oblivious routing scheme for the permutation routing on a d-ary de Bruijn network of $N = d^n$ nodes, in the worst case, will take $\Omega(\sqrt{N})$ steps under the *single-port* model. This lower bound improves the results shown in [2] and [5] provided d is not a constant. A matching upper bound is also given. We also present an efficient general sorting algorithm which runs in $O((\log d) \cdot d \cdot n^2)$ time for directed de Bruijn network with d^n nodes, degree d and diameter n.

1 Introduction

Message passing is undoubtedly a vital issue in parallel processing. During the course of parallel computation, individual processors need to communicate their partial results with other processors, and message passing problems thus arise. This can be handled by parallel sorting destination tags attached to each data item (packet), or by routing packets to their own destinations. Packet routing and parallel sorting are thus important due to their intrinsic significance in communication among Interconnection Networks (ICN). In this paper we consider the routing and sorting problems on the *de Bruijn network*, an ICN which has proven to be versatile for parallel processing (see e.g., [4] [10]). More specifically, the task of routing is to route a set of packets of information (a packet being an ⟨*source, destination*⟩ pair) to their correct destinations as quickly as possible such that at most one packet passes through any link of the network at any time. In particular, we consider *permutation routing* in which initially there is exactly one packet in each node of the network and the destinations form some permutation of the sources. Also, a routing scheme is said to be *oblivious* if the path taken by each packet depends only on its source and destination. An oblivious routing strategy is preferable since it will lead to a simple control structure for the individual processing elements. In this paper we are thus concerned with only oblivious routing strategies. An optimal randomized oblivious routing algorithm for the de Bruijn network has been obtained by Palis, Rajasekaran, and Wei in [8][1]. We therefore consider only deterministic routing problems on the de Bruijn network. We show that any deterministic oblivious routing scheme for the permutation routing on the d-ary de Bruijn network of $N = d^n$ nodes, in the worst case, will take $\Omega(\sqrt{N})$ steps provided a node in the network can process only one packet at a time. We then present a simple deterministic oblivious permutation routing algorithm whose running time matches the shown lower bound. We are also interested in finding efficient sorting algorithms for the de Bruijn network to make the network being more versatile. Because in addition to its importance in communication, a sorting algorithm is frequently invoked by various sorts of important applications, such as computational geometry, robotics, etc. In this paper we present an efficient general sorting algorithm for the de Bruijn network of arbitrary degree. To the best of our knowledge, this algorithm is so far the best known sorting algorithm which runs in $O((\log d) \cdot d \cdot n^2)$ time for directed de Bruijn network with d^n nodes, degree d and diameter n. As a corollary, on the binary de Bruijn network of $N = 2^n$ nodes, our sorting algorithm runs in $O(\log^2 N)$ time.

2 The de Bruijn Networks

A directed de Bruijn network $B(d, n)$ has $N = d^n$ nodes. A node v can be labelled as $d_n d_{n-1}...d_1$ where each d_i is a d-ary digit. Node $v = d_n d_{n-1}...d_1$ is adjacent to the nodes labelled $d_{n-1}...d_2 d_1 l$, denoted $SH(v, l)$, where l is an arbitrary d-ary digit. Note that a rotation of n digits of v, denoted $SH(v, d_n)$, represents an adjacent node of v. If v is also adjacent to the nodes labelled $l d_n d_{n-1}...d_2$, then the graph is referred to as undirected de Bruijn network after deleting self-loops and multiple edges. In this paper we focus on the study of directed de Bruijn networks, and henceforth, de Bruijn networks are referred to as such unless otherwise stated. Like many other static

---

[1]In [8], the de Bruijn network of degree d is named as d-way shuffle, a term borrowed from [11].

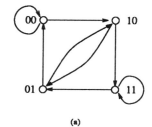

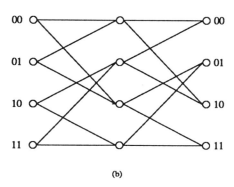

(b)

Figure 1: (a) B(2, 2) as a static network. (b) B(2, 2) as a multi-stage network.

networks, such as hypercube, CCC, shuffle-exchange, etc., the de Bruijn network can also be expressed as a multi-stage dynamic network[2]. Interestingly, the behavior of our routing and sorting algorithms can be easily observed and analyzed if we describe our algorithms using the multistage de Bruijn network. A $B(2,2)$ in both static and dynamic forms is depicted in Figure 1. Please note that throughout the paper, multistage networks are used only for the purpose of describing the behavior of our algorithms. The algorithms are basically designed for the static ones.

According to the definition of the network, it is obvious that the degree of the network is d. Besides, from the definition of the network, one can easily see that given any pair of nodes $x = x_n x_{n-1} \cdots x_1$ and $y = y_n y_{n-1} \cdots y_1$, by shifting x's address n times with appropriate digit, say y_{n-i+1}, as new digit of rightmost (first) bit at ith time, one can obtain a new address, say $y_n y_{n-1} \cdots y_1$, as desired. In other words, any node in the network is reachable from any other node in exact n steps although there might exist a shorter path. This shows that the diameter of the network is n, and there is a unique path of exact n links between any pair of nodes. We assume that the de Bruijn network is a MIMD machine in which at each step different nodes can perform different instructions.

---

[2]In [8], the multistage representation of a static network is named as leveled network, a term coined by Leighton, Maggs, and Rao in [7]

3   The Lower Bound

In [2], Borodin and Hopcroft have shown that for any graph of N nodes with degree d, the maximum delay, in the worst case, of any deterministic oblivious routing scheme is $\Omega(\sqrt{\frac{N}{d}})$ provided a node in the network can process only one packet at a time (referred to as single-port model henceforth). In this section we show that any deterministic oblivious permutation routing algorithm will take $O(\sqrt{N})$ time, in the worst case, for the N-node de Bruijn network of arbitrary degree d under the single-port model. This improves the one given in [2], provided the degree d of the network is not constant. This lower bound is in fact a tight one as we will construct, in the next section, a matching upper bound algorithm for the de Bruijn network of arbitrary degree.

Theorem 3.1 Any deterministic oblivious permutation routing scheme on the de Bruijn network $B(d, n)$ of $N = d^n$ nodes, in the worst case, takes at least $\Omega(\sqrt{N})$ steps under the *single-port* model.

Proof Sketch : In a permutation routing, each node in the network is a destination of one packet, and the packet can start from any node in the network. Therefore, for each destination node v, we define a *destination tree* consisting of all N nodes of the de Bruijn network in a recursive way as follows: (a) the root v connects to its adjacent nodes, (b) each new node in the tree connects to any adjacent nodes which are not already in the tree, (c) repeat (b) until all nodes in the network have been included in the tree. For each node v_0 in the destination tree of v, we define its *origins* as all its descendants in the tree. If a packet heading for node v is placed at any of these origins, it will go through node v_0. Accordingly, if one can prove that there exist at least a node, say v_0, which has at least ρ origins in each of ρ destination trees, it can be concluded that the delay must be at least ρ. Because we can start a packet at each of these destination trees at a different origin of v_0 such that the packets heading for distinct destinations will go through v_0. Thus the crux of the proof lies in finding some nodes which have at least $\sqrt{N}$ origins in each of $\sqrt{N}$ destination trees. In counting the number of origins of a given node, we must exclude those descendants which have appeared as the ancestors of the given node. It can be shown that for a node $y = y_1 y_2 \cdots y_n$, the condition for its descendant at distance j being identical to its ancestor at distance i (and thus will not be included as an origin) is that $y_1 = y_{i+j+1}, y_2 = y_{i+j+2}, \cdots$. Consider nodes whose addresses are $\underbrace{\alpha\alpha \cdots \alpha}_{n/2}\underbrace{\beta\beta \cdots \beta}_{n/2}, 0 \leq \alpha, \beta < d$, and $\alpha \neq \beta$ For each of these nodes, say $v_o = x_1 x_2 \cdots x_n$, the number of ancestors at distance

$\frac{n}{2}$ is $d^{n/2} = \sqrt{N}$ since $x_1 = \alpha \neq x_{1+n/2} = \beta$. This implies that node v_o is at level $\frac{n}{2}$ in $\sqrt{N}$ destination trees. The origins of v_o would be the nodes of the subtree rooted at v_o[3] excluding v_o itself and those which have appeared as ancestor of v_o. More specifically, let D be the number of origins of v_o, then $D = \sum_{i=1}^{n/2} d^i - \varphi$, where φ is the number of nodes which have appeared as ancestor of v_o. It is the case that $\varphi \leq 1$. Because $x_1 \neq x_{n/2+1}$ and then the possible descendants which can have appeared as an ancestor of v_o must be located at distance $\frac{n}{2}$, which are at level n, and the ancestor must be the root of the destination tree. There is only one node can be of this kind of descendant. Therefore, we have $D = \sum_{i=1}^{n/2} d^i - 1 \geq \sqrt{N}$ □

4 The Routing Scheme

Informally, our routing algorithm forces each packet to take the unique path of exactly n links between its source and destination. The routing time is thus n plus the delay of the latest packet. The delay that a packet suffers is the number of packets it meets during the entire course of the routing. The reason that we force each packet to travel exact n links along the unique path between its source and destination is that the behavior of the algorithm is easier to observe. This can be seen from the analysis follows. We analyze the performance of the algorithm by counting the maximum number of packets that any packet will meet during the entire course of the routing.

Lemma 4.1 (Queue line lemma) The number of steps a packet x is delayed is less than or equal to the number of packets that *overlap* with x provided the routing scheme is nonrepeating. A routing scheme is **nonrepeating** provided the following is true: if the paths taken by any two distinct packets share some links and then diverge, then the remainder of these two paths will never share any link again.[11].

Theorem 4.1 The above routing algorithm can realize any permutation routing in $O(\sqrt{N})$ time.

Proof Sketch : A permutation routing on a static de Bruijn network can be viewed as a permutation routing of N packets from the first column, one packet per node, to the last column on a multistage de Bruijn network. Consider any packet x, on the way to its destination, x will meet some other packets at each intermediate column. Based on Lemma 4.1 and the fact that our routing scheme on the de Bruijn network is nonrepeating, the maximum delay x may suffer is the summing up of the number of packets x meets at

---

[3]The subtree is of height $\frac{n}{2}$.

every column. We then count the number of packets which will be accumulated at a node in column $i, 0 \leq i < n$. It can be easily seen (from any multistage de Bruijn network) that for any node v in column $i, 0 \leq i \leq \lfloor \frac{n}{2} \rfloor$, the number of packets that have potential to go through v is d^i. However, this is not true for any node v' in column $j, j > \lfloor \frac{n}{2} \rfloor$ because in a permutation routing, only d^{n-j} packets will be destined for the nodes (in the last column) which are reachable from node v'. Therefore, the number of packets that can possibly go through a node in column i is $\min(d^i, d^{n-i})$. The total number of packets that x can possibly meet on the way to its destination is thus $\sum_{i=0}^{n-1} \min(d^i, d^{n-i}) \leq 3d^{\lfloor \frac{n}{2} \rfloor} = O(\sqrt{N})$ □

5 The Sorting Algorithm

The indexing scheme adopted in our sorting algorithm is lexicographic order, i.e. after sorting is done, node $x = x_n x_{n-1} \cdots x_1$ will accommodate the item whose rank is $\sum_{i=1}^{n} d^{i-1} \cdot x_i$ among sorted items. Our general sorting algorithm works for the de Bruijn network of arbitrary degree d. It's an adaptation of Batcher's bitonic sort [1]. Batcher's bitonic sorting network can sort any bitonic sequence (defined in Definition 5.1) into ascending or descending order. The bitonic sorting network (bitonic-sorter) is constructed by recursively combining *half-cleaners*. A half-cleaner with k inputs and k outputs can move $k/2$ smaller items to the upper $k/2$ outputs, $k/2$ larger items to the lower $k/2$ outputs, and each subsequence of $k/2$ items remains a bitonic sequence. However, our sorting algorithm is designed based on m-sorter modules [6] instead of 2-sorters adopted by most of the previous parallel sorting schemes. Simply because in the network of degree m, it's necessary to merge $\frac{m}{2}$ bitonic sequences of length $2L$ into one longer bitonic sequence of length mL. We use the m-sorter to construct an *m-way cleaner*. To prove the correctness of our sorter, we need only to prove Lemma 5.1. The proof of Lemma 5.1 could be simplified based on the zero-one principle (Theorem 5.1).

Definition 5.1 A sequence $X = \langle x_1, x_2, \cdots, x_N \rangle$ of N numbers is said to be bitonic if either $x_1 \leq x_2 \leq \cdots \leq x_i \geq x_{i+1} \geq \cdots \geq x_N$ or $x_1 \geq x_2 \geq \cdots \geq x_i \leq x_{i+1} \leq \cdots \leq x_N$ for some $i, 1 \leq i \leq N$.

Theorem 5.1 If a network with N inputs can sort all 2^N possible sequences of 0's and 1's into ascending (or descending) order, it will sort any sequence of arbitrary numbers into ascending (or descending) order [6].

Fact 5.1 In a de Bruijn network, if we simultaneously rotate the address of each node, the obtained new addresses form a new permutation, i.e. there will be no different nodes with the same address.

Lemma 5.1 Given an m-way cleaner of k m-sorters (denoted as $m-cleaner_k$) with $m \times k$ inputs and $m \times k$ outputs, which are grouped into m subsequences of k contiguous outputs, if the input to the cleaner is a bitonic sequence of 0's and 1's, then the output satisfies the following properties: each of m subsequences is a bitonic sequence; each item in ith group, $1 \leq i < m$, is at least as small as (or as large as) every item in $(i+1)$th group; each group of the output is clean[4] except jth group for a j, $1 \leq j \leq m$, which may be clean or dirty, but is still bitonic[9].

Informally, our general sorting algorithm for $B(d,n)$ consists of n stages. In the ith stage, there are d^{n-i+1} sorted sequences of length d^{i-1}, produced by $(i-1)$th stage, in which each pair of odd-even sequences forms a bitonic sequence. The task of ith stage is to merge these sequences into d^{n-i} bitonic sequences of length d^i. To do so, each stage consists of two phases, viz., merge phase and bitonic sort phase. The merge phase iterates for $\log\lceil\frac{d}{2}\rceil = \lceil\log d\rceil - 1$ times to merge $\frac{d}{2}$ bitonic sequences into a single one which can then be sorted in the bitonic sort phase. To implement a d-sorter, the algorithm routes all d items in the same sorter to a group of d nodes such that each node contains a copy of each item from the same sorter, i.e. items at distance d^i for some $i, 0 \leq i < n$, and then kth node, for all k, $1 \leq k \leq d$, in the group selects kth item from the set of items, which will result in a sorted sequence along the sorter. We use $\mathsf{SL}(k)$ to denote a selection which selects kth element from a given set of elements. We assume that each node in the network can perform a $\mathsf{SL}(k)$ function. The detail of the algorithm is given in Figure 2. For description convenience, in Algorithm A, we use $SH(x,[p,q]), 0 \leq p < q < d$, to denote a set of adjacent nodes of x, namely $SH(x,p), SH(x,p+1), \cdots, SH(x,q-1), SH(x,q)$, and $SH(x,[p,q]) \Leftarrow item$ represents that x sends item to all of the adjacent nodes in the set.

Theorem 5.2 Algorithm A can sort $N = d^n$ items stored one per node in $B(d,n)$ in ascending (or descending) order in $O((\log d) \cdot d \cdot n^2)$ steps under the single-port model.

Proof Sketch: Follows from Lemma 5.1, Fact 5.1, and the complexity analysis stated in what follows. Let T be number of steps taken by Algorithm A, we have $T \leq \sum_{i=1}^{n}(\alpha+\beta)$, where $\alpha = \sum_{m=1}^{\lceil\log d\rceil-1}(n-i+O(d)+$

---

[4]A subsequence is *clean* if it consists of either all 0's or all 1's, otherwise it is *dirty*

Algorithm A
{ *Item* denotes the element being processed by node x. Any item coming from any of incoming links is stored in the local memory of the node to be processed.}
for each node $x = x_n x_{n-1} \cdots x_1$ do in **parallel**
    for $i := 1$ to n do
    begin
        {Merge phase}
        for $m := 1$ to $\lceil\log d\rceil - 1$ do
        begin
            for $j := 1$ to $n - i$ do

1. $SH(x, x_n) \leftarrow item$;
2. $SH\left(x, [\lfloor\frac{x_n}{2^m}\rfloor \cdot 2^m, \min(\lfloor\frac{x_n}{2^m}\rfloor \cdot 2^m + 2^m - 1, d-1)]\right) \Leftarrow it$
            if $\lfloor\frac{x_1}{2^m}\rfloor$ is *even*
            then
3. $item \leftarrow \mathsf{SL}((x_1 \bmod 2^m) + 1))$
            else
4. $item \leftarrow \mathsf{SL}(2^m - (x_1 \bmod 2^m))$
            endif
            for $k := n - i + 2$ to n do
            begin
5. $SH(x, [0, d-1]) \Leftarrow item$;
            if $\lfloor\frac{x_i}{2^m}\rfloor$ is *even*
            then
6. $item \leftarrow \mathsf{SL}(x_1 + 1)$
            else
7. $item \leftarrow \mathsf{SL}(d - x_1)$
            endif
        end
        end
        {Bitonic sort phase}
        for $j := 1$ to $n - i$ do
8. $SH(x, x_n) \leftarrow item$;
         for $k := n - i + 1$ to n do
         begin
9. $SH(x, [0, d-1]) \Leftarrow item$;
         if $i < n$
         then
            if x_{k-j+1} is *even*
            then
10. $item \leftarrow \mathsf{SL}(x_1 + 1)$
            else
11. $item \leftarrow \mathsf{SL}(d - x_1)$
            endif
         else
12. $item \leftarrow \mathsf{SL}(x_1 + 1)$;
         endif
        end
    end

Figure 2: A sorting algorithm for $B(d,n)$.

$\mathsf{SL}(d) + \sum_{k=n-i+2}^{n}(O(d) + \mathsf{SL}(d)))$, and $\beta = n - i + \sum_{n-i+1}^{n}(O(d) + \mathsf{SL}(d))$
$= O(\log d \cdot (O(d) + \mathsf{SL}(d)) \cdot n^2)$, where $\mathsf{SL}(d)$ is the time needed for a node to select kth item from a set of d items. Let each node run a **linear time** selection procedure given in [3], we have $T = O(\log d \cdot (O(d) + O(d)) \cdot n^2) = O((\log d) \cdot d \cdot n^2)$.□

6   Conclusions

In this paper we have shown a lower bound of deterministic oblivious permutation routing for the de Bruijn network, and come up with a matching upper bound algorithm. We also present an efficient generalized sorting algorithm for the de Bruijn network of arbitrary degree d, which runs in $O((\log d) \cdot d \cdot n^2)$ time, where n is the diameter of the network. To the best of our knowledge, this algorithm is, so far, the best known sorting algorithm for the de Bruijn network of arbitrary degree. The routing and sorting problems we investigated are a *store and forward* one. Exploring problems of cut through routing and $k - k$ sorting on the de Bruijn networks would also be of interesting. From the experience we gained in designing the sorting and routing algorithms (both deterministic and randomized ones) for the de Bruijn networks and star graphs, we conjecture that increasing the degree of an interconnection network doesn't help with deterministic sorting, but even suffers a setback. However, increasing the degree of a network does help with the randomized routing. We also conjecture that m-sorter is inferior to 2-sorter in designing a sorting network, a question given in [6].

Acknowledgements

The authors would like to thank anonymous referees for their helpful comments and suggestions.

References

[1] K. Batcher, "Sorting Networks and Their Applications," Proc. AFIPS Spring Joint Comput. Conf., 1968, pp. 307-314.

[2] Borodin, A. and J. E. Hopcroft, "Routing, merging and sorting on parallel models of computation," Proc. Symposium on Theory of Computing, 1982, pp. 338-344.

[3] M. Blum, R.W. Floyd, V. Pratt, R.L. Rivest, and R.E. Tarjan, "Time bounds for selection," *Journal of Computer and System Sciences*, 7(4), 1973, pp. 448-461.

[4] J.C. Bermond, and C. Peyrat, "de Bruijn and Kautz Networks: A Competitor for the Hypercube?," *Hypercube and Distributed Computers*, F. André, and J.P. Verjus, eds., Elsevier Science Publishers, 1989, pp. 279-293.

[5] C. Kaklamanis, D. Krizanc and Th. Tsantilas, "Tight Bounds for Oblivious Routing in the Hypercube," Proc. ACM Symposium on Parallel Algorithms and Architectures, 1990, pp. 31-36.

[6] D.E. Knuth, *The Art of Computer Programming*, Vol. 3, Addison-Wesley, 1973.

[7] T. Leighton, B. Maggs, and S. Rao, "Universal packet routing algorithms," Proc. Symposium on Foundations of Computer Science, 1988, pp. 256-269.

[8] M. Palis, S. Rajasekaran, and D.S.L. Wei, "Packet Routing and PRAM Emulation on Star Graphs and Leveled Networks," *Journal of Parallel and Distributed Computing*, vol. 20, no. 2, Feb. 1994, pp. 145-157.

[9] S. Rajasekaran, and D.S.L. Wei, "Selection, Routing and Sorting on the Star Graph," *Technical Report 94-1-028*, Dept. of Computer Software, The U. of Aizu, Japan.

[10] M.R. Samatham, and D.K. Pradhan, "The De Bruijn Multiprocessor Network: A Versatile Parallel Processing and Sorting Network for VLSI," *IEEE Trans. on Computers*, Vol. 38, No. 4, April, 1989, pp. 567-581.

[11] L.G. Valiant, and G.J. Brebner, "Universal Schemes for Parallel Communication," Proc. Symposium on Theory of Computing, 1981, pp. 263-277.

SOFTWARE BASED FAULT-TOLERANT OBLIVIOUS ROUTING IN PIPELINED NETWORKS*

Young-Joo Suh[†], Binh Vien Dao[†], Jose Duato[¥], and Sudhakar Yalamanchili[†]

[†]Computer Systems Research Laboratory
School of Electrical and Computer Engineering
Georgia Institute of Technology
Atlanta, Georgia 30332-0250
e-mail: {suh, dao, sudha}@ee.gatech.edu
phone: (404) 894-2940 fax: (404) 853-9959

[¥]Facultad de Informatica
Universidad Politecnica de Valencia
P.O.B. 22012
46071 Valencia, Spain
e-mail: jduato@aii.upv.es

Abstract -- *This paper presents a software based approach to fault-tolerant routing in oblivious, wormhole routed networks. When a message encounters a faulty output link it is removed from the network by the local router and delivered to the messaging layer of the local node's operating system. The message passing software can re-route this message along a non-minimal oblivious path or via an intermediate node, which will forward the message to the destination. A message may encounter multiple faults and pass through multiple intermediate nodes. This paper discusses deadlock, livelock, and performance issues. Router designs are minimally impacted remaining compact, oblivious, and fast. Therefore this approach is a good candidate for incorporation into the next generation of wormhole routed multiprocessor networks.*

1.0 Introduction

Interconnection networks in modern machines make use of oblivious, fixed path wormhole routing to achieve high throughput and low message latency. However, fault tolerant communication requires the network to be able to dynamically route messages along alternative, possibly non-minimal paths. This paper proposes a software based approach for re-routing messages blocked by faults. The techniques reported here are motivated by several considerations. First it is targeted towards environments where the fault rates are relatively low, i.e., on the order of a maximum of 3 failed components between repair cycles. During this time we wish the machine to continue functioning, with possibly degraded communication performance. Solutions for higher fault rates have been addressed elsewhere [9]. Second, we wish to retain the features of existing oblivious router designs, i.e. compactness and speed. This implies that additional hardware complexity in the form of additional virtual channels should be avoided. Finally, we wish to make the common case fast. Therefore messages that do not encounter faults should be minimally impacted.

The basic idea is quite simple. When a message encounters faulty link, it is removed from the network by the local router and delivered to the messaging layer of the local node's operating system. The message passing software either i) modifies the header so the message may follow an alternative dimension order path, or ii) computes an intermediate node address. In either case, the message is re-injected into the network. In the case that the message is transmitted to the intermediate node,

* This research was supported in part by a grant from the National Science Foundation under grant CCR-9214244

it will be forwarded upon receipt to the final destination. A message may encounter multiple faults and pass through multiple intermediate nodes. The problem is distinct from adaptive packet routing in networks using packet switching or virtual cut through [14]. Routing is oblivious and is based on wormhole. Re-routing must consider dependencies across multiple routers caused by small buffers (< message size) and pipelined dataflow.

The proposed techniques do accommodate a range of fault patterns: more than previously proposed wormhole routing techniques. Only messages that encounter faults are affected, and degradation is largely proportional to the number of faults. If the mean time between failures is large, this approach is viable. Thus, we feel it is a good candidate for inclusion in the next generation of wormhole routed networks, and it targets commercial multiprocessors. This is particularly true when the application environment does not justify the use of expensive, custom, fault-tolerant backplanes.

2.0 Fault Model

This paper considers k-ary n-cube networks. Adjacent faulty links and faulty nodes are coalesced into fault regions. Fault regions may overlap forming a larger fault region. We also assume that the fault regions do not disconnect the network. Fault regions may be convex or L-shaped or U-shaped concave regions. Some example fault regions are illustrated in Figure 1. Messages are routed obliviously, and therefore cannot progress when the required output link at a node leads to a fault region. The message is removed from the network and delivered to the local node's message handling software. This message is said to be *absorbed* at the local node and is subsequently referred to as a *faulted* message. This status is recorded in the message header.

3.0 Fault Tolerant Oblivious Routing

This section describes the software-based fault-tolerant oblivious routing algorithm, *e-sft*. In the absence of faults, the e-sft routing algorithm is equivalent to the traditional dimension order routing algorithm, e-cube [8]. When the outgoing link at a node leads to a fault region, the message is absorbed and delivered to the messaging layer of the local node. The header may now be modified to reflect a different, non-minimal path around the faulty region. For example in Figure 1, a message from node P to node Q is blocked by a fault region. The message can be transmitted along the negative X-dimension across the wrap-around channel to node Q. Alternatively, an intermediate node may be selected, and the message is sent to this node. The intermediate node will receive the message and

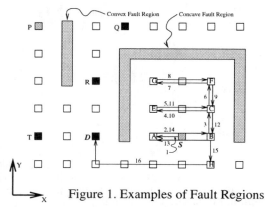

Figure 1. Examples of Fault Regions

X-Dest	Y-Dest	F	X-Final	Y-Final	RT	DF	PF

RT:Re-route Table   X-Dest:X-coordinate offset
F:Faulted Status Bit   Y-Dest:Y-coordinate offset
PF:Prevent Flag   X-Final:X-coordinate of final destination
DF:Direction Flag   Y-Final:Y-coordinate of final destination

Figure 2. Format of the message header

forward it to the final destination. For example in Figure 1, a message from node P to node R may be first sent to intermediate node T. Node T can then transmit the message in dimension order to node R. If other faults are encountered by the re-routed message, the process is repeated.

3.1   Routing Header

In order to keep track of the manner in which a message is re-routed, the header contains a 2 bit flag called the Direction Flag (DF). The DF is used to record the following information regarding the path that a message has taken.

00 - Traversing the shortest distance along the X-dimension

01 - Traversing the longer distance along the X-dimension

10 - Traversing the shortest distance along the Y-dimension

11 - Traversing the longer distance along the Y-dimension

For example, the message from P to Q in Figure 1 will have a DF value of <01>. Since fault-free routing is dimension order, a DF value of <10> would imply the message had attempted to traverse the X-dimension in both directions and is now trying to traverse the Y-dimension. The exception to this interpretation is when the source and destination nodes are in the same column. In this case the message will attempt the Y-dimension first and the interpretation of DF is reversed, i.e., <00> and <01> (<10> and <11>) refer to the Y-dimension (X-dimension). The DF value in the header is only modified when the message is absorbed at a node. It enables e-sft routing to keep track of the directions along each dimension that have been attempted and therefore aid in re-routing decisions.

There are three additional fields in the header. A *faulted* status bit (F) indicates that the message has encountered at least one fault and is being re-routed. This bit enables a node to distinguish between messages destined for the local node and messages which must be forwarded. A *prevent flag* (PF) status bit is used to prevent the occurrence of certain livelock situations. The role of the PF will be clear from the example described later. A two bit re-route table field (RT) specifies one of three tables to be used for re-routing decisions. Finally, since messages may be routed through intermediate nodes, the header must contain two sets of address fields. The first records the final destination address (X-Final, Y-Final). This is an absolute address. The second is used for routing the messages and is an offset within each dimension (X-

Dest, Y-Dest). The message header now appears as shown in Figure 2. Note that the routers only process the offset fields and set the F bit. All of the remaining header processing is done in software, and only when messages encounter faults. Thus, router operations are minimally impacted.

3.2   The Routing Algorithm

The network hardware routes messages using traditional dimension order routing based on the X-Dest and Y-Dest fields in the message header. If the outgoing channel at a router is faulty, the router sets the F bit, and routes the message to the local processor interface. This causes the message to be marked as a faulted message and ejected to the local messaging layer. If the F bit is set, the messaging software checks the X-Final and Y-Final fields to determine if the message is to be delivered locally. If not, a re-routing function is invoked. The X-Dest and Y-Dest fields are updated, and if necessary the DF and PF flags (as described below) are modified. The re-routing function depends *only* on the relative address of the first node where a fault is encountered and the final destination. Let the coordinates of the node where the message first encounters a faulty channel be (x_f, y_f) and the coordinates of the destination be (x_d, y_d). Let the offsets at (x_f, y_f) in each dimension be Δx (given by $x_d - x_f$) and Δy (given by $y_d - y_f$). There are three possible cases i) $\Delta y = 0$, ii) $\Delta x = 0$, and iii) $\Delta x \neq 0$ & $\Delta y \neq 0$. The re-routing decisions for each case are captured in Tables 1-3. The RT field identifies which of the three cases apply to the message and is set at (x_f, y_f). The notation $(x_d, y_d)_x^s$ signifies that the message header is modified to be transmitted to the node with coordinates (x_d, y_d), along the shortest path in the X-dimension. The notation $(x_d, y_d)_x^l$ signifies transmission along the longer path in the X-dimension. Note that re-routed messages still follow dimension order, though not shortest path within a dimension (due to faults). When re-routed messages are absorbed at nodes due to faults, the RT field identifies the table to be used in making the re-routing decision for the message. If the RT field is 0, then this is the first node at which the message encountered a fault and RT must be set to signify one of the three cases above.

The tables can be interpreted as follows. If a header to be re-routed has a DF value shown in the first column, it is re-routed to the node shown in the second column. The notation specifies the direction in the dimension the message is to be transmitted. The remarks column specifies the action that takes place if the message subsequently encounters a fault before reaching the node specified in column 2 or arrived at the intermediate node without meeting a fault. The Tables attempt to capture the following ideas. When a message encounters a fault, it is first re-routed in the same dimension in the opposite direction. If another fault is encountered, the message is routed in an orthogonal dimension in an attempt to route around the faulty regions. The DF keeps track of the directions attempted. The PF will be shown to prevent certain live-lock situations within concave fault regions.

DF	To/Via	Remarks
00	$(x_d, y_d)_x^s$	DF=<01> after meeting a fault
01	$(x_d, y_d)_x^l$	DF=<10> after meeting a fault if PF = 0 DF=<11> after meeting a fault if PF = 1
10	$(x_c, y_c + r)_y^s$	DF=<00> & PF is unchanged if no fault met DF=<11> and PF = 1 if fault met
11	$(x_c, y_c - r)_y^s$	DF=<00> after received by a node

Table 1: Re-routing decisions for the case $\Delta y = 0$

DF	To/Via	Remarks
00	$(x_c, y_d)_y^s$	DF=<01> after meeting a fault
01	$(x_c, y_d)_y^l$	DF=<10> after meeting a fault if PF = 0 DF=<11> after meeting a fault if PF = 1
10	$(x_c + r, y_c)_x^s$	DF=<00> & PF is unchanged if no fault met DF=<11> and PF = 1 if fault met
11	$(x_c - r, y_c)_x^s$	DF=<00> after received by a node

Table 2: Re-routing decisions for the case $\Delta x = 0$

DF	To/Via	Remarks
00	$(x_d, y_d)_x^s$	DF=<01> after meeting a fault
01	$(x_d, y_d)_x^l$	DF=<10> after meeting a fault
10	$(x_c, y_d)_y^s$	DF=<11> after meeting a fault DF=<00> if not meet a fault
11	$(x_c, y_d)_y^l$	DF=<00> after received by a node

Table 3: Re-routing decisions for the case $\Delta x \neq 0$ & $\Delta y \neq 0$

Table 1 shows the routing decisions for the case $\Delta y = 0$, i.e., the first node at which the message was absorbed, and destination node are on the same row. The message is injected into the network with DF=<00> and PF=0 along the shortest path in the X-dimension. This is captured in the first row of the table. The remarks column indicates the action taken when the message encounters a faulty link. In this case, DF is changed to <01> to signify re-routing along the longer path in the X-dimension and the message is re-injected. If the message encounters another faulty channel before reaching (x_d, y_d), DF is changed to <10> (PF=0) or <11> (PF=1) as denoted in the remarks column of row 2. Let us assume DF is set to <10>. In this case from row 3 we see that the message is re-routed r hops along positive direction in the Y-dimension in an attempt to find a path around the fault region.

The message is explicitly sent to an intermediate node $(x_c, y_c + r)$, where (x_c, y_c) is the current node. When the message reaches this intermediate node, it is absorbed and DF is reset. However, if instead the message with DF=<10> encounters a faulty channel, DF is set to <11> and PF set to 1, and the message is re-routed r hops in the negative direction in the Y-dimension. The choice of r is arbitrary. The right choice depends upon the expected height of the fault region. The safest (and most expensive) choice would be to use $r = 1$ hop.

Table 2 illustrates the routing decisions for the case of $\Delta x = 0$. These are nearly similar to the case for $\Delta y = 0$. But, when the offset in Y-dimension is eliminated at an intermediate node, RT field is changed for Table 1. Table 3 captures decisions for the case of $\Delta x \neq 0$ & $\Delta y \neq 0$. In this case the message must traverse both the X- and Y-dimensions, and, after the message encounters a faulty channel along the longer path in X-dimension, e-sft attempts to eliminate the offset in one of the dimensions reducing subsequent re-routing decisions to those captured in Table 1 and Table 2. When one of the dimensions has been eliminated, the RT field is changed to reflect the choice of Table 1 or Table 2.

Figure 1 illustrates an example where $\Delta y = 0$. In the figure S denotes the source node and D the destination node. The path taken by the message is numbered. A message sent with DF=<00> from S meets a faulty channel, and is absorbed by node A (1). Since $\Delta y = 0$, DF is set to <01> and the message is transmitted to $(x_d, y_d)_x^l$, where it meets a faulty channel at node B. Now DF is changed to <10> and the message is to be transmitted along the Y-dimension. However, since $\Delta y = 0$, the purpose now is to traverse the Y-dimension just enough to be able to be routed around the faulty region. In this example $r = 1$ so, the message is sent to node C, which is located one hop away from node B (3). Since the message is received rather than being ejected due to a fault, DF is reset to <00>. The message tries the shortest path in the X-dimension and fails (4) and the process is repeated until step 8. After step 8, DF is set to <10> and node F tries to send the message to $(x_c, y_c + 1)_y^s$. However, the Y-dimension channel is faulty. Therefore DF is changed to <11>, which indicates that the message should be sent in the opposite path in the Y-dimension. After some more failures in the X-dimension, the message is arrived at node H. In the next step, the message is delivered to the final destination. Note that in step 11 and step 5 (or step 2 and step 14) the DF value is <01>, and after both steps the message meets a faulty channel at node C in the X-dimension. If node C changes DF to <10> after both steps 5 and 11, the message passes through nodes E-C-F-G-F-C-E-C-F......, infinitely, resulting in livelock. Therefore we require DF to be set to <11> after step 11 to force the message to traverse the negative Y-dimension. This distinction is realized by the PF bit in the header. The PF is initially 0. When DF makes a transition from <10> to <11>, PF is set to 1 and remains at 1. The value of PF is used to prevent cycles in the course of the message transmission (which corresponds to a cycle of DF values) and thus avoid certain livelock situations.

3.3 Deadlock and Livelock Freedom

A few observations can be made about the behavior of e-sft routing. Re-routed messages still follow dimension order, which is deadlock-free [8], though not necessarily the shortest path in each dimension. When intermediate nodes are utilized, these nodes are selected to be in the same column or row as the current node. Finally, absorbed messages use dynamically allocated buffers in node memory (rather than router buffers) to prevent introducing dependencies between consumption channels and injection channels. It has been shown that these latter dependencies could lead to deadlock [2]. Thus we have the following theorem.

***Theorem 1.** e-sft is deadlock-free.*

Due to space limitations, the proof of deadlock freedom is provided in a detailed technical report [15]. While e-sft is deadlock-free it is not necessarily livelock-free for arbitrarily large fault patterns. However, when the number of faults is limited to a small number, e.g., 3, livelock freedom can also be guaranteed [15]. Under the current fault model, the PF flag enables a message to exit certain types of concave regions, and routing does proceed around convex regions. These observations and experience with simulations are encouraging. However, the issue of livelock-freedom for the current fault model is still under study.

4.0 Performance Evaluation

The performance of e-sft was evaluated with flit-level simulation studies of message passing in a 16-ary 2-cube with 16 flit long messages and a single flit routing header. We use a congestion control mechanism (similar to [1]) by placing a limit on the size of the buffer on the injection channels. If the input buffers are filled, messages cannot be injected into the network until a message in the buffer has been routed. Injection rates are specified as the number of 16 flit messages injected each 5000 cycle period. Thus, injection rate of 20 corresponds to 0.064 flits/node/cycle. Note that these are 32 bit flits.

A 32 bit header enables routing within a 64×64 torus. A flit crosses a channel in a single cycle, and traverses a router from input to output in a single cycle. Routing decisions are assumed to take a single cycle with the network operating with a 50 Mhz clock, and 20 ns cycle time. The software cost for absorbing and re-injecting a message is derived from measured times on an Intel Paragon and reported work with active message implementations [11]. Based on these studies we assess this cost at 25μs per absorption/injection or 50μs each time a message must be processed by the messaging software at an intermediate node. If the message encounters busy injection buffers when it is being re-injected, it is re-queued for re-injection at a later time. Absorbed messages have priority over new messages to prevent starvation. Relative to existing router designs [4], the only additional functionality required is in the Routing Arbitration Block [4].

One side effect of the increased header size is a possible increase in virtual channel buffer size and the width of the internal datapaths, although 32 bit datapaths appear to be reasonable for the next generation of routers. The remaining required functionality of e-sft is implemented in the messaging layer software.

4.1 Simulation Results

In a fault-free network the behavior of e-cube and e-sft is identical. Simulation experiments placed a single fault region of varying size within the network. Performance of e-sft is shown in Figure 3 for three different sized concave fault regions (5, 8, and 11 nodes) and for a 9 node convex fault region. Due to the greater difficulty in "entering" and "exiting" a concave fault region, the average message latency for e-sft is greater in the presence of concave fault regions rather than for equivalent sized convex fault regions. The curves also show that for each of the particular fault configurations, the latency remains relatively constant as the throughput increases. As the throughput increases, the number of messages each node injects and receives increases, but the percentage of messages that encounter the fault region remains relatively constant. Therefore, the latency remains relatively flat. Another factor is that the high latencies of re-routed messages

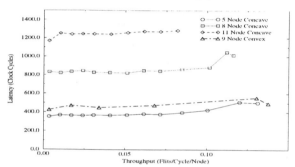

Figure 3. Latency-throughput curves

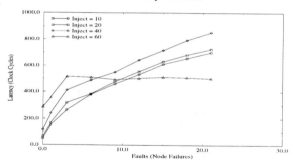

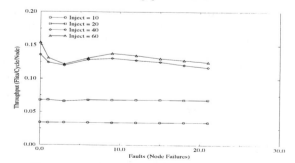

Figure 4. Latency-throughput vs. node faults

masks some of the growth in the latency of messages that do not encounter faults, though a close inspection of the graphs reveals a small but steady growth in average latency.

Figure 4 shows the performance of e-sft in the presence of a single convex fault region ranging in size from 1 failed router node to 21 failed router nodes. The latency plot indicates that when the network is below saturation traffic, the increase in the size of the fault block causes significant increases in the average message latency. This is due to the increase in the number of messages encountering larger fault regions (an 1 node fault region represents 0.4% of the total number of nodes in the network, while a 21 node fault region represents 8.2% of the total number of nodes). The latency and throughput curves for high injection rates (60) represent an interesting case. Throughput and latency appear to remain relatively constant. At high rates and larger fault regions, more messages become absorbed and re-routed. However, the limited buffer size provides a natural throttling mechanism for both new messages as well as absorbed messages waiting to be re-injected. As a result, active faulted messages in the network form a smaller percentage of the traffic and both the latency and throughput characteristics are dominated by the steady state values of traffic unaffected by faults. The initial drop in throughput for small number of faults is due to the fact that a higher percentage of faulted messages are delivered reducing throughput. These results suggest that sufficient buffering of faulted messages and priorities in re-injecting them have a significant impact on the performance of faulted messages.

At lower injection rates the throughput of the network remains relatively constant independent of the size of the fault blocks since fault blocks only increase the latency of the messages. Since messages are guaranteed delivery, when operating well below saturation the network quickly reaches the steady state throughput.

The effect of the overhead on the message latencies can be significant. Message latency histograms show peaks at intervals of 2500 cycles (corresponding to the 50µ s software overhead each time a message passes through the messaging layer software at a node). Among messages that do encounter faults, it appears that on the average the majority of faulted messages do not require more than three re-routing steps to be routed around the fault region.

In general, the results demonstrate good performance with messages being re-routed a few times. In practice, we find that the probability of multiple router failures before repair to be very low. Therefore we expect that large majority of faulted messages will not have to pass through more than one node. This would make these techniques attractive for next generation wormhole routed networks.

5.0 Concluding Remarks

We find that performance of re-routed messages is significantly affected by the techniques for buffering and re-injection. While the large majority of traffic is unaffected by faults, reliable message delivery and improving the latency of faulted messages will require better understanding of how re-routed messages should be handled. Finally, it appears that this approach can be extended nat-

urally to networks employing adaptive routing. For example, many fully adaptive routing protocols rely on dimension order routing over a subset of channels [10,7] to avoid deadlock. Messages which are blocked waiting on these channels, and experience faults on these channels can be absorbed and re-routed. Furthermore, partially adaptive routing protocols such as those based on the Turn Model [13] can also be adapted in a similar fashion. These issues and extensions are the subject of ongoing research.

REFERENCES

[1] R. Boppana and S. Chalasani. "A comparison of adaptive wormhole routing algorithms," *Proc. of Int. Symp. on Computer Architecture*, pp. 351-360, 1993.
[2] R. Boppana and S. Chalasani, "Fault-tolerant routing with non-adaptive wormhole algorithms in mesh networks," *Proc. of Supercomputing*, pp. 693-702, 1994.
[3] S. Chalasani and R. Boppana, "Fault-tolerant wormhole routing in tori," *Proc. of Int. Conf. on Supercomputing,* 1994.
[4] A. Chien, "A cost and speed model for k-ary n-cube wormhole routers," *Proc. of Hot Interconnects Workshop*, Aug. 1993.
[5] A. Chien and J. H. Kim, "Planar-adaptive routing: Low-cost adaptive networks for multiprocessors," *Proc. of Int. Symp. on Computer Architecture*, pp. 268–277, 1992.
[6] W. J. Dally, "Virtual-channel flow control," *IEEE Trans. on Parallel and Distributed Systems*, vol. 3, pp. 194-205, March 1992.
[7] W. J. Dally and H. Aoki, "Deadlock-free adaptive routing in multicomputer networks using virtual channels," *IEEE Trans. on Parallel and Distributed Systems,* vol. 4, pp. 466-475, April 1993.
[8] W. J. Dally and C. L. Seitz, "Deadlock-free message routing in multiprocessor interconnection networks," *IEEE Trans. on Computers*, C-36, pp. q547–553, May 1987.
[9] B. V. Dao, J. Duato, and S. Yalamanchili, "Configurable flow control mechanisms for fault tolerant routing," *Proc. of Int. Symp. on Computer Architecture*, June 1995.
[10] J. Duato, "A new theory of deadlock-free adaptive routing in wormhole networks," *IEEE Trans. on Parallel and Distributed Systems,* vol. 4, pp. 1320-1331, Dec. 1993.
[11] DT. von Eicken, D. E. Culler, S. C. Goldstein, and K. E. Schauser, "Active messages: a mechanism for integrated communication and computation," *Proc. of Int. Symp. on Computer Architecture*, pp. 256-266, 1992.
[12] C. J. Glass and L. Ni, "Fault-tolerant wormhole routing in meshes," *Proc. of the Fault Tolerant Computing Symposium,* 1993.
[13] C. J. Glass and L. Ni, "The turn model for adaptive routing," *Proc. of Int. Symp. on Computer Architecture,* pp. 278-287, 1992.
[14] S. Konstantinidou and L. Snyder, "Chaos router: architecture and performance," *Proc. of Int. Symp. on Computer Architecture*, pp. 212-221, 1991.
[15] Y. J. Suh, B. V. Dao, J. Duato, and S. Yalamanchili, "Software based fault tolerant routing," *Technical Report TR-GIT/CSRL-95/04*, Georgia Institute of Technology, Atlanta, Georgia 30332-0250, April 1995.

Fault-tolerant Routing in Mesh Networks *

Younes M. Boura and Chita R. Das
Department of Computer Science and Engineering
The Pennsylvania State University
University Park, PA 16802
E-mail: {boura | das}@cse.psu.edu

Abstract

A deadlock-free fault-tolerant routing algorithm for n-dimensional meshes is proposed in this paper. The algorithm provides full adaptivity and fault-tolerance at a modest cost of 3 virtual channels per physical channel. A node labeling mechanism is utilized to identify nodes that cause routing difficulties. Messages are routed adaptively in healthy regions of the network. Once a message faces a faulty region, it is routed around it using a nonminimal path. The advantage of the proposed algorithm is its ability to tolerate any number of faults and fault patterns.

1 Introduction

A class of direct networks, known as n-dimensional meshes, has received increasing attention recently due to its scalability and high performance. A network with many components experiences various types of component failure, which can be classified as node and link failure at the network level. A communication network should tolerate these faults gracefully and operate in a gradual degraded mode. The objective of this paper is to develop a fault-tolerant routing algorithm for n-dimensional meshes. The proposed routing algorithm is based on the wormhole switching paradigm, and builds upon adaptive routing [2] and virtual channel flow control [6].

The fault-tolerant routing algorithms proposed in the literature exhibit a trade-off between the degree of adaptivity, degree of fault-tolerance, and the number of virtual channels required to guarantee deadlock-freedom. For example, the algorithm proposed in [4] tolerates any number of faults, but it is partially adaptive and turns off a lot functional nodes to avoid routing difficulties. The algorithm in [5] has a degree of adaptivity and fault tolerance that is dependent on the number of virtual channels. On the other hand, Glass and Ni's algorithm requires only one virtual channel, but it is partially adaptive and tolerates a small number of faults [9]. Finally, Boppana and Chalasani's algorithm tolerates any number of faulty nodes and requires 4 virtual channels [1]. However, it does not offer any adaptivity. In this paper, an algorithm that tolerates any number of faults, requires only 3 virtual

channels per physical channel, and is fully adaptive is presented. The proposed algorithm extends an adaptive routing algorithm that is described in [2].

The rest of the paper is organized as follows. Necessary notations and definitions are described in Section 2. Section 3 describes the fault tolerant routing algorithm. Simulation based performance evaluation of the proposed algorithm is conducted in Section 4. Finally, Conclusions are drawn in Section 5.

2 Preliminaries

Definition 1 An n-dimensional mesh is defined formally as an interconnection structure that has $K_0 \times K_1 \times \ldots \times K_{n-1}$ nodes where n is the number of dimensions of the network, and K_i is the radix of dimension i. Each node is identified by an n-coordinate vector $(x_0, \ldots, x_{n-1})$, where $0 \leq x_i \leq K_i - 1$. Two nodes, $X(x_0, \ldots, x_{n-1})$ and $Y(y_0, \ldots, y_{n-1})$ are connected if and only if there exists an i such that $x_i = y_i \pm 1$, and $x_j = y_j$ for all $j \neq i$.

Definition 2 A channel along dimension i is termed a positive channel if its source node $X(x_0, \ldots, x_{n-1})$ and sink node $Y(y_0, \ldots, y_{n-1})$ differ in the ith coordinate such that $x_i = y_i - 1$.

Definition 3 A channel along dimension i is termed a negative channel if its source node $X(x_0, \ldots, x_{n-1})$ and sink node $Y(y_0, \ldots, y_{n-1})$ differ in the ith coordinate such that $x_i = y_i + 1$.

3 Fault-tolerant routing

A fault-tolerant routing algorithm should guarantee the delivery of messages in the presence of faulty nodes/links. Bypassing faults forces messages to use nonminimal paths in some cases. Hence, fault-tolerant routing algorithms are nonminimal by nature. In this section, a deadlock-free fault-tolerant routing algorithm is presented. The algorithm is based on a node fault model, and link faults are treated as node faults. The algorithm uses a node labeling scheme that identifies nodes that cause routing difficulties [10]. Faulty regions are converted to rectangular regions to facilitate routing decisions at each node. The node labeling strategy is based on the following two rules.

Node deactivation rule A *healthy* node connected to 2 *faulty* nodes is deactivated and marked *faulty*.

*This research was supported in part by the National Science Foundation under Grant No. MIP-9104485.

Node activation rule A deactivated node connected to a *healthy* node is activated and marked *unsafe*.

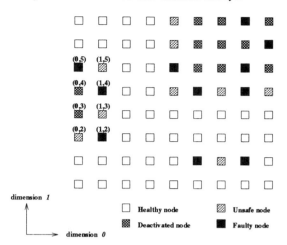

Figure 1: **Application of the node deactivation and activation rules in a 9×9 2-dimensional mesh.**

First, faulty regions are transformed to rectangular regions by applying the **node deactivation rule**. Nodes that present potential routing problems are deactivated. Second, the **node activation rule** is applied to activate deactivated nodes residing at the boundaries of faulty regions. Figure 1 demonstrates the application of the node labeling scheme. The following notation is used for describing the routing algorithm.

- $U(N)$ = set of *unsafe* nodes connected to node N.

- $F(N)$ = set of *faulty* nodes connected to node N.

- $S(c)$ = sink node of channel c.

- $L(c)$ = label of channel c.

- $L(N)$ = label of node N *(healthy, unsafe, faulty)*.

- c_i = channel along dimension i.

- $c_{class_{x\,i}}$ = virtual channel of class x along dimension i.

- $c_{class_{x\,i}+}$ = positive virtual channel of class x along dimension i.

- $c_{class_{x\,i}-}$ = negative virtual channel of class x along dimension i.

The proposed algorithm routes messages adaptively in healthy regions of the network. Once a message faces a faulty region, it is routed around it using a nonminimal path to get closer to its destination. *Healthy* nodes connected to only *healthy* nodes have two classes of virtual channels (i.e. $class_1$ and $class_2$) per physical channel. *Healthy* nodes connected to either *unsafe* or *faulty* nodes have three classes of virtual channels (i.e. $class_2$, $class_3$, and $class_4$) per physical channel. Hence, *healthy* nodes connected to faulty regions divide $class_1$ channels into two

classes ($class_3$ and $class_4$). Therefore, at least 2 $class_1$ channels and 1 $class_2$ channel are needed resulting in 3 virtual channels per physical channel. In healthy regions of the network, a message considers any $class_1$ channel that gets it closer to its destination. If no available $class_1$ channel is found, a message examines the state of $class_2$ channels. $class_2$ channels are traversed in two phases. In the first phase, positive $class_2+$ channels are traversed one dimension at a time in an increasing order. In the second phase, negative $class_2-$ channels are traversed in any order. The algorithm $ADAPTIVE\text{-}ROUTE()$ presents a formal description of adaptive routing in healthy regions.

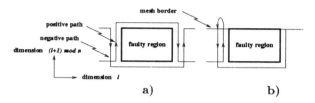

Figure 2: **A message reaching a faulty region routes around it. a) Dimension $(i + 1)$ *mod* n and dimension i virtual channels are utilized to misroute a message around a faulty region and get it closer to its destination along dimension i. b) A message utilizes a positive path first. When a message reaches the mesh border, it utilizes a negative path.**

Once a message faces a faulty region, it is routed around it to get closer to its destination following the steps described in the algorithm $MISROUTE()$. In phase one (two), dimension $(i+1)$ *mod* n $class_3$ ($class_4$) and dimension i $class_2+$ ($class_2-$) virtual channels are utilized to get a message closer to its destination along dimension i, where i is the lowest dimension that a message needs to traverse. Figure 2a shows graphically the available nonminimal paths that a message uses to avoid faulty regions. A message utilizes the positive path first. If a message reaches the system boundary, then it utilizes the negative path as illustrated in Figure 2b.

The main algorithm $FAULT\text{-}TOLERANT\text{-}ROUTE()$ is composed of four major parts. First, if the destination node is *unsafe*, *healthy* nodes are used until a message becomes one hop away from its destination where any available virtual channel is used. Second, if the source node is *unsafe*, a message is routed to a *healthy* node at the first hop using any available virtual channel. Third, if a message needs to traverse a channel along dimension i where the sink node of that channel is either *unsafe* or *faulty*, then the message is misrouted according to the algorithm $MISROUTE()$. Finally, if a message resides at a *healthy* node that is connected to only *healthy* nodes, a message is routed according to the algorithm $ADAPTIVE\text{-}ROUTE()$. A procedure-like description of the proposed fault-tolerant algorithm is given below.

FAULT-TOLERANT-ROUTE (routing_tag)
begin
 /* $routing\_tag[i] = destination[i] - source[i]$ */
 if $(\not\exists\ i\ |\ routing\_tag[i] \neq 0)$ **then**
  /* message reached its destination */
  return;
 else
  if $((\exists\ i\ |\ routing\_tag[i] == \pm 1)\&\&(\forall j \neq$
   $i,\ routing\_tag[j] == 0)\&\&$
   $(L(S(c_i)) == unsafe))$ **then**
   /* deliver message to an *unsafe* destination */
   route message along a virtual channel
   $(c_i\ |\ ((c_i \in (C_{class2_i} \cup C_{class3_i} \cup C_{class4_i}))$
   $\&\&(c_i\ is\ free)));$
   return;
  else
   if $(L(N) == unsafe)$ **then**
    /* forward message from an *unsafe* node
    to a *healthy* node */
    $C = \{c_i | ((L(S(c_i)) == healthy)\&\&$
     $(c_i \in (C_{class2_i} \cup C_{class3_i} \cup C_{class4_i}))\&\&$
     $(c_i\ is\ free)\ for\ (0 \leq i \leq n-1))\};$
    route message along a virtual channel
    $(c_i | c_i \in C);$
    return;
   else
    /* $L(N) == healthy$ */
    if $(\exists\ i\ |\ routing\_tag[i] > 0)$ **then**
     $i := min_{0 \leq j \leq n-1}(j\ |\ routing\_tag[j] > 0);$
     /* message is in phase one */
     if $(S(c_i) \in (F(N) \cup U(N)))$ **then**
      $MISROUTE(i, routing\_tag);$
     else
      $ADAPTIVE\text{-}ROUTE(routing\_tag);$
     end ifelse;
    else
     /* $\forall\ i,\ routing\_tag[i] \leq 0$ */
     /* message is in phase two */
     $i := min_{0 \leq j \leq n-1}(j\ |\ routing\_tag[j] \neq 0);$
     if $(S(c_i) \in (F(N) \cup U(N)))$ **then**
      $MISROUTE(i, routing\_tag);$
     else
      $ADAPTIVE\text{-}ROUTE(routing\_tag);$
     end ifelse;
    end ifelse;
   end ifelse;
  end ifelse;
 end ifelse;
end;

ADAPTIVE-ROUTE (routing_tag)
begin
 /* $routing\_tag[i] = destination[i] - source[i]$ */
 /* C_{class1} is a set of free virtual channels */
 $C_{class1} = \{c_{class1_i} | (routing\_tag[i] \neq 0)$
  $\&\&(c_{class1_i}\ is\ free)\};$
 if $(C_{class1} \neq \emptyset)$
  $routing\_tag[i] = routing\_tag[i] \pm 1;$
  route message along channel $(c_{class1_i} | c_{class1_i}$
  $\in C_{class1});$
  return;
 else
  if $(\exists\ i\ |\ routing\_tag[i] > 0)$ **then**
   /* message is in phase one */
   /* $C_{class2+}$ is a set of free positive virtual
    channels */
   $C_{class2+} = \{c_{class2_i} | (i := min_{0 \leq j \leq n-1}[j\ |$

$routing\_tag[j] > 0])\&\&(c_{class2_i}\ is\ free)\}$
$routing\_tag[i] = routing\_tag[i] - 1;$
route message along channel $(c_{class2_i} | c_{class2_i}$
$\in C_{class2+});$
return;
else
 /* $\forall\ i,\ routing\_tag[i] \leq 0$ */
 /* message is in phase two */
 /* $C_{class2-}$ is a set of free negative virtual
  channels */
 $C_{class2-} = \{c_{class2_i} | (routing\_tag[i] \neq 0)\&\&$
  $(c_{class2_i}\ is\ free)\};$
 $routing\_tag[i] = routing\_tag[i] + 1;$
 route message along channel $(c_{class2_i} | c_{class2_i}$
 $\in C_{class2-});$
 return;
end ifelse;
end ifelse;
end;

MISROUTE (i, routing_tag)
begin
 if $(routing\_tag[i] > 0)$ **then**
  /* message is in phase one */
  misroute around faulty region along a positive
  path using available $c_{class3_{(i+1) mod\ n}}$ and
  c_{class2_i+} virtual channels.
  if (message is blocked by mesh border) **then**
   misroute around faulty region along a negative
   path using available $c_{class3_{(i+1) mod\ n}}$ and
   c_{class2_i+} virtual channels.
  end if;
 else
  /* message is in phase two */
  misroute around faulty region along a positive
  path using available $c_{class4_{(i+1) mod\ n}}$ and
  c_{class2_i-} virtual channels.
  if (message is blocked by mesh border) **then**
   misroute around faulty region along a negative
   path using available $c_{class4_{(i+1) mod\ n}}$ and
   c_{class2_i-} virtual channels.
  end if;
 end ifelse;
end;

Theorem 1 *FAULT-TOLERANT-ROUTE() is dead-lock free (for proof see [3]).*

4   Experimental results

Extensive simulations are conducted to analyze the performance of the proposed fault-tolerant routing algorithm. Two unidirectional channels are assumed to exist between two connected nodes. Wormhole switching is assumed to be the underlying switching mechanism. All virtual channels are assumed to be a single buffer deep. *Healthy* and *unsafe* nodes generate messages at time intervals selected from a negative exponential distribution. Message destinations are uniformly distributed among *healthy* and *unsafe* nodes. Messages are assumed to be 20 flits long. Conflicts of requests for the same output channel by multiple messages are resolved randomly.

When a message has a number of valid and available output channels, the first channel along the lowest dimension is selected. Each simulation is run for a total of 30000 cycles. Performance data are not collected in the first 10000 cycles to allow the system to stabilize.

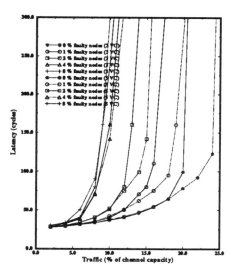

Figure 3: **Message latency versus traffic rate in a 10 × 10 2-dimensional mesh for a varying number of faulty nodes and for 3 and 5 virtual channels.**

A 10 × 10 2-dimensional mesh is selected as the underlying network topology. Experiments with different numbers of faulty nodes are performed. In order to evaluate the effect of different fault patterns for a given number of faulty nodes, 10 different experiments having the same number of faults and different fault patterns are conducted. The collected data are averaged to obtain average performance measures. Figure 3 shows the performance of the proposed fault-tolerant routing algorithm in various faulty environments for 3 and 5 virtual channels per physical channel. The figure suggests that as the number of faulty nodes increases, the network saturates at a lower traffic rate. The degradation in performance is primarily due to the nonuniformity in traffic rates on different channels. The traffic rate on channels surrounding a faulty region is higher than the traffic rate on channels distant from faulty regions. This nonuniformity in traffic rates is attributed to the loss in network connectivity. Consequently, channels surrounding faulty regions become heavily congested and a source of bottleneck. As the number of virtual channels is increased from 3 to 5, network performance improves when the number of faults is small. As the number of faults increases, the increase in the number of virtual channels has no effect on network performance.

5 Conclusions

Fault-tolerant routing is very essential for guaranteeing the delivery of messages in the presence of faulty nodes/links. In this paper, a deadlock-free fault-tolerant routing algorithm for n-dimensional meshes is proposed. 3 virtual channels per physical channel are required to guarantee deadlock-freedom. Using a node labeling scheme, faulty regions are converted to rectangular regions. A message is routed adaptively until it reaches its destination or a faulty region. Once a message faces a faulty region, it is routed around it using a nonminimal path.

References

[1] R. Boppana and S. Chalasani, "Fault-tolerant routing with non-adaptive wormhole algorithms in mesh networks," Supercomputing 94, pp. 693-702, 1994.

[2] Y. M. Boura and C. R. Das, "Efficient fully adaptive wormhole routing in n-dimensional meshes," *Proc. The 14th International Conference on Distributed Computing Systems*, June 1994.

[3] Y. M. Boura, "Design and analysis of routing algorithms and communication switches for mesh-connected architectures," The Pennsylvania State University, Ph.D. dissertation 1995.

[4] A. A. Chien and J. H. Kim, " Planar-adaptive Routing: Low-cost Adaptive Networks for Multiprocessors," *Proc. The 19th Annual International Symposium on Computer Architecture*, pp. 268-277, May 1992.

[5] W. J. Dally and H. Aoki, "Deadlock-free adaptive routing in multicomputer networks using virtual channels," *IEEE Transactions on Parallel and Distributed Systems*, Vol. 4, pp. 466-475, April 1993.

[6] W. J. Dally, "Virtual channel flow control," *IEEE Transactions on Parallel and Distributed Systems*, Vol. 3, No. 2, pp. 194-205, March 1992.

[7] W. J. Dally and C. L. Seitz, "Deadlock-free message routing in multiprocessor interconnection networks," *IEEE Transactions on Computers*, Vol. C-36, pp. 547-553, May 1987.

[8] J. Duato, "A new theory of deadlock-free adaptive routing in wormhole networks," *IEEE Transactions on Parallel and Distributed Systems*, pp. 1320-1331, December 1993.

[9] C. J. Glass and L. M. Ni, " Fault-tolerant wormhole routing in meshes," *Proc. The 23rd Annual International Symposium on Fault-Tolerant Computing*, pp. 240-249, 1993.

[10] T. C. Lee and J. P. Hayes, "A fault-tolerant communication scheme for hypercube computers," *IEEE Transactions on Computers*, pp. 1242-1256, October 1992.

INCORPORATING FAULT TOLERANCE IN THE MULTISCALAR FINE-GRAIN PARALLEL PROCESSOR

Manoj Franklin

Department of Electrical & Computer Engineering
Clemson University
102 Riggs Hall . Clemson, SC 29634-0915, USA
Email: mfrankl@blessing.eng.clemson.edu

Abstract -- *The multiscalar processor, proposed recently for exploiting fine-grain parallelism, is a collection of sequential processors that are connected together in a logically decoupled and physically decentralized manner using a ring-type network. The central idea behind this processing paradigm is to divide a sequential instruction stream into tasks, and then exploit parallelism by allocating these tasks to multiple processing elements, which can execute code in parallel. This paper investigates the issue of incorporating fault tolerance in the multiscalar processor. Fault tolerance is achieved by replicating run-time software tasks and executing them on multiple execution units, which are already provided for performance reasons. Two re-execution schemes (both using the same replication technique) are described. The paper also presents the results of extensive simulation studies that show the performance loss due to the introduction of fault tolerance. These results suggest that exploiting spare resources is an excellent choice for providing fault tolerance in the multiscalar processor.*

1. INTRODUCTION

Recent advances in VLSI technology have enabled millions of transistors to be placed in a single chip. Current processors incorporate 7-9 million transistors in a chip, and technology projections predict this number to reach about 100 million by A.D. 2000. One way these resources are being used, and will continue to be used, is for incorporating performance-enhancing hardware structures. Whether we like it or not, a bludgeoning fact of life that has to be reckoned with, is that most of these hardware resources are going to be idle a substantial part of the time. Common examples of such hardware resources, with not-so-high utilization, are different stages of a pipelined functional unit, multiple functional units, and substantial portions of the instruction cache array and data cache array. Any hardware resource with low utilization is an ideal target for incorporating low-overhead fault tolerance schemes. Several proposals for exploiting this phenomenon can be found in the literature [2, 4, 9, 11-13].

This paper investigates the use of idle resources to incorporate fault tolerance in the multiscalar processor [6-8]. Introduced recently for exploiting instruction level parallelism (ILP), the multiscalar processor is replete with repetitive hardware. It exploits ILP by executing in parallel multiple blocks of code with the help of multiple processing elements that are connected together in a decoupled and decentralized manner using a unidirectional ring-type network. The parallelly executed code blocks can have both data and control inter-dependencies between them (which makes the multiscalar processor entirely different from a multiprocessor). We selected the multiscalar processor for this study for the following reasons:

(1)   It is a novel architecture, and has great potential to become the processing paradigm of choice for a circa 2000 general-purpose ILP processor.

(2)   Because it has a decentralized design, it is amenable to fault tolerance.

(3)   A detailed simulator is available for experiments.

The rest of this paper is organized as follows. Section 2 unifies the existing body of knowledge in processor fault tolerance. Section 3 describes the multiscalar processor, and explains how time-redundant techniques can be applied in the multiscalar processor to achieve fault tolerance. Section 4 describes two schemes for incorporating fault tolerance in the multiscalar processor. Section 5 gives the results of a simulation study that quantitatively measures the performance overhead due to implementing these fault tolerance schemes in the multiscalar processor. Section 6 presents the conclusions.

2. RELATED WORK

In this section, we attempt to put the existing state of the art in processor fault tolerance techniques in perspective, in order to provide a motivation for our approach to incorporate fault tolerance in the multiscalar processor. The basic idea behind any fault tolerance scheme is to have some form of redundancy. This redundancy can be in the form of (i) hardware redundancy, (ii) software redundancy, (iii) information redundancy, or (iv) time redundancy. Of course, real-life fault tolerance schemes invariably use combinations of the different types of redundancy, and at different levels of the system.

2.1. Hardware Redundancy

Hardware redundancy techniques involve replication of hardware units, and executing multiple copies (or different versions) of the software on these hardware units. The results produced by the multiple units are compared with the help of comparators/voters, which determine the correct result and identify the faulty units, if any. Hardware-redundant schemes have not been popular for commercial processors, because the extra silicon area needed for incorporating hardware redundancy could potentially be used for implementing hardware that improves the performance.

2.2. Information Redundancy

Information redundancy techniques involve adding additional information to existing information. Examples are (i) error-correcting code (ECC), (ii) control flow based signatures, and (iii) algorithm-based checksums. The principle behind ECC is that in a set of all possible combinations of symbols, only a subset of combinations, called *code words*, are allowed to be valid combinations so that the occurrence of an error most likely changes it into a non-code word.

In control flow checking, additional information on program control flow is stored in the program by means of signatures or checksums. At run time, these signatures are recalculated based on the run-time control flow, and compared against the compile-time calculated signatures. This technique detects control flow errors that cause the CPU to take a path that is not present in the control flow graph (CFG) of the executed program. However, control flow errors that cause the CPU to take an incorrect path through the CFG are not detected.

2.3. Time Redundancy

Time-redundancy techniques involve re-executing a piece of code or an operation using the same piece of hardware, and comparing the results. Examples for techniques at the subroutine execution level are: (i) rollback and recovery schemes [3, 10] and (ii) N version programming [1]. Examples for techniques at the instruction execution and data transmission level are instruction re-execution, data retransmission, and recomputation with shifted operands (RESO).

2.4. Exploiting Existing Hardware Redundancy

In high-performance processors that exploit parallelism, hardware redundancy in some form is already provided for exploiting concurrency. For example, vector processors routinely pipeline each functional unit heavily. VLIW processors even provide multiple copies of the same functional unit so as to issue multiple operations per cycle. Multiprocessors provide multiple copies of the processing element to exploit parallelism at a coarse level. Any time extra hardware is thrown in for improving the performance, there is a good chance that its utilization becomes low. For instance, in multiprocessors, not all processors may be active all the time, and this redundancy can be used for executing identical copies of programs [4, 5, 13]. Similarly, in ILP processors, not all stages of the functional unit pipelines may be busy during every clock cycle (as evidenced by the substantial difference between the peak and sustained issue rates of these processors). This makes it particularly attractive to employ time redundancy in these processors by making use of the redundant hardware [2, 9, 11, 12].

In [12], the authors investigated the application of time redundancy to make use of empty pipeline slots in the Cray-1. In their scheme, the hardware replicates instructions that use the (pipelined) functional units, and executes them in the same functional units, but the duplicate instructions use shifted versions of the operands (RESO).

In [9], Holm and Banerjee proposed a way of exploiting the functional unit redundancy in VLIW processors, to detect errors occurring due to faults in the functional units. In their scheme, the compiler replicates instructions that use functional units, generates code for comparing the results of regular and duplicate instructions, and schedules them. The overall performance degradation for their scheme is fairly high (ranging from 1.7% to 187% for one method and from 0.2% to 161% for another method).

Blough and Nicolau also proposed a similar technique for incorporating fault tolerance in the functional units of VLIW and superscalar processors [2]. Again, the error coverage is low, because only the errors occurring due to faults in the functional units are detected.

3. THE MULTISCALAR PROCESSOR AND REDUNDANCY

The multiscalar processor is a recent processing paradigm introduced for exploiting ILP. References [6, 8] describe different facets of the multiscalar processor in great detail. A sound understanding of the multiscalar processor is essential for a good appreciation of the fault tolerance techniques presented in this paper.

3.1. Multiscalar Processor

Figure 1 gives the block diagram of an 8-unit multiscalar processor. It consists of multiple execution units that are connected by a unidirectional ring-type network such that information flows in one direction around the circular queue. This provision, along with hardware head (H) and tail (T) pointers, imposes a sequential order among the units, the head pointer indicating the oldest active unit (when the queue is not empty). The order between the head and tail is implicit in the queue.

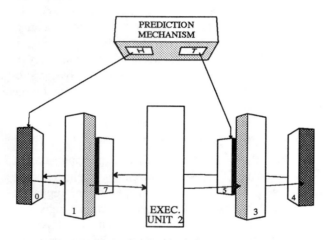

Figure 1: An 8-Unit Multiscalar Processor

A program executes on the multiscalar processor as follows. Each cycle, if the tail execution unit is idle, a prediction is made by the prediction mechanism to determine the next task in the dynamic instruction stream; a task is a subgraph of the control flow graph of the executed program. This predicted task is invoked on the tail unit. Upon a successful invocation, the tail pointer is advanced, and the invocation process continues at the new tail in the next cycle. Each active unit executes the instructions in its task, maintaining the appearance of sequential semantics within the tasks and between the tasks. When the head unit completes its task, it commits the instructions in its task. Upon a successful commit, the head pointer is advanced, causing that unit to become idle. In the event an incorrect execution (due to incorrect

task prediction) is detected, all units between the incorrect execution point and the tail are discarded in what is known as a *squash*. The tail pointer is adjusted to the unit at the point of incorrect execution, and the invocation process resumes at the correct target. execution maintains sequential semantics (although the tasks are executed in parallel) because the sequence of instructions that are committed by the units corresponds directly to the instructions which would be processed in a serial execution of the tasks.

The tasks being executed in parallel can have both control and data dependencies between them. Inter-task register data dependencies are taken care of with the help of busy bits and by forwarding the last update of each register in a task to the subsequent tasks, as and when the last updates are generated. Potential inter-task data dependencies through memory, occurring through loads and stores, are handled by allowing the loads and stores of multiple tasks to execute in any order, with special hardware provided to detect any violations to sequential program semantics, and to initiate recovery actions.

We do not dwell into further details of the multiscalar processor here, as they have been described elsewhere [6-8]. In passing, it is worth pointing out that *decentralization of critical resources, which was the philosophical backbone of the multiscalar processor, aids not only in expandability, but also in the incorporation of testing and fault tolerance.* Testing the execution units during program execution is relatively straightforward, because it is possible to logically disconnect an execution unit temporarily and test it while the remaining units are busy executing the tasks assigned to them. Notice that in other ILP processing paradigms such as the superscalar and VLIW processors, the entire processor is more or less a single monolithic unit, which makes it difficult to do concurrent testing or fault tolerance.

3.2. Redundancy in the Multiscalar Processor

For the multiscalar processor to be functional and be executing the tasks in a program, strictly speaking, a single execution unit is sufficient. Multiple execution units are provided only to boost the performance, and are not essential to the functionality of the processor. In that sense, the multiple execution units are redundant hardware. If there are N execution units, they can generally execute a total of N instructions in every clock cycle. However, it is rare that this execution rate is achieved in a continuous manner, because of data and control dependencies inherent in the program. This fact is evident from the performance results presented in [6, 8] for the multiscalar processor.

When instructions are not executed at the peak rate, *stall cycles* exist in the execution units in which no instruction execution was initiated. Since an execution unit is not used to initiate any useful computation during a stall cycle, that cycle can potentially be used for initiating fault tolerance operations. By making effective use of the stall cycles, the overall performance overhead for performing fault tolerance operations can be significantly reduced. Let us next consider ways of doing this.

4. FAULT TOLERANCE SCHEMES

4.1. What to Re-Execute?
(Unit of Re-Execution)

The first issue to be considered is the unit of re-execution. The unit of redundancy in the multiscalar processor is an execution unit, which is responsible for executing an entire task. Therefore, it is convenient to incorporate fault tolerance at the task execution level, and to consider a task as the unit of re-execution.

The re-execution can be carried out as follows: for each dynamic task T, generate a duplicate task T_d and execute it in one of the execution units. Of the two tasks assigned to an execution unit — an *original task* and a *duplicate task* — give preference to the execution of instructions from the original task so as to minimize the impact of fault tolerance on performance. That is, in any cycle, an execution unit first attempts to execute ready-to-execute instructions from the original task allocated to it, and if there are any vacant slots, then it executes ready-to-execute instructions from the duplicate task allocated to it. If there are many stall cycles in the execution of the original task, then the execution of the duplicate task is not likely to take many extra cycles, thereby reducing the overall performance overhead.

There are 2 additional important questions while contemplating the re-execution of each task. They are: (i) *where* to execute the duplicate task, and (ii) *when* to execute the duplicate task. We shall next address them.

4.2. Where to Execute the Duplicate Task?

This is an important question as it concerns 3 issues: (i) error coverage, (ii) communication of operands for the duplicate task, and (iii) communication of results for comparison purposes. If the duplicate task is re-executed in the same execution unit in which the original task is executed, then the only errors that are detected are those occurring due to transient faults in that execution unit; errors due to permanent faults may go undetected because the same error might occur while executing T as

well as T_d. Therefore, for better error coverage, it is desirable to execute the duplicate task in a different execution unit. However, if the duplicate task is executed in a distant execution unit, then significant global communication is introduced for two reasons: (1) the results produced by $T-1$ are the operands for T_d, and therefore need to be communicated to T_d, and (2) the values produced by T_d need to be compared with the values produced by T. Even if point 1 can be taken care of by using the results of $(T-1)_d$ as the operands of T_d, point 2 still remains valid. Therefore, a good choice is to execute T_d in an execution unit adjacent to the one in which T is executed. Thus, if T is executed in unit E, then executing T_d in unit $E_d = (E \pm 1) \bmod N$, where N is the number of execution units in the multiscalar processor, provides both good error coverage and localized communication.

Of the two choices in $(E \pm 1) \bmod N$, let us choose $(E-1) \bmod N$. That is, execute T_d in the execution unit preceding the one in which T is executed. In that case, because tasks $(T-1)$ and T_d are executed in the same execution unit, the external operands for T_d (i.e., the operands of T_d produced by preceding tasks) are available in the same execution unit, and need not be communicated across the register forwarding network. The results produced by T_d are forwarded to the succeeding unit, where task T is executed, and compared with the results generated by T to detect any errors.

4.3. When to Execute the Duplicate Task?
(Unit of Recovery)

Having decided to execute the duplicate task in the unit preceding the unit in which the corresponding original task is executed, let us next address the issue of when to do the re-execution. Re-execution delay refers to the time difference between the execution of an original task and the corresponding duplicate task, and is an indication of the *unit of recovery* in times of error detection. As the re-execution delay increases, the unit of recovery, and hence the size of the storage (for storing the uncommitted values produced by the original tasks and the operands for the execution of the duplicate tasks) also increases.

4.3.1. Zero-Delay Scheme

A straightforward choice is to do the re-execution more or less at the same time as the execution of the corresponding original task. That is, when execution unit E executes task T, execution unit $(E-1) \bmod N$ executes task T_d (in addition to task to $T-1$). We shall call this scheme a *zero-delay* scheme, as the execution of tasks T and T_d may overlap. Figure 2 shows conceptually where and when duplicate tasks are executed in the zero-delay scheme. It shows 3 execution units, numbered $(E-1)$ mod N, E, and $(E+1)$ mod N. They have been assigned the original tasks $T-1$, T, and $T+1$, respectively. The duplicate tasks assigned to them at the same time are T_d, $(T+1)_d$, and $(T+2)_d$, respectively.

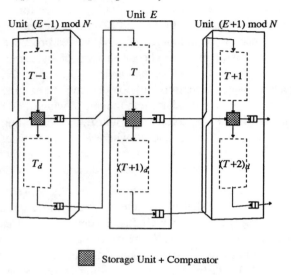

Figure 2: **Conceptual Schematic of Zero-Delay Scheme**

The major advantage of the zero-delay scheme is that the comparison of the results take place not long after a result is produced, and therefore only limited storage is needed to store the results. Notice that if a result from a duplicate task T_d reaches the storage unit of unit E earlier than the corresponding result from task T, it can update the storage unit. The comparison in that case is done when T produces the corresponding result. Recovery actions are simple with the zero-delay scheme, as the error latency is small.

Another advantage of the zero-delay scheme is that the duplicate task being executed in a unit is the logical successor of the original task being executed in that unit, and so the two tasks can be considered to be a single task for the purposes of instruction fetching, decoding, execution, and intra-task memory disambiguation; the instructions need to be treated differently only for the purpose of forwarding of results.

4.3.2. Unit-Delay Scheme

Although additional hardware requirements are minimal for the zero-delay scheme, it can result in degraded performance. This is because in a given cycle, both $T-1$ and T_d, the two tasks that are being executed in execution unit $(E-1)$ mod N, may be waiting for their operands to be ready, with the result that unit $(E-1)$ mod N may not be able to execute any instructions in that

cycle. This happens because T_d and $T-1$ (which produces most of the external operands of T_d) are executed at the same time. To overcome this problem, we can execute T_d later, along with original task $T+N-1$, *when all of T_d's external operands are available*. We call this scheme as a ***unit-delay*** scheme, as task T_d is executed in unit $(E-1)$ mod N along with the next regular task executed in $(E-1)$ mod N. Figure 3 shows conceptually where and when the duplicate tasks are executed in the unit-delay scheme. Again, there are 3 execution units in the figure, numbered $(E-1)$ mod N, E, and $(E+1)$ mod N as before. They have been assigned the original tasks $T+N-1$, $T+N$, and $T+N+1$, respectively. But, notice that the duplicate tasks assigned to them at the same time are T_d, $(T+1)_d$, and $(T+2)_d$. In this case, task T_d has an instruction ready for execution every cycle (except when there are data dependencies with other long latency instructions of T_d), as there are no inter-task dependencies to be taken care of among the active duplicate tasks.

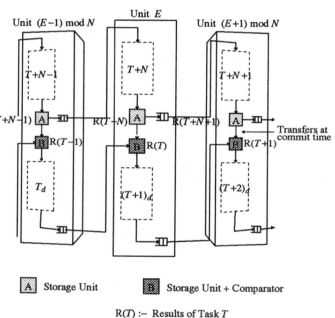

$R(T) :-$ Results of Task T

Figure 3: **Conceptual Schematic of Unit-Delay Scheme**

Now consider the hardware for storing the results while they are waiting to be compared. For the unit-delay scheme, in addition to storing the results of T, the operands for executing T_d also need to be stored. A comparison of Figures 2 and 3 shows that an extra storage unit is required for storing the results produced by the previous original task executed in a unit. Among the two storage units (marked A and B in Figure 3) in an execution unit, storage unit A is used to store the results produced by the current original task being executed in that unit, and storage unit B is used to store the results produced by the previous original task executed in that unit. When the execution of an original task is completed and the task is committed, the results produced by the task are transferred from storage unit A to storage unit B. Thus, when a duplicate task T_d starts execution, all its external operands are available in storage unit B. An important aspect to be noted here is that to transfer results from A to B, data needs to be moved only logically, which can be easily implemented by toggling a flip-flop that identifies which storage unit is A and which is B.

4.4. Other Details

4.4.1. Forwarding of Register Values

In the zero-delay scheme, a register value produced by a duplicate task is forwarded to the next stage, provided the value is the last instance for that architectural register in that duplicate task. For performing this forwarding, the same forwarding network used for forwarding the regular results is used. The one difference in the forwarding of these values is that they are not forwarded beyond the subsequent execution unit.

4.4.2. Recovery Due to Incorrect Task Prediction

When a task prediction (say, the prediction made after allocating a task to unit E) is found to be incorrect, recovery actions are initiated by the processor. These actions involve squashing the tasks being executed in the execution units beyond E. In the zero-delay scheme, the recovery mechanism also discards the duplicate tasks being executed in these units. Further, the duplicate task being executed in unit E is also discarded (as it belongs to an incorrect execution path), and the next duplicate task from the correct execution path is allocated to unit E.

In the unit-delay scheme, the recovery process does not affect any of the duplicate tasks being executed in the processor. This is because, in the unit-delay scheme, duplicate tasks assigned to the execution units are guaranteed to be along the correct execution path. This is evident from the fact that their corresponding original tasks have completed their execution. Thus, in the unit-delay scheme, duplicate tasks are never squashed because of incorrect task prediction. This is another reason to expect better performance for this scheme.

4.4.3. Committing an Execution Unit

Before an execution unit can be committed, both the original task and the duplicate task being assigned to the unit should complete their execution. Although it is not mandatory that the duplicate task should have

finished its execution, committing an execution unit before the completion of the duplicate task will require yet another level of physical storage — to temporarily store the results produced by the new original task that is assigned to the unit after committing the previous one.

4.5. Hardware Cost

Let us evaluate the hardware cost for the proposed fault tolerance schemes. Introduction of these fault tolerance schemes requires additions to the control unit and the datapath within each execution unit. Whereas the additions to the datapath can be easily quantified, the additions to the control unit complexity can be determined only with the help of a detailed hardware design.

First let us consider the datapath. An additional register file is required in each execution unit to be used as a working file by the duplicate task being executed in that unit. (For the unit-delay scheme, the number of additional register files required is two.) Furthermore, n comparators, where n is the cross-section bandwidth of the register forwarding network (n is typically 1 or 2), are required in each execution unit to do the comparisons. If out-of-order execution is performed in each execution unit, then the hardware execution window in each execution unit has to be made large enough to hold instructions from both the original task and the duplicate task.

The control unit has to be modified to fetch instructions from the duplicate task as well. It has to check for additional conditions before committing an execution unit. Finally, it has to initiate appropriate recovery actions when it detects an error.

4.6. Error Coverage

It is worthwhile to study the error coverage of the proposed fault tolerance schemes. Because code duplication and re-execution are performed at the task level, and because the two sets of tasks are executed in different execution units, any error occurring in an execution unit gets detected. This includes errors due to fetching incorrect instructions (within an execution unit), errors due to improper control mechanism within an execution unit, errors due to improper register access, errors due to faults in functional units, etc. The only errors that are not directly covered by this scheme are those occurring due to faults in the global instruction cache, the data cache, and the prediction mechanism. Of these, both the instruction cache and the data cache can be protected by ECC check bits. The decisions made by the global control unit (task predictions) are counter-checked by the execution units when they execute tasks.

5. EXPERIMENTS AND RESULTS

To determine the actual degradation in program execution time due to the introduction of fault tolerance techniques, we carried out a detailed performance evaluation study.

5.1. Experimental Framework

Our methodology of experimentation is simulation, and the experimental framework is same as the one used for obtaining the results given in [8]. The multiscalar simulator was modified to model the two fault tolerance techniques discussed in section 4. Cache sizes, and other parameters (such as number of instructions simulated) were also fixed at the same values as in [8]. However, a task was restricted to have at most one basic block. While executing instructions from an active execution unit, preference was given each cycle to the execution of instructions from the original task. Furthermore, instructions of the duplicate task are fetched only after fetching all the instructions of the original task being executed in the same unit.

5.1.1. Benchmarks and Performance Metrics

For benchmarks, we used integer programs from the SPEC '92 benchmark suite, and tycho, a cache simulator program. These programs were compiled for a DECstation 3100 using the MIPS C compiler. Notice that this code has not been scheduled for a multiscalar processor, as a compiler for this processor is still under development. Because of this, the results presented in this paper should be considered as optimistic.

For measuring performance, we use instruction completion rate as the metric, with the following steps adopted to make the metric as reliable as possible: (i) nops are not counted while calculating the instruction completion rate, (ii) speculatively executed instructions whose execution was not required are not counted while calculating the instruction completion rate, and (iii) duplicate instructions, executed for fault tolerance purposes, are also not counted. Thus, our metric can be called as *useful instruction completion rate (UICR)*.

5.2. Performance Results with 4 Execution Units

Figure 4 presents the results of our simulations done with 4 execution units. Each cycle, a maximum of two instructions are fetched in each of the active execution units. In the figure, the light shade corresponds to the results with no fault tolerance incorporated; the next dark shade corresponds to the results with the zero-delay scheme. The dark shaded region corresponds to the

results with unit-delay scheme. Let us consider the results of Figure 4 in some detail.

First, consider the zero-delay scheme. Because the duplicate tasks are executed more or less at the same time their corresponding original tasks are executed, this fault tolerance technique is not able to make full use of the stall cycles that are made available to it, because the duplicate instructions themselves stall for their data dependencies to be resolved. The performance degradation can be as high as 19% (for *tycho*). For the remaining programs, the performance degradation is not so high. The zero-delay scheme suffers the worst performance degradation; however, the worst degradation is still less than a respectable 19% (for code compiled for a single-issue processor).

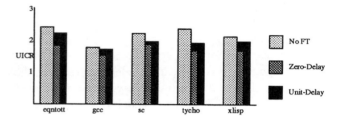

Figure 4: UICR with 4 Execution Units

With a unit-delay scheme, the performance degradation is less because there are no inter-task dependencies affecting the execution of the duplicate tasks, and the dependence paths in the re-execution part are limited to within each duplicate task.

5.3. Performance Results with 8 Execution Units

Next let us see how the results are if 8 execution units are used instead of 4. Figure 5 presents the same set of results obtained with 8 execution units. Rest of the parameters are same as those used to obtain the results with 4 execution units. The results of Figure 5 are an eye-opener. Almost no performance penalty is being paid for introducing the unit-delay fault tolerance scheme!

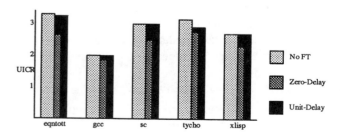

Figure 5: UICR with 8 Execution Units

6. CONCLUSIONS

We have developed a strategy for implementing fault tolerance in the multiscalar processor, and our strategy is to re-execute each task in an execution unit different from the one in which it was originally executed. We proposed two schemes for carrying out this re-execution. Both re-execute a task in the unit preceding the one in which it was originally executed. Whereas the first scheme re-executes a task in parallel to the execution of its original task, the second scheme performs the re-execution later, when all of its operands are available. We also evaluated the performance degradation due to the two schemes using a simulation study. The two schemes were found to degrade performance by at most 19% and 16% for 4 execution units. For 8 execution units, the unit-delay scheme was found to have almost zero performance impact. Based on these results, we believe that time-redundant fault tolerance techniques are an excellent choice for use in the multiscalar processor. This is because the architectural paradigm used in the multiscalar processor — multiple execution units — is replete with redundant hardware, and is well suited to time-redundant fault tolerance schemes.

ACKNOWLEDGEMENTS

This work was supported by the NSF Research Initiation Award grant CCR 9410706.

REFERENCES

[1] A. Avizienis and L. Chen, "On the Implementation of N-Version programming for Software Fault-Tolerance during Execution," *Proceedings of Compsac 77*, pp. 149-155, November 1977.

[2] D. M. Blough and A. Nicolau, "Fault Tolerance in Super-Scalar and VLIW Processors," *Proceedings of IEEE Workshop on Fault-Tolerant Parallel and Distributed Systems*, pp. 193-200, 1992.

[3] K. M. Chandy and C. V. Ramamoorthy, "Rollback and Recovery Strategies for Computer Programs," *IEEE Transactions on Computers*, vol. C-21, pp. 546-556, June 1972.

[4] A. T. Dahbura, K. K. Sabnani, and W. J. Hery, "Spare Capacity as a Means of Fault Detection and Diagnosis in Multiprocessor Systems," *IEEE Transactions on Computers*, vol. 38, pp. 881-891, June 1989.

[5] J.-C. Fabre, Yves Deswarte, and J.-C. Laprie, "Saturation: Reduced Idleness for Improved Fault Tolerance," *Proceedings of 18th International Symposium on Fault-Tolerant Computing (FTCS-18)*, pp. 200-205, 1988.

[6] M. Franklin and G. S. Sohi, "The Expandable Split Window Paradigm for Exploiting Fine-Grain Parallelism," *Proceedings of 19th Annual International Symposium on Computer Architecture*, pp. 58-67, 1992.

[7] M. Franklin and G. S. Sohi, "Register Traffic Analysis for Streamlining Inter-Operation Communication in Fine-Grain Parallel Processors," *Proceedings of 25th Annual International Symposium on Microarchitecture (MICRO-25)*, pp. 236-145, 1992.

[8] M. Franklin, "The Multiscalar Architecture," Ph.D. Thesis, Technical Report TR 1196, Computer Sciences Department, University of Wisconsin-Madison, 1993.

[9] J. G. Holm and P. Banerjee, "Low Cost Concurrent Error Detection in a VLIW Architecture using Replicated Instructions," *Proceedings of International Conference on Parallel Processing*, 1992.

[10] K. H. Kim, "Programmer-Transparent Coordination of Recovering Concurrent Processes: Philosophy and Rules for Efficient implementation," *IEEE Transactions on Software Engineering*, vol. 14, pp. 810-821, June 1988.

[11] M. A. Schuette and J. P. Shen, "Exploiting Instruction-Level Parallelism for Integrated Control-Flow Monitoring," *IEEE Transactions on Computers*, vol. 43, pp. 129-140, February 1994.

[12] G. S. Sohi, M. Franklin, and K. K. Saluja, "A Study of Time-Redundant Fault Tolerance Techniques for High-Performance Pipelined Processors," *Digest of Papers, 19th International Symposium on Fault-Tolerant Computing (FTCS-19)*, pp. 436-443, 1989.

[13] S. Tridandapani and A. K. Somani, "Efficient Utilization of Spare Capacity for Fault Detection and Location in Multiprocessor Systems," *Proceedings of 22nd International Symposium on Fault-Tolerant Computing (FTCS-22)*, pp. 440-447, 1992.

Fault-Tolerant Multicast Communication in Multicomputers*

Rajendra V. Boppana
Div. of Computer Science
The Univ. of Texas at San Antonio
San Antonio, TX 78249-0664
boppana@runner.utsa.edu

Suresh Chalasani
ECE Department
University of Wisconsin
Madison, WI 53706-1691
suresh@ece.wisc.edu

Abstract. *We describe fault-tolerant routing of multicast messages in mesh-based wormhole-switched multicomputers. With the proposed techniques, multiple convex faults can be tolerated. The fault information is kept locally—each fault-free processor needs to know the status of the links incident on it only. Furthermore, the proposed techniques are deadlock- and livelock-free and guarantee delivery of messages. In particular, we show that the previously proposed column-path and Hamilton-path based algorithms can be made tolerant to multiple faults using two or three virtual channels per physical channel.*

Keywords: block faults, fault-tolerant routing, Hamilton path routing, multicast routing, wormhole routing.

1 Introduction

Many commercially available parallel computers use mesh or grid based networks for interprocessor communication with a processor and router module at each node [5, 10]. The interprocessor communication functions in a multicomputer are usually handled by a router which receives data from incoming channels and transmits data to outgoing channels using a suitable channel assignment protocol. The channel assignment is specified as a routing algorithm and is implemented in distributed manner such that each router routes messages from its input channels to its output channels based on its local information only.

Wormhole switching [7], a form of pipelined communication, is the most commonly used switching technique in multicomputers; store-and-forward and virtual cut-through are alternatives to wormhole switching.

Several methods are available for unicast routing, where each message is between a pair of nodes. An excellent survey of wormhole routing methods can be found in [12]. The issue of routing becomes complicated when there are faults in the parallel computer or when multicast communication should be supported. In multicast communication, a node sends a message to multiple destinations. Multicast communication is complicated, since a multicast message, in general, requires more resources than a unicast message.

In this paper, we study the problem of fault-tolerant multicast communication for wormhole switched multi-

computers. This is an important problem, since multicast communication is a natural communication primitive to handle synchronizations, invalidations and updates of cache lines in distributed shared memory computers, and since parallel computers must be used efficiently even in the presence of faults.

There are some recent results on fault-tolerant wormhole routing [6, 4, 9, 8, 1, 2] and multicast wormhole routing [13, 11, 3], but very few results exist on fault-tolerant multicast routing (for a result on hypercubes with limited number of faults see [14]).

In this paper, we show how to provide fault-tolerant communication using two recently proposed multicast routing algorithms for mesh based multicomputers. We consider the convex or block fault model used in literature [4, 1] with no global knowledge of fault information. If the network is connected, our techniques provide deadlock- and livelock-free delivery of messages to all of their destinations.

Section 2 describes the key multicast algorithms used in this paper and the fault model. Section 3 describes fault-tolerant routing with the column path algorithm. Section 4 describes fault-tolerant routing with Hamilton path based routing algorithm. Section 5 concludes the paper.

2 Fault model and routing algorithms

We consider k-radix, 2-dimensional meshes. But all the results and discussions can be applied to multidimensional tori and meshes with suitable modifications.

The two dimensions of the mesh are denoted as DIM_1 and DIM_0. The rows of a 2D mesh are numbered from top to bottom $0, 1, \ldots, k-1$, and the columns are numbered from left to right $0, 1, \ldots, k-1$. Node x, $0 \leq x < k^2$, in a 2D mesh may be represented by a two-tuple (x_1, x_0), where x_1 is the node's row number and x_0 the node's column number in the 2D grid. The hops taken by a message in a row correspond to hops through processors in DIM_0 and hops in a column correspond to hops in DIM_1. In addition, a hop may be a '+' or a '−' hop depending on the indices of the current node and the next node in the dimension of travel. For example, DIM_{1+} hops correspond to column hops from top to bottom. A communication channel from node x to y is denoted by $< x, y >$. Each node is a combination of processor, local memory, and router. The router handles all the communication and is connected to its processor through injection and consumption (delivery) channels, and connected to other nodes (routers)

*Rajendra Boppana's research has been supported by NSF Grant CCR-9208784. Suresh Chalasani's research has been supported in part by a grant from the Graduate School of UW-Madison and the NSF grants CCR-9308966 & ECS-9216308.

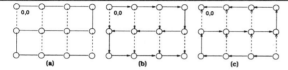

Figure 1: Example of (a) an undirected Hamilton path and the corresponding (b) H_u and (c) H_l directed networks of a mesh. The solid lines indicates the Hamilton path and dashed lines indicate the links that could be used to reduce path lengths in message routing.

through communication links. Each communication link is a full-duplex channel implemented using two unidirectional physical channels.

2.1 Multicast routing algorithms

First we describe a few recent multicast algorithms: the Hamilton path based algorithms [11] and the e-cube based *column path* algorithm [3].

2.1.1 Multicast routing based on Hamilton paths

First an undirected Hamilton path, which goes through each node exactly once, is constructed. An example of an undirected Hamilton path, with node $(0,0)$ as an end node, is given in Figure 1. From this two directed Hamilton paths can be constructed: one starts at $(0,0)$, the H_u network, and another ends at $(0,0)$, the H_l network. The links that are not part of the Hamilton path may be used with appropriate direction to reduce path length.

Dual-path algorithm. Due to the construction of the Hamilton paths H_u and H_l, the paths from any node to any other node are acyclic. In particular, some nodes are reached from a given node via H_u network only and the rest via H_l network only.

In the dual-path algorithm, multicast messages from a node are transmitted on appropriate parts of the H_u and H_l networks. Figure 2.1.1(a) illustrates the portions of H_u and H_l networks used by node $(3,2)$ to send its multicast messages. Hence, the destinations of a multicast message are placed into two groups. One group has all the destinations that can be reached from the source node using the H_u network, and the other has the remaining destinations, which can be reached using the H_l network.

Thus each source of a multicast message, depending on its location and the locations of the destinations, transmits either one or two copies of the message. For example, if $(3,2)$ needs to send a message to destinations $(0,5)$ and $(5,0)$, it will send two copies in opposite directions—one to $(0,5)$ and another to $(5,0)$. However, a multicast message from $(3,2)$ to destinations $(5,5)$ and $(5,0)$ will be sent as a single message. For shorter paths, vertical channels that are not part of the Hamilton path may be used appropriately. The routing of a multicast message from $(3,2)$ to $(0,5)$, $(1,4)$, $(5,0)$, and $(5,5)$ is indicated in Figure 2.1.1(a).

Multipath algorithm. The dual-path algorithm uses at most two copies of the message for multicast communication. This may increase the latency for some multicast messages. The multipath algorithm attempts to reduce long latencies by using up to four copies ($2n$ for

n-dimensional meshes) of the original multicast message. As per the multipath routing algorithm, all the destinations of the multicast message are grouped into four disjoint subsets. Each subset of destinations are serviced by one copy of the multicast message [11]. Figure 2.1.1(b) indicates the routing of a multicast message from $(3,2)$ to $(0,5)$, $(1,4)$, $(5,0)$, and $(5,5)$ using three copies.

The dual-path and multipath schemes provide deadlock-free routing of multicast messages. Further, they also provide minimal routing of unicast messages, since vertical links are used for shortcuts. Therefore, either scheme can be used to route unicast and multicast messages simultaneously in a common framework.

2.1.2 Column-path routing algorithm

The dual-path and multipath schemes are not compatible with the well-known e-cube routing algorithm. Since the e-cube is the most commonly used routing method, it is of interest to develop multicast techniques that can take advantage of implementation techniques and methods developed for e-cube. One such example is the column-path algorithm given in [3]. The e-cube routing is specified for unicast messages as follows: each message is routed in DIM_0 until it exhausts all its row hops, at which point it is in the same column as its destination; the message is then routed in DIM_1 until the destination is reached.

The column-path algorithm partitions the set of destinations of a multicast message into at most $2k$ subsets such that there are at most 2 messages directed to each column. Only one message is sent to a column if all destinations in that column are either below or above the source node; otherwise, two messages are sent to that column. In the example of Figure 2.1.1(c), the destinations for the multicast message with source $(2,2)$ are $(1,4)$, $(3,3)$, $(3,4)$, $(4,4)$, $(1,5)$ and $(0,5)$. In all four copies of the message are sent; one copy to $(3,3)$, one to $(1,4)$, another to $(3,4)$ and $(4,4)$, and yet another to $(1,5)$ and $(0,5)$. Each of these messages is routed using the e-cube (or, row-column) routing algorithm. Hence, the column-path routing is compatible with the unicast routing method used in the current parallel computers. A similar but more general method has been independently developed by Panda *et al.* [13].

2.2 The fault model

We consider both node and link faults. All the links incident on a faulty node are considered faulty. We assume that failed components simply cease to work and that messages are generated among nonfaulty processors only.

We model multiple simultaneous faults, which could be connected or disjoint. We assume that the mean time to repair faults is quite large, a few hours to many days, and that the existing fault-free processors are still connected and thus should be used for computations in the mean time. We assume that each non-faulty processor knows only the status of its neighbors.

A set F of faulty nodes and links indicates a (rectangular) fault block, or f-region, if there is a rectangle connecting various nodes of the mesh such that (a) the boundary of the rectangle has only fault-free nodes and links and (b) the interior of the rectangle contains all and only the components given by F. A fault set that includes a component from one of the four boundaries—top and

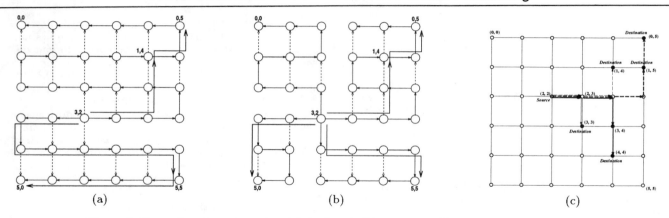

Figure 2: Examples of routing using (a) dual-path (b) multipath (c) column-path algorithms.

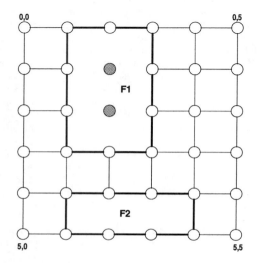

Figure 3: Block faults in a 2-D mesh. Faulty nodes are shown as filled circles, and faulty links are not shown.

bottom rows, left most and right most columns—of a 2D mesh denotes a rectangular fault block, if the above definition is satisfied when the mesh is extended with nonfaulty *virtual* rows and columns on all four sides. However, fault blocks abutting on network boundary are not convex [4]. Therefore, we do not consider faults on network edges. Figure 3 indicates two fault blocks: $F_1 = \{(1, 2), (2, 2)\}$ and $F_2 = \{< (4, 2), (5, 2) >, < (4, 3), (5, 3) >\}$.

We use the *block-fault* model, in which each fault belongs to exactly one fault block. Under the block-fault model, the complete set of faults in a 2D mesh is the union of multiple fault blocks (e.g., $F_1 \cup F_2$ in Figure 3). If faults disconnect the network, our results can still be applied to each subnetwork.

For each fault region, there is a ring of fault-free nodes and links such that it encloses the fault-free region. Such a ring with minimal number of links is called the fault-ring (f-ring) for that fault region. The f-ring of a block fault is rectangular. The f-rings associated with the two fault blocks in Figure 3 are indicated by thick lines. The four sides of an f-ring are classified into left and right columns and and top and bottom rows. If two f-rings have common

physical channels, then they are said to overlap.

When a fault occurs, the corresponding f-ring can be formed in a distributed manner using a two-step process. In the first step, each processor that detected a faulty link sends this message to its neighbors in other dimensions. Using the set of messages received, each node determines its position and neighbors on the f-ring. There are eight possible positions for a processor to be in an f-ring: four corner positions, two row positions and two column positions. For more details on forming f-rings, see [2].

An f-ring represents a two-lane path to a message that needs to go through the f-region contained by the f-ring. Routing messages on fault-rings creates new dependencies among resources acquired by messages and, hence, additional possibilities for deadlocks.

3 Fault-tolerant column-path routing

Because of its simple routing logic, the column-path algorithm can be easily enhanced to handle faults. When there are no faults, the original column-path algorithm is used. Even when there are faults in the network, each message is routed using the original column-path algorithm as much as possible. When a message arrives at a node, the next hop for that message is specified by the column-path algorithm. If that hop is on a faulty link, then the message is blocked by the fault. The routing logic is enhanced to handle such situations so that the message is routed around faults. Once the message is routed around faults, the column-path algorithm is used to route the message until it reaches all of its destinations or is blocked again. It is noteworthy that the modifications to routing are used only when a message is blocked by a fault. First, we consider nonoverlapping f-rings.

Modifications to the routing logic. The path of a message in column-path algorithm consists of two parts: the first part is on row (DIM$_0$) channels and the second part is on column (DIM$_1$) channels. Therefore, a message may be blocked by a fault while traveling in a row or in a column. A message blocked from taking its DIM$_0$ hop travels on two sides of the f-ring; a message blocked from taking its DIM$_1$ hop travels on three sides of the f-ring.

If a message is blocked from taking a DIM$_{0+}$ hop, then it touches the corresponding f-ring on the left column of the f-ring. This message travels on the the f-ring in clockwise orientation (up and right) if its first destination is in a row below its current row; otherwise it travels the f-ring in counter clockwise orientation (down and right). Similarly, if a message is blocked from taking its DIM_0- hop, then it travels on the f-ring in clockwise orientation if its first destination is in a row above the current row or counter clockwise orientation otherwise (see Figure 4). A DIM_0 message that is in the same column as its first destination becomes a DIM_1 message.

If a message is blocked from taking a DIM_{1+} hop, then it is blocked at a node in the top row of the f-ring. This message travels on three sides of the f-ring in clockwise orientation such that it reaches the same column at the bottom row of the f-ring. Finally, if a message is blocked from taking a DIM_{1-} hop, then it travels on the counter clockwise orientation on the f-ring starting from the bottom row of the f-ring to the top row of the f-ring such that it is in the same column as it was before being blocked by the fault. See Figure 4 for an illustration.

Because of this misrouting of messages when blocked by faults, some physical channels around f-rings are used by multiple messages. This creates cyclic dependencies. To break these new dependencies, we use a general technique given in [7] and simulate multiple virtual channels on each physical channel. The bandwidth of a physical channel is demand time-multiplexed among the virtual channels. When faults are such that only nonoverlapping f-rings occur, just two virtual channels per physical channel are sufficient to provide deadlock free routing even when there are multiple faults in the network.

We use two classes of virtual channels: c_0 and c_1. On each physical channel, a virtual channel of each class is simulated. The channel allocation is such that any new dependencies among the four classes of messages caused by sharing of the physical channels on f-rings are broken. For DIM_0 messages, virtual channels of class c_0 are used and for DIM_1 messages c_1 virtual channels are used.

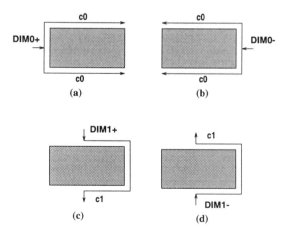

Figure 4: Routing of four different message types around a fault. The shaded area represents a faulty block, and directed lines indicate the paths of messages on the f-ring.

The complete routing logic and channel allocation are given in Figure 5. An example of fault-tolerant routing with the proposed method is shown in Figure 6.

Table 1: Orientations and virtual channels used by the Fault-Tolerant-Column-Path-Routing algorithm.

Message Type	Position of Next Destination	F-Ring Orientation (Virtual Channel)
DIM_0+	In a row above its row of travel	Counter Clockwise (c_0)
DIM_0+	In a row below its row of travel	Clockwise (c_0)
DIM_0-	In a row above its row of travel	Clockwise (c_0)
DIM_0-	In a row below its row of travel	Counter Clockwise (c_0)
DIM_1+	(don't care)	Clockwise (c_1)
DIM_1-	(don't care)	Counter Clockwise (c_1)

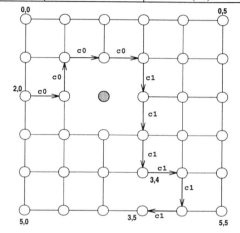

Figure 6: Example of fault-tolerant routing with the column-path algorithm. There is one faulty (shaded) node and one faulty link. The path of the multicast message from $(2,0)$ to $(3,4)$ and $(3,5)$ is shown by directed lines.

Proof of deadlock free routing. In multicast wormhole routing, deadlocks can arise from dependencies on communication channels between nodes and consumption channels between a router and its processor in a node [3]. First let us consider deadlocks on communication channels. The proof technique is similar to the one we have given in [1] for fault-tolerant e-cube routing, since the column-path algorithm is similar to the e-cube routing. Therefore, a sketch of the proof is given below.

For the deadlock to occur, there has to be a cyclic dependency on the virtual channels acquired by the messages involved in the deadlock. For the purpose of the following discussion, we define row messages as messages that need to take DIM_0 hops when normal and column messages are messages that have completed all their row hops and need to take DIM_1 hops only when normal.

Row messages may turn into column messages after a few hops, but column messages never turn into row messages. Since row messages use only c_0 virtual channels and column messages use only c_1 virtual channels, there cannot be a deadlock cycle involving both row and column

---

Procedure Fault-Tolerant-Column-Path-Routing(Message M)
/* M has a status field to indicate if it is currently misrouted or normal. Initially M's status is normal. */

1. As long as M is not currently misrouted and the hop specified by the original column-path algorithm is not blocked, route M accordingly.

2. If M is currently not misrouted but the hop specified by the column-path is blocked by a fault, then

   (a) set the status of M to misrouted, and

   (b) route it using the orientation and virtual channels specified in Table 1.

The misrouting of M is completed and M's status is set to normal if one of the following occurs:

   (a) M is a DIM_0 message and reached a corner node of the f-ring it is traversing.

   (b) M is a DIM_1 message and it is in the same column as its destination and is on the other side of the faulty block that caused the misrouting.

Figure 5: Modifications to the column-path routing to handle faults.

messages. Conceptually, the network may be considered as a union of two planes, plane 0 with virtual channels of c_0, and plane 1 with virtual channels of c_1. A message may move from plane 0 to plane 1 but never in the opposite direction. Therefore, if there is a deadlock, then it is among the channels of c_0 or c_1 only.

Class 0 channels are used by two types of row messages: DIM_{0+} and DIM_{0-} messages. The DIM_{0+} messages use virtual channels of c_0 only on DIM_{0+} physical channels, and virtual channels of c_0 on left columns of the f-rings in the network. The DIM_{0-} messages use virtual channels of c_0 only on DIM_{0-} physical channels, and virtual channels of c_0 on right columns of the f-rings in the network. The sets of physical channels and, hence, the set of virtual channels used by DIM_{0+} and DIM_{0-} are disjoint. Therefore, there cannot be deadlocks among row messages. A similar argument can be used to show that DIM_{1+} and DIM_{1-} messages use disjoint sets of physical channels.

3.1 Handling overlapping fault rings

The above routing method can be easily extended to handle overlapping f-rings, when the sets of physical channels of a pair of f-rings are not disjoint. The routing logic remains the same. Because of increased sharing of physical channels more classes of virtual channels are needed to ensure deadlock-free routing.

When two f-rings overlap along a column (respectively, row), some physical channels in that column (respectively, row) belong to the left column (respectively, top row) of one f-ring and to the right (respectively, bottom row) of another f-ring. Take DIM_0 messages: our arguments for DIM_0 messages for the nonoverlapping case are based on the fact that they use disjoint sets of physical channels. This is no longer true when f-rings overlap in a column. Therefore, DIM_{0+} and DIM_{0-} messages should use disjoint sets of virtual channels to break the new dependencies and preserve deadlock freedom. Similarly, when two f-rings overlap in a row, DIM_{1+} and DIM_{1-} messages share the physical channels of the row. Once again, new dependencies, this time among DIM_1 messages, occur. To break these dependencies we require three classes of virtual channels: c_0, c_1, c_2. Virtual channels used by different types of messages are given Table 2. Extending the above

Table 2: Use of virtual channels for misrouting messages by the column-path algorithm for overlapping fault rings.

Message type	Channel	Used for
DIM_{0-}	c_0	all hops
DIM_{1+}	c_1	all hops
DIM_{1-}	c_2	all hops
DIM_{0+}	c_0	DIM_0 hops
	c_1	DIM_{1-} hops
	c_2	DIM_{1+} hops

arguments, it can be shown the resulting routing is still deadlock free.

3.2 Deadlocks on consumption channels

Another source of deadlocks unique to multicast wormhole routing is the dependency among messages waiting for consumption channels. A unicast message upon reaching its destination does not compete for any additional communication resources. In multicast routing, however, a message may hold consumption channel (from router to processor) in one node and wait for other resources (communication/consumption channels) elsewhere [3].

Figure 7 illustrates the routing of two multicast messages in a row of a two-dimensional mesh. The first message, m_1, originates at node a and has destinations b, c; the second message, m_2, originates at node d and has destinations b, c. Furthermore, nodes a, b, c are left neighbors to b, c, d, respectively. The following scenario is shown in Figure 7. The message m_1 obtains the communication channel from a to b, denoted $[a, b]$, consumption channel in b, denoted $Cons_b$, and communication channel $[b, c]$; m_2 obtains the communication channel $[d, c]$, consumption channel in c, $Cons_c$, and communication channel $[c, b]$. The reservation of the consumption channel is shown by labeling the (flit) buffer associated to it with the name of the message that has reserved it.

Due to flit-level flow control in wormhole routing, node b can accept only the header flit (and may be a few data flits if more than one flit is sent between nodes at a time) of $m1$. Though the consumption channel in b is free, $m1$ cannot be consumed at b until it acquires the consumption

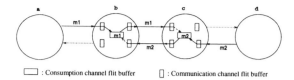

: Consumption channel flit buffer   : Communication channel flit buffer

Figure 7: Example of deadlocks on consumption channels.

channel in c also. A similar condition occurs with the consumption channel in c and message $m2$. This causes a circular wait on consumption channels between $m1$ and $m2$, which leads to deadlock.

One solution is to provide multiple classes of consumption channels and allocate them to messages using specific rules. For the fault-free column-path algorithm, two classes of consumption channels are sufficient [3]. One class of consumption channels are used by messages that travel on DIM_{1-} channels and the other class by messages that travel on DIM_{1+} channels. Messages that do not need to take any DIM_1 hops can be treated as DIM_{1+} or DIM_{1-} messages.

Since we use two virtual channels on each physical channel for fault-tolerant routing, more messages of each type compete for consumption channels in routers. Fortunately, the dependencies are still acyclic and therefore do not cause deadlocks if two classes of consumption channels are used as in the original algorithm. To see this, consider DIM_{1+} messages. Let $c+$ be the class of consumption channels they use. Also let the rank of the $c+$ consumption channel in node (i, j), $0 \leq i, j < k - 1$ be j. A DIM_{1+} message needs to deliver data to nodes in a single column only, and never takes DIM_{1-} hops, even when misrouted. So, DIM_{1+} messages do not cause cyclic dependencies on the consumption channels of its class. A similar argument can be used for DIM_{1-} messages. Therefore, two consumption channels per router are sufficient to eliminate deadlocks on consumption channels.

4   Fault-tolerant Hamilton path routing

In this section, we show that the Hamilton path based dual-path and multi-path algorithms can be made fault-tolerant, using two virtual channels per physical channel. We address the dual-path algorithm specifically, since all the dependencies that occur in multipath routing also occur in the dual path algorithm. In this section, Hamilton path and dual-path are used synonymously.

As before, our approach is to leave the original routing logic as it is and add new logic to help messages get around faults when they are blocked. There is one worst case scenario for the dual-path algorithm. This is shown in Figure 8. The source of a multicast message is $(0,0)$ and its destinations are $(1,4),(2,1),(3,4),(3,1)$, in the order to be visited. Because of fault block $\{(1, 2), (2, 2), (3, 2)\}$, however, the message travels on the f-ring many times, to preserve the order in which destinations are visited if there are no faults. This example shows that many classes of virtual channels are required just to keep this message from deadlocking itself. This can be avoided only when the message travels on the f-ring only a few times (typically, once or twice). Thus, faulty blocks spanning multiple rows cause severe problems for Hamilton path algorithms. Therefore,

for Hamilton path algorithms, we restrict the fault model to faults with f-rings of height (the number of links in a column of the f-ring) of two or less. Examples of fault-blocks with f-ring height two are less include (i) a block of node-faults in a row, (ii) isolated DIM_0 link faults, and (iii) all possible block faults involving DIM_1 links. Of these, DIM_1 faults do not affect Hamilton path routing, since the routing logic depends on fault-free DIM_1 links only on network boundaries, which, in our fault model, are fault-free.

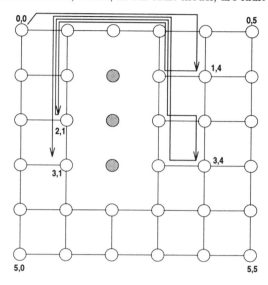

Figure 8: Worst case block fault for Hamilton path algorithms. Directed lines indicate the path of a multicast message from node $(0,0)$ to nodes $\{(1, 4), (2, 1), (3, 4), (3, 1)\}$, visited in the order given.

Now, we present our fault-tolerance technique to handle block faults with f-rings of height two or less. We directly present our method for overlapping f-rings case.

Let M_l (respectively, M_u) be the set of messages that use the Hamilton path H_l (respectively, H_u) in the fault-free network. We use two virtual channels for fault-tolerant routing: c_0 is used exclusively by messages in class M_l and c_1 is used by messages in class M_u. The fault-tolerant routing algorithm is shown in Figure 9.

An M_l message, if blocked by a faulty component in a row, takes the clockwise orientation if the faulty component is to the left of the message, and the counter-clockwise orientation otherwise. An M_u message, if blocked by a faulty component in a row, takes the counter clockwise orientation if the faulty component is to the left of the message, and the clockwise orientation otherwise. A message can be blocked by a fault, while taking a column-hop, only if it is trying to take a short-cut, since no component on the network boundary is faulty. Hence, a message, if blocked on a column-hop, simply continues to the next node in the same row.

Theorem 1 *Procedure Fault-Tolerant-Dual-Path-Routing tolerates multiple block faults of height two or less.*

Proof. First, we observe that messages in class M_l use c_0, while those in M_u use c_1. Hence, deadlocks, if occur, can be only among either M_l messages or among M_u messages. Let us consider M_l messages. Let H_{lp} (respectively,

---

Procedure Fault-Tolerant-Dual-Path-Routing(Message M)
/* Messages in class M_l use c_0 channels for all hops and those in class M_u use c_1 for all hops. */
If the next hop of M is not blocked by a fault
  route it as per the fault-free dual-path algorithm.

If the next hop of M is a row-hop which is blocked by a fault

  1 Let x be the node at which M is blocked and let y be the node in the same row as x at the other end of the f-ring. M is misrouted on the f-ring from x to y using the orientation given in Table 3.

If the next hop of M is a column-hop which is blocked by a fault

  1 The column-hop is being used by M to take a short-cut.
  2 M foregoes the opportunity to take the short-cut and takes a row-hop.

Figure 9: Modifications to the dual-path routing to handle faults.

Table 3: Orientations and virtual channels used Fault-Tolerant-Dual-Path-Routing algorithm.

Message Type	Current Direction of Travel	F-Ring Orientation (Virtual Channel)
M_l	DIM_0+	Counter Clockwise (c_0)
	DIM_0-	Clockwise (c_0)
M_u	DIM_0+	Clockwise (c_1)
	DIM_0-	Counter Clockwise (c_1)

H_{up}) be the H_l (respectively, H_u) network consisting only of c_p virtual channels, for $p = 0, 1$. For fault-tolerant routing, M_l messages use network H_{l0} (which is also used for the fault-free case) and a part of the network H_{u0}. Only those channels in H_{u0} around the f-rings are used for fault-tolerant routing of messages in M_l. Figure 10 illustrates this case for three overlapping f-rings in a 6×6 mesh. In this figure, messages in M_l use channels in H_{l0} (shown using thin lines with arrows) and a few channels in H_{u0} around the f-rings (shown using dashed lines with arrows).

If an M_l message is blocked at node x, then it is misrouted on the corresponding f-ring to node y, which is in the same row as x and at the other side of the faulty block. Thus, in Figure 10, a message blocked at x must use links marked A, B, C and D to reach y. As per our fault-tolerant logic, a message, once it uses link A, must use links B, C and D to reach y. Hence, the path from X to Y can be replaced with a single link from X to Y as far as dependencies are concerned. The resulting dependency graph is acyclic, and hence there cannot be deadlocks among messages in class M_l. A similar argument holds for deadlock-freedom of messages in class M_u. ∎

Example. Consider a message from $(5, 2)$ with destinations in $\{(4, 2), (3, 3), (2, 0), (1, 2), (0, 1)\}$. This message takes the path indicated by thick lines with arrows in Figure 11. This message takes a short-cut from $(5, 2)$ to $(4, 2)$. It tries to take a short-cut from $(4, 2)$ to $(3, 2)$. However, since $(3, 2)$ is faulty, it is routed to $(4, 1)$, where it takes a short-cut to $(3, 1)$. At $(3, 1)$, it is blocked by $(3, 2)$, and hence is misrouted to $(3, 3)$ as shown in Figure 11.

This example illustrates several possibilities for opti-

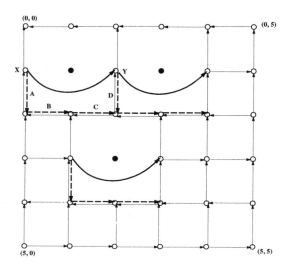

Figure 10: Links used by messages in M_l for fault-tolerant routing. Dashed lines with arrows indicate c_0 channels in the network H_{u0}, and thin lines with arrows indicate c_0 channels in H_{l0}. Thick (curved) lines indicate the net effect of misrouting.

mizing the paths taken by messages. For example, $(4, 2)$ can route the message to $(4, 3)$ based on its knowledge that node $(3, 2)$ is faulty; this allows the message to skip the path from $(4, 2)$ to $(3, 1)$ and then back to $(4, 2)$. Using a similar argument, the journey from $(2, 0)$ to $(1, 0)$ and then back to $(2, 0)$ can also be avoided. Such optimizations do not introduce deadlocks, and can be used to improve the latency of messages.

Consumption channel requirements of fault tolerant dual-path algorithm. It has been shown before that, for dual-path and multi-path algorithms, two consumption channels per node are sufficient to avoid deadlocks on consumption channels [3]. One consumption channel is used by messages in class M_l (traveling along H_l) and the other is used by messages in class M_u. Even for the fault-tolerant dual-path routing algorithm, two consumption channels are sufficient to avoid deadlocks on con-

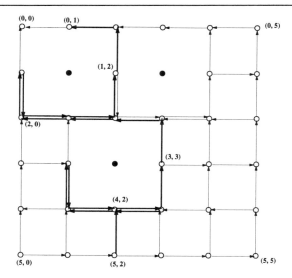

Figure 11: Example of fault-tolerant dual-path routing. Thick directed lines indicate the routing of a message from node $(5, 2)$ to nodes $(4, 2), (3, 3), (2, 0), (1, 2), (0, 1)$.

sumption channels: one consumption channel is used by M_l messages and the other by M_u messages, just as in the fault-free case.

The proof of deadlock-freedom is based on the fact that a message, while it is being misrouted, is not delivered to a destination. In other words, no message waits for consumption channels while it is being misrouted. To illustrate this, let us consider the message M in Figure 11. This message is misrouted from $(1, 0)$ to $(1, 2)$ via $(2, 0)$. Though $(2, 0)$ is a destination for this message, M would have acquired the consumption in $(2, 0)$ during its journey from $(2, 2)$ to $(1, 0)$. Hence, during misrouting M does not need to wait for consumption channels. Since no message waits for consumption channels while being misrouted, the dependencies on consumption channels are the same as those in the fault-free Hamilton path algorithms. Thus, two channels are sufficient to avoid consumption channel deadlocks for fault-tolerant Hamilton-path algorithms.

5 Summary and concluding remarks

In this paper, we have addressed the issue of reliable multicast communication in wormhole-switched multicomputers. Our techniques handle multiple convex faults for meshes. When a fault occurs, only the fault-free nodes around the faulty components need to know the information. The concept central to our approach is fault rings, which are formed around each fault and are of rectangular shape for 2-D meshes.

We have specifically considered two recently proposed multicast algorithms: Hamilton path and column-path. With two virtual channels per physical channel, multiple convex faults with nonoverlapping f-rings can be handled by the column-path algorithm. Overlapping f-rings can be handled by the column-path with three virtual channels.

For Hamilton path based algorithms, providing fault-tolerance is more complicated, since a message uses longer paths and visits destinations in a predetermined sequence.

However, when convex faults are such that the height of fault rings is two, two virtual channels per physical channel are sufficient to provide fault-tolerant routing.

The column-path and Hamilton path algorithms have different strengths. The column-path is compatible with e-cube and can be easily implemented in the next generation e-cube routers. The column-path is especially attractive when the number of destination is small [3]. The Hamilton path algorithm is a specialized algorithm and attempts to minimize congestion at sources of multicasts by limiting the number of copies of a message to 2 or 4, independent of the number of destinations. For large networks, a combination of these two algorithms may provide more efficient communication with less source congestion. In future, we plan to simulate the fault-tolerant versions of the column-path and Hamilton path algorithms and estimate performance degradation due to faults.

References

[1] R. V. Boppana and S. Chalasani, "Fault-tolerant routing with non-adaptive wormhole algorithms in mesh networks," in *Proc. Supercomputing '94*, Nov. 1994.

[2] R. V. Boppana and S. Chalasani, "Fault-tolerant wormhole routing algorithms for mesh networks," Tech. Rep. CS-94-2, Div. of Math and CS, U. Texas at San Antonio.

[3] R. V. Boppana, S. Chalasani, and C. S. Raghavendra, "On multicast wormhole routing in multicomputers," in *Proc. 1994 IEEE Symp. on Par. and Distr. Processing*.

[4] A. A. Chien and J. H. Kim, "Planar-adaptive routing: Low-cost adaptive networks for multiprocessors," in *Proc. 19th Ann. Int. Symp. on Comput. Arch.*, pp. 268–277, 1992.

[5] Cray Research Inc., *Cray T3D Architectural Summary*, Oct. 1993.

[6] W. J. Dally and H. Aoki, "Deadlock-free adaptive routing in multicomputer networks using virtual channels," *IEEE Trans. on Parallel and Distributed Systems*, vol. 4, pp. 466–475, April 1993.

[7] W. J. Dally and C. L. Seitz, "Deadlock-free message routing in multiprocessor interconnection networks," *IEEE Trans. on Computers*, vol. C-36, no. 5, pp. 547–553, 1987.

[8] J. Duato, "A theory to increase the effective redundancy in wormhole networks," *Parallel Processing Letters*. To appear.

[9] C. J. Glass and L. M. Ni, "Fault-tolerant wormhole routing in meshes," in *Twenty-Third Annual Int. Symp. on Fault-Tolerant Computing*, pp. 240–249, 1993.

[10] Intel Corporation, *Paragon XP/S Product Overview*, 1991.

[11] X. Lin, P. K. McKinley, and L. M. Ni, "Deadlock-free multicast wormhole routing in 2-d mesh multicomputers," *IEEE Trans. on Parallel and Distributed Systems*, vol. 5, pp. 793–804, Aug. 1994.

[12] L. M. Ni and P. K. McKinley, "A survey of wormhole routing techniques in direct networks," *IEEE Computer*, vol. 26, pp. 62–76, Feb. 1993.

[13] D. K. Panda, S. Singhal, and P. Prabhakaran, "Multidestination message passing mechanism conforming to base wormhole routing scheme," in *Proc. of Parallel Routing and Communication Workshop*, May 1994.

[14] Y.-C. Tseng and D. K. Panda, "A trip-based multicasting model for wormhole-routed networks with virtual channels," in *Proc. 1993 Int. Parallel Processing Symp.*

DESIGN AND ANALYSIS OF HIGHLY AVAILABLE AND SCALABLE COHERENCE PROTOCOLS FOR DISTRIBUTED SHARED MEMORY SYSTEMS USING STOCHASTIC MODELING*

Oliver E. Theel[†] and Brett D. Fleisch

Department of Computer Science
University of California
Riverside, CA 92521-0304, U.S.A.
E-Mail: {oliver | brett}@cs.ucr.edu

Abstract – We present a new coherence protocol class for DSM systems whose instances offer highly available access to shared data at low operation costs. The protocols proposed scale well; an increase in the number of client sites does not increase the operation costs after a certain threshold has been reached. The results presented give strong guidelines for the overall design of DSM systems which offer highly available, uninterrupted services.

INTRODUCTION

DSM systems have been extensively investigated during the past several years. Although a large number of approaches has been developed only a few of these solutions have taken aspects of scalability and fault-tolerance into consideration (e.g. [6, 4]). Earlier work emphasized finding a solution to the DSM coherence problem with good performance for a relatively small network consisting of only a few sites. But as networks become larger, the need for DSM coherence protocols which scale well and provide highly available services becomes increasingly important. *Scalability* of a DSM coherence protocol is required for a larger network with an increasing number of users that are expected to make use of DSM. *Fault-tolerance*, in terms of high availability, is required for uninterrupted DSM service even in a large-scale environments with a greater number of potentially malfunctioning components.

MODEL

In this section, we give definitions which we use in the paper. Additionally, we describe the functional model of the DSM system under consideration.

Definitions

We assume a networking environment consisting of m sites $R_1, \ldots, R_m$ with independent failure rates. The single sites do not behave in a Byzantinian manner. The sites are arbitrarily connected by communication links which may also fail independently from each other and independently from the sites. This may result in network partitions with several independently operating subnets. In addition, we assume that

within the network, a DSM service is available which functions in the manner described in the next section. The DSM system maintains one or more *segments*. A segment consists of one or more *pages*, and finally all of those pages consists of a certain number of basic memory units, e.g. bytes or words. Let $\mathbf{C} \subset \{R_1, \ldots, R_m\}$ be the set consisting of all potential client sites of the DSM system, with $|\mathbf{C}| = n < m$ giving the cardinality of this set. $S \subseteq \{R_1, \ldots, R_m\}$ is the DSM server site. Only sites included in $\mathbf{C}$ are allowed to submit requests to the DSM server.

Functional Model of a DSM System

Besides operations for creating and destroying segments, the most important operations that a (segmented) DSM system must support are *read* and *write operations* to a particular page of a segment. These operations are carried out using the client/server paradigm: the client site which desires to read (write) a specific page currently not available in the local memory submits a *read request (write request)* to the *DSM server* of the segment. The DSM server[1] is a designated site of the network owning data structures and algorithms which enable it to carry out the requested operations. In particular, the DSM server maintains a request queue and keeps directory information for the page copies distributed in the network. Once a client has submitted its request (i.e. it received an acknowledgment from the DSM server), it blocks and waits for the *result* delivered from the server. The result of a request typically consists of the requested page's data and a *mode*. In both cases, the requested operation is regarded as being *successful*, since the client received or achieved what it wanted. Due to the state of the network or the state of the DSM system itself, it might not always be possible to carry out the requested operation. In such a case, an error message is returned and the request is deemed *unsuccessful*. It is, among others, the aim of this work to construct DSM systems where the vast majority of requests are carried out successfully for all points in time. This leads to highly available DSM.

---

*This research was partially sponsored by NSF CCR-9209405 and the DEC External Research Program.

[†]also with the University of Darmstadt, Germany

[1]We assume a single, non-replicated DSM server. In case of a server failure, a new server is selected using the standard election algorithms for distributed systems.

Once a client received the results of a successful request, which includes the data of a page, it maps the page data into its memory and uses it only according to the granted mode. The page is regarded as being *cached* in a particular mode with respect to a particular operation. The mode determines how subsequent read and write operations operate. Figure 1 gives an overview of the modes used by the functional model.

Mode	Description
(local read, local write)	the page data can be read and modified locally, i.e. without the need to submit read or write requests to the DSM server
(local read, global write)	the page data can be read locally, but write operations have to be submitted to the DSM server in order to execute a global write operation
(global read, global write)	neither a read nor a write operation can be executed locally. This mode is implicitly assumed for all pages which are not cached at a client site

Figure 1: *Modes associated with DSM pages at the client sites*

A *mode* is a two-dimensional tuple consisting of a *read attribute* in the first dimension and a *write attribute* in the second dimension. The mode instances are stored at the client sites. The read (write) attribute defines how a read (write) operation originating at a particular client site must be handled: If the client wants to read (write) the data of a page cached in a mode containing a local read (local write) attribute, then this results in a *local read operation* (*local write operation*) carried out locally on the client site. If a particular page is not cached at all or cached in a mode including a global read (global write) attribute, then a read (write) request must be submitted to the DSM server, triggering a global read operation (global write operation) performed by the server, as described earlier in this section.

NEW APPROACH

The key concept for achieving fault-tolerance in terms of high availability is *redundancy*. The higher the degree of redundancy the more malfunctioning system components can be tolerated. Applying this concept to DSM systems means increasing the number of page copies cached (also called *replicas*) at sites in the network. Clearly, there is a price to be paid: the more replicas, the more management overhead. Options for management include keeping the multiple copies either up-to-date, or preventing the client from reading replicas when they have become out-of-date. Although these two approaches appear straightforward, it is not easy to decide which approach is best. The choice has a strong impact not only on the costs of subsequent operations but also on the data availability. For instance,

when updating *all* existing copies of a page as part of a write operation to a DSM page, then the number of up-to-date (or *actual*) replicas of a page remains constant. Thus, the data availability is not affected. The same holds true for the costs: since the number of replicas and their storage locations has not changed, subsequent read and write operations may be carried out at the same costs. This is not the case when a write operation on a page alters the number of replicas: deleting replicas has a severe impact on the data availability and the costs of subsequent operations. If, for example, a write operation decreased the number of replicas from 5 to 1, and the only site caching the most recent copy subsequently fails, then the page is not available any more. On the other hand, reducing the number of replicas from 5 to 1 reduces the costs of the next write operation, since only one replica must be updated instead of five. This discussion shows that there is a strong correlation between operation costs and data availability.

Design Goals

As suggested earlier, operation costs affect data availability and vice versa. It is important to understand factors that have an impact on those parameters. Clearly, the coherence protocol used by the DSM system strongly influences operation costs and data availability. A suitable coherence protocol for a large, error-prone networking environment must have the following properties:

• (P1) *Limited Workload Dependability*: the number of replicas should, only to a certain degree, be dependent on the workload, i.e. the mixture of read and write operations

• (P2) *Lower Bound in the Number of Cached Copies*: situations where only a single replica exists should be avoided. In this case, the availability of the page is extremely vulnerable to even single component failures

• (P3) *Upper Bound in the Number of Cached Copies*: situations where all the client sites cache a copy of a page and all those pages must be updated as part of a global write operation must be avoided. This is required when there is a substantial number of global write operations inherent in the workload.

A coherence protocol class based on the functional model which takes these observations into account is given in the following section.

New Coherence Protocol Class

A DSM server's workload can be divided into two phases. The first phase starts with the execution of the first global read operation after a global write operation and ends with the next global write operation. The second phase starts with the first global write operation following a global read operation and continues until the execution of the next global read operation.[2] We call the former phase *read phase* since it consists entirely of read operations and the latter phase analogously *write phase*. In order to control the number of page copies present in the critical situations (P1) – (P3), we establish the following constraints.

Protocol Invariants   The coherence protocol class guarantees that during a read phase, copies of a page are exclusively cached in (local read, global write)-mode, and that the number of replicas N_r during this phase always satisfies the constraint:

$$r_{min} \leq N_r \leq r_{max} \text{ with } r_{min}, N_r, r_{max} \in \mathbf{N} \quad (1)$$

with r_{min} (r_{max}) representing the minimum (maximum) number of replicas in the read phase. $\mathbf{N}$ is the set of positive integers excluding zero. For the number of replicas present and modified as part of the execution of a local or global write operation N_w in the write phase, the following always holds:

$$w_{min} \leq N_w \leq w_{max} \text{ with } w_{min}, N_w, w_{max} \in \mathbf{N} \quad (2)$$

With these invariants, Figure 2 shows three cases which correspond to coherence protocols known from the literature. Part a) of the figure shows the bound-

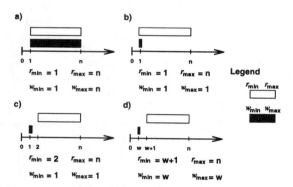

Figure 2: *Boundary Settings for three coherence protocols: a) WB, b) WI, and c) WIwD. d) shows the boundary settings for the BR coherence protocols*

ary setting for the *Write-Broadcast* (WB) coherence protocol [3], part b) for the *Write-Invalidation* (WI) coherence protocol [3], and part c) shows the boundary setting for a protocol named *Write-Invalidation with Downgrading* (WIwD) [1]. The WIwD coherence protocol is similar to the WI protocol, except when a global read occurs and the requesting site gets an actual copy, the copy of the former writer's page is

[2]Note that local read and local write operations cannot be monitored by a DSM server since they are not resulting in a request submitted to the server.

downgraded from mode (local read, local write) to (local read, global write). Since the page is not deleted, it increases the *minimum* number of replicas in the read phase to two. Unfortunately, even this approach suffers from too few copies present in the write phase. In order to address this serious drawback, we have studied a design modification that increases the minimal and maximum number of replicas modified by a write operation (i.e. required in the write phase) together with the minimum number of replicas in the read phase by $w-1$ with $w \in \{2, 3, \ldots\}$. This leads to boundary settings given in part d) of Figure 2. These constraints are enforced by the DSM server while serving a read or write request, and as part of the initial configuration when a segment is created (see [5] for a description of the protocol operations). The values for r_{min}, r_{max}, w_{min}, and w_{max} can ideally be different for every page, leading to a different degree of availability for the pages' data and to different operation costs.

We call the coherence protocol class with boundary settings as specified in Figure 2 *Boundary-Restricted* (BR) coherence protocol class, since it restricts the number of cached copies to lie between w_{min} and r_{max}. Furthermore, because $w_{min} = w_{max}$ and $r_{min} = w_{max} + 1$, it suffices to instantiate w_{min} and r_{max}. For the ease of discussion, we abbreviate w_{min} with w and set r_{max} to n, the number of potential client sites. As a convention, we refer to an instance of the BR coherence protocol class with given values for w and n as $BR(w, n)$ coherence protocol.

ANALYSIS

The properties of DSM coherence protocols which are of primary interest in the current context are the read and write *costs*, the *availability* of the user data managed by the DSM system (and therefore managed by the coherence protocol), and the impact of an increasing number of sites using a DSM system with a particular coherence protocol, in other words, the *scalability* of the coherence protocol.

Definitions

The number of sites contacted by the DSM server using the communication infrastructure of the network in order to carry out a global read (write) operation are called *read (write) costs*. The probability that at an arbitrary point in time at least one functioning site caches an actual copy is called *data availability*. □

Stochastic Model

Figure 3 gives the Markov chain used for analyzing the BR coherence protocol class. The description of the transistion probabilities is given in Figure 4.

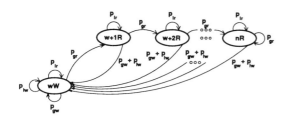

Figure 3: *Markov chain used for deriving the steady state distribution of the BR coherence protocol*

Prob.	Description
p_{lr}	Probability that a site already caching a page re-reads the data. It depends on the read attribute as to whether this can be done locally (local read attribute) or whether the site must issue a read request (global read attribute).
p_{gr}	Probability that a site currently not caching a copy of a page requests the data.
p_{lw}	Probability that a site already caching a copy of a page wants to modify it. In this case, if a local write attribute is present, this can be done locally. If the global write attribute is set, then the site must submit a write request to the DSM server.
p_{gw}	Probability that a site currently not caching a copy of a page wants to modify the page's data.

Figure 4: *Transition probabilities used for the Markov chain analysis*

After processing "a lot" of requests, for example while doing a parallel matrix calculation, the system reaches a *steady state*. For this steady state, the probability that an observer finds the DSM system in a particular state, can be computed. This so-called *steady state distribution* is the basis for calculating the mean operation costs and data availability.

Costs

Figure 5 shows a characteristic behavior of the mean read, mean write, and mean operation costs of BR coherence protocols when the upper boundary n is fixed to a certain value (here 100) and the lower boundary w is varied. Additionally, the mean number of cached copies is given. Obviously, an increase of the lower boundary w results in an increase of the mean costs and of the mean number of copies cached. This corresponds to our expectation: an increase of the minimum number of copies cached must result in an increase of the mean number of cached copies. The mean write costs must increase, since for every write operation initiating a write phase, a higher number of sites must be contacted when w grows. The mean read costs grow because at the beginning of every read phase, an increasing number of copies is cached at the client sites. Finally, because the mean read and write costs increase with a larger w, the mean operation costs must also grow. A remarkable fact is, that the

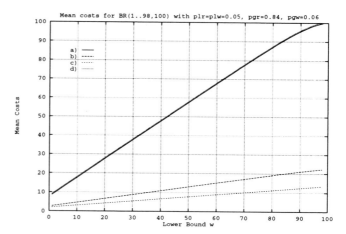

Figure 5: *Mean costs and mean number of cached copies of the BR(w,100) coherence protocol with w varying from $w = 1, \ldots, 98$. Graph a) represents the mean number of cached copies, b) the mean operation costs, and graphs c) and d) the mean read costs and the mean write costs respectively*

behavior of all those cost measures is almost linear. Except in those cases, where the difference between the lower and the upper boundary, i.e. $n - w$, is relatively small, the mean number of cached copies and the mean write costs behave non-linearly. (The same holds for the mean operation costs, since it is directly impacted by the mean write costs.) The mean read costs always behaves linearly. They are *de facto* not affected by the size of $n - w$.

As a consequence, any further increase of these values does not have an impact on the steady state distribution; therefore, it has no impact (or only a very limited one) on the mean costs and mean number of copies. These properties have significant consequences with respect to availability and scalability issues.

We performed analyses with different values for p_{lr}, p_{gr}, p_{lw}, p_{gw}, n, and w. Results given in Figure 5 are characteristic for the BR coherence protocol class. The higher the ratio of $(p_{lr} + p_{gr})/(p_{lw} + p_{gw})$ (referred to as *read/write ratio or rwr* hereafter), the more elliptical are the graphs for the mean write costs and the mean number of cached copies when w is almost as large as n. In other words, those are the cases where the approximations are imprecise for small values of $n - w$. Additionally, the higher rwr, the higher is the offset of the mean write costs and the mean number of cached copies. That is because the system returns often to the "wW" state, causing heavy costs. The mean read costs vary insignificantly with respect to a variation of the ratio of local to global operation, e.g. $(p_{lr} + p_{lw})/(p_{gr} + p_{gw})$ (referred to as *local/global ratio* hereafter). On the other side, their steepness is severely governed by rwr: the smaller rwr, the higher the increase of the mean read costs when w approaches n.

Availability

In this section, we compare the *data availability* provided by the protocols under investigation. The mean number of cached copies is used to determine with what probability the DSM system is able to deliver the most recent value of a particular page. As already pointed out, the higher the mean number of replicas present in the networking environment, the higher is the probability that such a copy can be accessed by the DSM server. In Figure 6 selected mean numbers of cached copies are given for all three coherence protocols discussed so far (Markov chain analyses for the WI and WB coherence protocols can be found in [5]). The local/global ratio is 1/9.

Mean Number of Cached Copies										
n	rwr	p_{lr}	p_{gr}	p_{lw}	p_{gw}	WI	BR $(1,n)$	BR $(2,n)$	BR $(3,n)$	BR $(4,n)$	WB
10	8	0.05	0.84	0.05	0.06	5.5	6.1	6.8	7.4	8.0	10
10	4	0.05	0.75	0.05	0.15	3.6	4.3	5.2	6.0	6.8	10
10	2	0.05	0.62	0.05	0.28	2.2	2.8	3.8	4.8	5.7	10
10	1	0.05	0.45	0.05	0.45	1.4	1.9	2.9	3.9	4.9	10
10	0.5	0.05	0.28	0.05	0.62	1.1	1.4	2.4	3.4	4.4	10
25	8	0.05	0.84	0.05	0.06	7.3	8.1	9.1	10.1	11.0	25
25	4	0.05	0.75	0.05	0.15	3.9	4.7	5.7	6.7	7.7	25
25	2	0.05	0.62	0.05	0.28	2.2	2.8	3.8	4.8	5.8	25
25	1	0.05	0.45	0.05	0.45	1.4	1.9	2.9	3.9	4.9	25
25	0.5	0.05	0.28	0.05	0.62	1.1	1.4	2.4	3.4	4.4	25
50	8	0.05	0.84	0.05	0.06	7.7	8.5	9.5	10.5	11.5	50
50	4	0.05	0.75	0.05	0.15	3.9	4.7	5.8	6.8	7.8	50
50	2	0.05	0.62	0.05	0.28	2.2	2.8	3.9	4.9	5.9	50
50	1	0.05	0.45	0.05	0.45	1.4	1.9	2.9	3.9	4.9	50
50	0.5	0.05	0.28	0.05	0.62	1.1	1.4	2.4	3.4	4.4	50

Figure 6: *Mean number of cached copies in the WI, the WB, and the BR coherence protocol*

Obviously, the WB protocol caches the maximum number of copies in all those cases. This is generally true, but unfortunately, the WB coherence protocol achieves this advantage at a high price: the mean operation costs for any reasonable value of p_{lw} are extremely high. The price for achieving this degree of data availability is not only too high but also not necessary. As expected, the mean number of cached copies of the BR(w,n) protocol is essentially equal to the corresponding value of the BR$(w+1,n)$ protocol plus one for $w=1,2,3$. Only for small values of n is the difference between BR(w,n) and BR$(w+1,n)$ not exactly one.

Next, we focus on how many replicas are actually necessary to achieve a certain degree of data availability in a particular network with failure-bound components. We assume that all sites of the network – or at least those caching copies – are functioning with a uniform probability of p. This does not only imply that the individual sites are up and running but also that the software needed to manage the copy is in working order [2]. Since a page data is available in the network as long as at least one of $a > 0$ copies is accessible, the data availability is computed according to a binomial distribution as $A_{data}(p) = 1 - (1-p)^a$. For realistic values of $p > 0.75$ [2], four cached copies are sufficient to provide an almost perfect degree of page data availability. Thus, those coherence protocols which offer a mean number of cached copies of at least four at the lowest cost possible are of particular interest. Using the approximation of the BR coherence protocols, those protocols can easily be identified.

CONCLUSION

We presented a new coherence protocol family for DSM systems, whose instances offer highly available access to the shared data at low operation costs. Results show that the protocols proposed scale well since increasing the number of client sites does not increase the operation costs beyond a certain boundary. The analysis with respect to data availability suggests that it is not sensible to maintain more than four copies in the network.

References

[1] B. D. Fleisch and G. J. Popek. Mirage: A coherent distributed shared memory design. In Proc. of the 12th ACM Symp. on OS Principles, published in *OS Review* 23(5), pages 211–223, The Wigham, AZ, Dec. 1989.

[2] D. D. E. Long, A. Muir, and R. Golding. A Longitudinal Survey of Internet Host Reliability. Technical Report UCSC–CRL–95–16, Dept. of Computer and Information Sciences, University of California, Santa Cruz, CA 95064, Mar. 1995.

[3] A. Mohindra and U. Ramachandran. A survey of distributed shared memory in loosely-coupled systems. Technical Report GIT-CC-91/01, Georgia Institute of Technology, Atlanta, GA, Jan. 1991.

[4] G. G. Richard, III and M. Singhal. Using logging and asynchronous checkpointing to implement recoverable distributed shared memory. In *Proc. of the 12th Symp. on Reliable Distributed Systems*, pages 86–95, Princeton, NJ, Oct. 1993.

[5] O. E. Theel and B. D. Fleisch. Design and Analysis of Highly Available and Scalable Coherence Protocols for Distributed Shared Memory Systems Using Stochastic Modeling. Technical Report UCR–CS–95–1, Dept. of Computer Science, University of California, Riverside, CA 92521, Apr. 1995.

[6] K.-L. Wu and W. K. Fuchs. Recoverable Distributed Shared Virtual Memory. *IEEE Trans. on Computers*, 39(4):460–469, Apr. 1990.

DIAGNOSING MULTIPLE BRIDGE FAULTS IN

BASELINE MULTISTAGE INTERCONNECTION NETWORKS*

V. Purohit+, F. Lombardi+, S. Horiguchi++ and J. H. Kim++

+Texas A&M University
Department of Computer Science
College Station, USA

++ Japan Advanced Institute of Science and Technology
School of Information Science
Tasunokuchi, Japan

Abstract.

This paper proposes new approaches for diagnosing baseline multistage interconnection networks (MIN) in the presence of single and multiple faults under a new fault model. This model referred to as the geometric fault model, considers defective crossing connections (bride faults) which are located between adjacent stages.

1. Introduction.

Multistage interconnection networks (MIN) have been extensively analyzed in the literature due to their wide applicability in today's computer systems. The correct operation of a MIN is *indispensable* for the proper functioning of a computer/communication system; hence, diagnosis of MINs (fault detection and location) has been analyzed extensively [1,2,3]. It has been proved [1] that under a single fault assumption, a baseline interconnection network with $\log_2 N$ stages can be diagnosed using $\max\{12, 6+2\lceil \log_2(\log_2 N)\rceil\}$ test vectors. Few papers have addressed fault diagnosis of MINs under multiple faults [1,2]. In [1], it has been proved that $2(1+\log_2 N)$ tests may detect in most cases multiple faults under an unrestricted combinatorial fault model for the switching elements and a stuck-at model for the links. [5] has presented an algorithm for multiple fault diagnosis in interconnection networks made of baseline switching elements. This approach requires the execution of two phases, each made of $1+\log_2 N$ tests.

If multiple bridge faults are considered, fault diagnosis can not be accomplished using previous approaches because either only detection is pursued within a different fault model [3] or the execution of the diagnostic procedure is directed to multiple faults mainly in the SE's and only a single stuck-at fault in the links is assumed [1,5]. Also, it is not possible to collapse a bridge fault to the SE's to which the lines

are directly connected because the effects of this type of fault can not be related to a specific fault in the switching characteristics of a SE as the faulty output is a function of two SE's in a stage, i.e. for diagnosing a bridge fault, tests must be generated to account for this dependency between SE's. The objective of this paper is to present a new fault model for MIN; this fault model is based on a geometric characterization of the links between stages in a MIN.

2. Preliminaries.

In this paper, it is assumed that for simplicity, the fanin (the number of input links of a switching element) and the fanout (the number of output links of a switching element) are both equal to 2. A switching element (SE) of a MIN with two inputs (denoted as I_1 and I_2) and two outputs (O_1 and O_2) can be considered as a 2×2 crosspoint switching matrix with a maximum of 16 possible states [1]. The valid states of a SE in a baseline interconnection network are S_5 (cross) and S_{10} (straight). The fault model of [1] assumes that a faulty SE can be in any one of the 16 states from a given valid state. In the fault model of [1], only permanent faults are considered; there are two types of faults: link-stuck fault and switching element fault. Fault diagnosis under the combinatorial fault model of [1] can be accomplished using a number of very simple test vectors. Two types of test vectors are permitted [1]. These are denoted as t (of value 01) and t' (the complement value of t).

In this paper, the stages of the MIN (each made of $\frac{N}{2}$ SE's) are numbered from 0 to $\log_2 N - 1$, where N is the number of primary input lines to the network. The primary input lines and the primary output lines (out of the last stage) are numbered from 0 to $N-1$. The set of links (or lines) which connect the SE's in stage i to the SE's in stage $i+1$ constitutes the

* This research is supported in part by grants from the Texas Advanced Technology Program and the Ministry of Science and Culture of Japan.

intermediate stage i. Thus, for a MIN consisting of N primary input lines, there are $\log_2 N$ stages (made of SE's) and $\log_2 N - 1$ internal intermediate stages (numbered from 0 to $\log_2 N - 2$). Hereafter, the generic term of "stage" refers to a stage made of SE's, while a stage made of connections is specifically referred to as an intermediate stage. The following definition must be introduced to understand the proposed approaches. Let the *order* of a MIN be given by the number of stages in the MIN. A MIN can be defined recursively as follows (provided a SE is assumed to have a fanin/fanout of 2): (1) A SE is a MIN of order 1. (2) For two MINs A and B (each of order M) which are connected to 2^M SE's such that all upper output lines of the SE's are connected to the input lines of A and all lower lines of the SE's are connected to the input lines of B, then these MINs constitute a MIN or order $M+1$. For convenience, the two MINs (each of order M) which make up the MIN of order $M+1$, are referred to as the sub_MINs of the latter, i.e a MIN of order M has 2 sub-MINs of order $M-1$, 4 sub-MINs of order $M-2$ .... 2^{M-1} sub-MINs of order 1. For a MIN of order M, sub_min(i,j) (for $0 \le i \le M-1$ and $0 \le j \le 2^{i+1}-1$) denotes the jth sub_min starting at stage i of the MIN.

The following assumptions are applicable to the analysis presented in the next sections: (1) All SE's are assumed to be fault free. (2) Multiple link faults (of the type stuck-at-0 or stuck-at-1) are not considered (these faults can be diagnosed using [1,5]).

3. Basic Properties of the Proposed Approach.

Initially, some basic functional relationship in the lines of a MIN [1] must be established. A *thread* is defined as a line which connects a primary input line of a SE (i.e. in stage 0) to a primary output line of a SE (i.e. in the last stage of the MIN) when all switches are in S_{10}. The *number of a thread* is defined as the number of the primary output line (i.e. in the last stage of the network) which the thread passes through. *reverse*(i) is the number which is obtained by reversing the representation of i (the subscript denotes the base); if $i_{10} = (P_{M-1}P_{M-2}....P_0)_2$, then *reverse*($i_{10}$)$=(P_0 P_1 .... P_{M-1})_2$, where $M \ge 0$. *complement*(i) is the number which is obtained by complementing the bits of the binary representation of i; if $i_{10} = (P_{M-1}P_{M-2}....P_0)_2$, then *complement*($i_{10}$) $= (\overline{P_{M-1}P_{M-2}....P_0})_2$, where $M \ge 0$.

The above definitions yield the following observations (they establish a relationship between thread number and the number of the primary input line for each of the two states of a baseline SE). *Observation 1*. If a primary input line in stage 0 of thread i has a number j, then $i = reverse(j)$ provided the SE's are in S_{10} and M is equal to the number of stages in the MIN (as according to the definition of *reverse* given

previously). *Observation 2*. If the number of the primary input line at stage 0 of thread i is given by j, then $i = complement(reverse(j))$ provided the SE's are in S_5 and M is equal to the number of stages in the definition of *complement* and *reverse*.

The above analysis can be also extended to two threads. Therefore, the following definitions are applicable. Let i and j be the two threads. Threads i and j are said to *intersect* if in the fault free topology of the MIN, these two threads intersect between two adjacent stages (provided the SE's are in S_{10}). The point at which the two threads intersect, is referred to as the *intersection point*. The *location of an intersection point* is defined to be i if it occurs in intermediate stage i. The following observations are applicable. *Observation 3*. If i and j are two thread numbers, then these two threads intersect if and only if the following condition holds: $(i > j) \rightarrow reverse(i) < reverse(j)$ AND $(i < j) \rightarrow (reverse(i) > reverse(j))$, where $\rightarrow$ means logical implication, i.e. *intersect*(i,j)=(not ($i > j$) OR (*reverse*(i)<*reverse*(j))) AND (not ($i < j$) OR (*reverse*(i) > *reverse*(j))). *Observation 4*. If i and j are two threads, then *intersect*(i,j) implies that the intersection point of these two lines occurs at a location given by the position of the first 1 starting from the most significant bit (MSB) position in (i EXOR j). The position of the MSB in the representation of the number is defined to be the one whose index is 0. *Observation 5*. No two intersecting threads can ever pass through the same SE or conversely, two threads which pass through a common SE, can never intersect.

4. Fault Model.

A bridge fault is commonly defined to be a faulty contact between two lines (otherwise unconnected). In this paper, the faulty contact occurs between two lines in adjacent stages of a MIN. In this paper, an approach based on geometry is proposed for identifying the fault free topology of the MIN (hereafter, referred to as the correct topology). A *geometric intersection* is defined as the intersection which occurs in the (two-dimensional) representation of the network. This model is based on the two-dimensional layout of a MIN (as a shuffle-exchange network) as given in [4]. As an example, consider Figure (1) and the two threads numbered 7 and 10. These two threads intersect (geometrically) in intermediate stage 0.

The following assumptions are used in this paper. (1) A fault may occur between a pair of lines which connect two stages of SE's, if the following conditions hold (examples are given later in this section): two lines *intersect geometrically* in between the two stages which they connect. This type of fault is referred to as a *type A fault*; two lines are *adjacent* to each other and

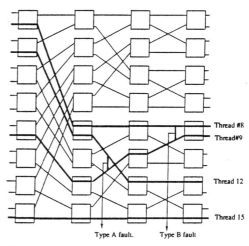

Figure 1: Types of Bridge Faults under the Geometric Fault Model.

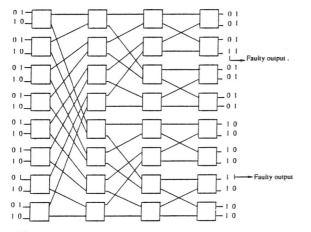

Figure 2: Tests to Detect Type B Faults at Input Lines and Type A faults in Intermediate Stage 0; All Switches are in the Parallel Mode.

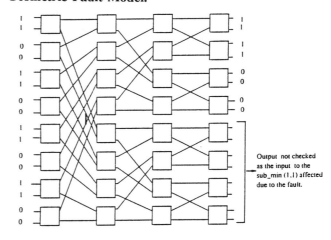

Figure 3: Tests to Detect Type B Faults at Input Lines of Stage 1 and Type A faults in Intermediate Stage 1.

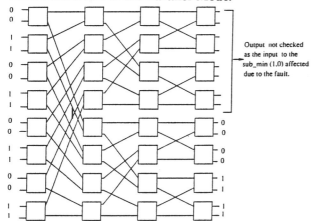

Figure 4: Complemented Vector to Detect Type B Faults at Input Lines of Stage 1 and Type A Faults in Intermediate Stage 1.

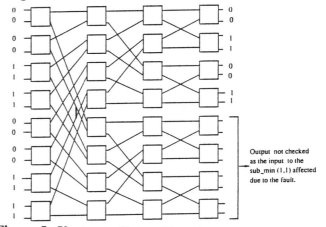

Figure 5: Vector to Detect Type B Faults at Input Lines of Stage 2 and Type A Faults in Intermediate Stage 2.

do not have any other line lying completely in between them. This type of fault is referred to as a *type B fault* (as shown in Figure (1)). (2) Faults are assumed to occur in intermediate stages only and not in the *SE*'s, i.e. *SE*'s are assumed to be fault free. (3) If a bridge fault occurs between two threads, then the value of the bits being carried beyond the bridge fault (unidirectional fault) by these threads is the OR of the bits being brought into the bridge fault (wired-OR). (4) Only a pair (two) of lines can have a bridge between them. The possibility of three simultaneous lines with a bridge in between them, is remote in practice and is not considered.

5. Fault Detection.

In this section, an approach for detecting bridge faults in a baseline MIN is presented. The main principle behind the proposed approach is to test each intermediate stage of the MIN separately and distinguishing the fault type and observe its effects at the primary output lines of the MIN. If there exists a fault of type A between two lines in the intermediate stage under test, or a fault of type B in the previous intermediate stage, then the fault is detected because at least an erroneous value will appear at the primary output lines of the MIN. If no erroneous value appears, then no fault (under the assumed fault model) exists. Detection of at least a bridge fault requires the proposed approach to execute a different phase for testing subsequent stages. In these circumstances, the main objective of the algorithm is to establish if there are any additional bridge faults in the MIN. Input test vectors are generated such that the effects of an already detected bridge fault are masked and testing of each of the subsequent intermediate stages is possible. This is accomplished in two phases. This process terminates when the following scenarios occur: (1) The total number of bridge faults which has been detected, is greater than one, thus indicating that multiple bridge faults are present in the MIN (multiple fault detection). (2) The entire MIN has been tested stage by stage, thus resulting in the detection of either only a single bridge fault or none (single fault detection or fault free status).

Test generation is based on generating input vectors with alternate groups of 0's and 1's, each of size 2^{int_stage}, where int_stage is the intermediate stage under test. The intermediate stage under test receives alternate 0's and 1's as inputs provided int_stage is the intermediate stage under test and all intermediate stages prior to int_stage in the MIN are fault free. This implies that if the first $i-1$ intermediate stages are fault free and the intermediate stage under test is i, then every two geometrically intersecting lines in intermediate stage i receive different bits as inputs, i.e. if there exists a fault of type A in intermediate stage i, then it is detected as at least an erroneous value

appears at a primary output line of the MIN.

Theorem 1. If there exist no faults up to the intermediate stage under test i (starting from intermediate stage 0), then the input vector for $i+1$, detects whether there exists either a bridge fault of type A in intermediate stage $i+1$, or a type B fault in intermediate stage i.

Note that if no erroneous output is observed during this test, then it is easy to show that no type A fault exists in intermediate stage $i+1$ and no type B fault exists in intermediate stage i. Hence, the fault free status of the MIN can be established by testing each intermediate stage, i.e. in $O(\log_2 N)$.

6. Bridge Fault Location.

The process which finds the location of a bridge fault in an intermediate stage (given the thread numbers with erroneous values at the primary outputs), can be presented. This approach is executed provided that single fault detection has been accomplished. The following steps must be be executed to establish whether the bridge fault exists in intermediate stages int_stage-1 or int_stage. (1) Put the *SE*'s given by (out_line1)/2 and (out_line2)/2 of int_stage in S_5. (2) Generate the input vector in_vect using the proposed test generation procedure for int_stage. (3) Find the numbers of the input lines to the first stage which provide the input bits to the switches which were set in S_5. Let these *SE*'s be in1, in2, in3 and in4. (4) Complement the bits int_vect[in1], int_vect[in2], int_vect[in3],int_vect[in4]. This occurs because the two *SE*'s in S_5 have been found and the input to subsequent stages must not be altered. (5) Check if the faulty bits in the output vector are the same as the ones obtained when all *SE*'s were in S_5. (6) If the fault exists in intermediate stage int_stage, then the positions of the faulty bits in the output vector do not change. However, if the faulty output occurs at a new position, then a fault exists in int_stage - 1. Set faulty_stage = int_stage-1 (for the latter case) or equal to int_stage (in the former case). (7) If the faulty bit occurs at its original position as well as at a new position, then it is due to the occurrence of a bridge fault in both stages. Set faulty_stage as int_stage. The occurrence of two these faults does not hinder the continued execution of the above process.

The following theorems can be introduced.

Theorem 2. Let the intermediate stage under test be i and no erroneous outputs be detected till the $(i-1)$th intermediate stage has been tested. If it is found that there are n (n $\geq$ 1) primary output lines in the last stage of the MIN which have erroneous values, then the following conditions are applicable: (1) There exist at least n bridge faults in the intermediate stages beyond and including the $(i-1)$th intermediate stage.

(2) All faults in intermediate stage i-1 are of type B.
(3) There exists at least one fault of type B in intermediate stage i-1, or at least one fault of type A in intermediate stage i.

Theorem 3. Let a single bridge fault exist in intermediate stage i. Also, let the two lines which have the bridge between them, be the output lines numbered l and m of intermediate stage i and j ($j > i$) be the number of the intermediate stage under test, $dist = 2^{j-i-1}$. If $((\frac{l}{dist}) \bmod 2) = ((\frac{m}{dist}) \bmod 2)$, then the presence of this bridge fault does not affect the inputs to intermediate stage j, i.e. the test vector of alternating 0's and 1's. Else, the sub_min at stage i+1 (which receives the faulty input test vector) is given by: (1) $(l \times \frac{2^i}{N}) \times 2 + 1$ if odd(l) and even($\frac{l \times 2^i}{N}$). (2) $(\frac{l \times 2^i}{N}) \times 2$ if even(l) and even($\frac{l \times 2^i}{N}$). (3) $(\frac{m \times 2^i}{N}) \times 2 + 1$ if odd(m) and even($\frac{m \times 2^i}{N}$). (4). $(\frac{m \times 2^i}{N}) \times 2$ if even(m) and even($\frac{m \times 2^i}{N}$).

The process for single bridge fault location is as follows: first, the fault type must be discriminated; the exact location of the fault is established by finding whether the fault occurs at the input or at the output lines of a SE in the stage under test. Figures (2) through (5) show the diagnosis process and the test vectors for faults in intermediate stage 1.

Theorem 4. Fault detection and location require at most ($2\log_2 N$) test vectors.

This process will also locate multiple faults; the only case under which the process fails to fully locate a fault, is when a bridge fault exists between two lines which act as input lines (or output lines) to a common SE in the MIN. In both these cases, it is not possible to fully locate it, i.e. whether the bridge fault exists at the inputs or outputs of the SE (even though the location of its stage can be identified).

7. References.

[1] Wu, C. and T. Feng, "Fault Diagnosis for a Class of Multistage Interconnection Networks," *IEEE Trans. on Comput.*, Vol. C30, No. 10, pp. 743-758, 1981.

[2] Falavarjani, K.M. and D.K. Pradhan, "Fault-Diagnosis of Parallel Processor Interconnection Networks," *Proc. FTCS*, pp. 209-212, 1981.

[3] Mourad, A., B. Ozden and M. Malek, "Comprehensive Testing of Multistage Interconnection Networks," *IEEE Trans. on Comput.*, Vol. C40, No. 8, pp. 935-951, 1991.

[4] Ullman, J.D., "Computational Aspects of VLSI," Computer Science Press, Rockville, Md., 1984.

[5] Feng, C., F. Lombardi and W.-K. Huang, "Detection and Location of Multiple Faults in Baseline Interconnection Networks," *IEEE Transactions on Computers*, Vol. C41, No. 11, pp. 1340-1344, 1992.

Track Piggybacking: An Improved Rebuild Algorithm for RAID5 Disk Arrays

Robert Y. Hou, Yale N. Patt
Department of Electrical Engineering and Computer Science
University of Michigan, Ann Arbor 48109-2122

Abstract

When a disk in a RAID5 disk array has failed, requests to that disk can only be serviced by reading data from all surviving disks and regenerating the missing data. Performance is dramatically reduced until the missing data is rebuilt to a spare disk. Several algorithms have been suggested for reducing the rebuild time and thus minimizing this disruption in service. We propose a new scheme, track piggybacking. We have found that track piggybacking reduces rebuild times by up to 29% over previously published rebuild algorithms, and for some workloads, can reduce response times by 93% or more.

1 Introduction

Dramatic improvements in microprocessor and main memory speeds have far outpaced improvements in disk access time. The resulting disparity has caused the storage subsystem, and the disk drive hardware in particular, to become a serious bottleneck for many applications. This will also pose a problem for applications that are compute-bound as they are moved to faster machines and become I/O-bound.

Transaction processing is one application that has strong requirements of its disk subsystem. One important requirement is that the array must have high availability.[1] RAID5 disk arrays have been proposed as a hardware solution for providing high data availability. When a disk in a RAID5 array fails, the data on that disk can still be accessed, albeit less efficiently, by reading all surviving disks and regenerating the data via the exclusive-OR operation. The array cannot, however, survive a second disk failure. Therefore it is necessary that the data on the failed disk be restored to a spare disk in a timely manner.

The simplest algorithm for rebuilding the failed disk, *Minimal Update* [Holl92], sequentially reads data

---

[1]Availability is based on the fraction of time during which the array can service requests [Gray91].

from all surviving disks to regenerate the contents of the failed disk to the spare disk. Minimal Update implements a general rebuild process. Several more sophisticated algorithms have been suggested for enhancing simple rebuild algorithms like Minimal Update by reducing the time required to rebuild an array. For example, when an application wishes to read a block of data from the failed disk, where a block is a small fraction of a track, all surviving disks must be accessed to satisfy the request. It is then a simple matter to also write the reconstructed block to the spare disk. As a result, this block does not have to be rebuilt at some later point using Minimal Update. Muntz and Lui [Munt90] called this rebuild algorithm *Piggybacking*. The drawback of this algorithm is that it is unlikely an entire track will be rebuilt to the spare disk via Piggybacking [Holl92]. If only a partial track is rebuilt, the remaining blocks on the track will still have to be regenerated later using Minimal Update, incurring a major fraction of the time required to rebuild the entire track. Thus, the reduction in rebuild time is small.

We propose a new rebuild algorithm, *Track Piggybacking*, which eliminates the partial track problem. Rather than reading just one block from each surviving disk to reconstruct the requested block, Track Piggybacking reads a track of data and rebuilds a full track of data to the spare disk. The result is that a track can be rebuilt to the spare disk for each application request accessing non-rebuilt data on the failed disk. There are no missing blocks and the track does not have to be regenerated later using Minimal Update, thus reducing the overall rebuild time. In addition, Track Piggybacking is invoked on write requests as well as read requests. The original Piggybacking algorithm does not do this, although there is no reason why it cannot.

The remainder of this paper is organized in seven sections. Section 2 acknowledges previous work in the field. Section 3 discusses the rebuild algorithms evaluated in this paper. Section 4 describes the simulator and workloads used. Sections 5 and 6 compare rebuild times and response times for Track Piggybacking

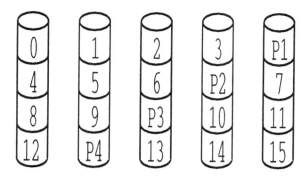

Figure 1: RAID5 disk array

and the other rebuild algorithms using a synthetically generated workload of uniformly distributed requests. Section 7 compares these algorithms using two traces of disk requests. Section 8 presents our conclusions and some topics for future research.

2   Previous Work

Increasing the number of concurrently active disk drives in the disk subsystem is a common technique for improving disk subsystem performance. Striping, or interleaving, data across disks can further increase performance [Kim86, Ng88, Redd89]. Disk striping allows the data for large requests to be accessed in parallel from multiple disks. Disk striping with a large stripe unit can also be used to form one logical disk capable of performing many concurrent accesses [Livn87].

Unfortunately, increasing the number of disks also increases the probability of data loss due to disk failure. Several hardware redundancy schemes have been devised to improve reliability, including mirroring [Bate85, Tera85, Bitt88] and disk arrays with parity [Kim86, Patt88, Gray90, Holl92]. The RAID5 disk array has become popular recently since it requires fewer disks than a mirrored disk array while ensuring data availability in the event of a single disk failure. Figure 1 illustrates how data can be laid out for a RAID5 disk array.

A disk array can operate in three different modes [Meno92]. It operates in Normal Mode when all disks are available and the disk array is providing its highest performance. It moves to Degraded Mode when one of the disks has failed. When the failed disk is accessed in a RAID5 disk array, corresponding blocks from all surviving disks must be read and exclusive-ORed to re-

generate the missing data. Response times during Degraded Mode are higher than in Normal Mode, making it more difficult to deliver high throughput while providing low response times. A RAID5 disk array in Degraded Mode also cannot survive a second disk failure. Thus, it is important for the disk array to enter Rebuild Mode and rebuild the failed disk as quickly as possible.

It is also important for the disk array to continue servicing the application's disk requests. Much of the research into the Rebuild Mode performance of disk arrays assumes the application prefers its requests be serviced, albeit with longer response times, and is willing to take the chance that a second disk failure may result in data loss. Menon and Mattson [Meno92] analyzed response times and rebuild times for a RAID5 disk array assuming application requests are given priority over rebuild requests. They used a rebuild algorithm called Baseline Copy [Munt90] which is similar to Minimal Update.

Several researchers have proposed and investigated solutions for more efficiently rebuilding the failed disk. Muntz and Lui [Munt90] compared three rebuild algorithms using an analytical model. Their *Baseline Copy* algorithm simply reads blocks sequentially from each of the surviving disks, regenerates the missing data and writes it to the spare disk. All write requests to the failed disk are also serviced by the spare disk.

Their second rebuild algorithm, *Redirection of Reads*, augments Baseline Copy by using the spare disk to service application read requests to the failed disk if the requested data has already been rebuilt to the spare disk. The third rebuild algorithm, *Piggybacking*, enhances Redirection of Reads by updating the spare disk with any data blocks regenerated due to application read requests to the failed disk. If a data block on the failed disk is accessed and has not yet been rebuilt to the spare disk, the data block is regenerated by reading the corresponding surviving disks. It is then a simple matter to also write this data block to the spare disk. As a result, this data block does not have to be regenerated later by the rebuild process. Muntz and Lui assumed that the rebuild time was reduced by a fraction proportional to the number of blocks rebuilt via Piggybacking.

Holland and Gibson [Holl92] extended Muntz and Lui's work by performing simulations of the rebuild algorithms and comparing response times and rebuild times. They also evaluated a rebuild algorithm that is simpler than Baseline Copy, called *Minimal Update*. Unlike Baseline Copy, Minimal Update does not up-

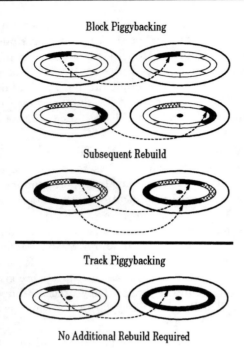

Figure 2: Block Piggybacking vs. Track Piggybacking

date the spare disk with all writes that access the failed disk. Only writes which access data already rebuilt to the spare disk will update the spare disk. Holland and Gibson were able to more precisely determine the benefits of Redirection of Reads and Piggybacking. In particular, they found that Piggybacking provided little if any benefit over Redirection of Reads.

3 Methodology

We have evaluated several algorithms for rebuilding a degraded RAID5 disk array. The first one is Minimal Update [Holl92]. The others are Redirection of Reads and Piggybacking [Munt90], which are used to enhance Minimal Update. Minimal Update is used to rebuild any blocks not rebuilt via these enhancements. Piggybacking, referred to as *Block Piggybacking* in the remainder of this paper, provides little improvement over Redirection of Reads. This can be understood by referring to figure 2. There are two disk surfaces shown in each row of figure 2. The left disk surface represents the failed disk while the right disk surface represents the spare disk. One track on each disk surface is shown with six blocks of data per track. The blackened blocks are the blocks being accessed while

the cross-hatched blocks have been rebuilt to the spare disk.

As observed by Holland and Gibson, Block Piggybacking only rebuilds one block on the track to the spare disk for each read request to the failed disk. A second block on a track rebuilt via Block Piggybacking still leaves the rest of the blocks on the track untouched. If only a partial track is rebuilt, the remaining blocks will have to be subsequently rebuilt using Minimal Update, incurring a major fraction of the time required to rebuild the entire track. Furthermore, parity blocks are not rebuilt by Block Piggybacking since they are not read by any disk request.

We propose a new scheme, *Track Piggybacking*, where an entire track is read from all surviving disks each time a block on the failed disk needs to be regenerated, as shown in the bottom of figure 2. The resulting track of regenerated data is then written to the spare disk. Since this track will not have to be rebuilt later via Minimal Update, the rebuild time for the array is significantly reduced. Also, Track Piggybacking is invoked for both read and write requests while Block Piggybacking is only invoked for read requests.

In addition to reducing rebuild time, Track Piggybacking can improve response times for many applications. If an application reads disk blocks sequentially, the first block read triggers a track piggyback request and takes a little longer to complete since a single block read becomes a full track read. A request that reads the following block on the same track, however, now finds that block available on the spare disk. Reconstructing that block is not required as it is for the other rebuild algorithms. Applications that exhibit this type of behavior include database sorts and various file system applications.

For some transaction processing applications, however, there is no sequentiality exhibited by the stream of disk requests. Each read to the failed disk is converted into a full track read and response times are increased. Workloads containing uniformly distributed requests therefore give worst case response times for Track Piggybacking. The rebuild time for the array, however, is still reduced.

4 Workload and Simulator

Trace-driven simulation is used to evaluate the rebuild times and response times for the rebuild algorithms. All experiments assume a nine-disk RAID5 array. This array size was chosen to ensure that par-

bytes per block	4096
blocks per track	6
surfaces	14
cylinders	949
rotational speed	4318 RPM
seek time (ms)	$2.0 +$ $0.01\ (\#of\,cylinders) +$ $0.46\ \sqrt{\#of\,cylinders}$
single track seek	2.5 ms
average seek time	12.7 ms
maximum seek time	25.5 ms

Table 1: Disk Drive Parameters

ity consumed a relatively small amount of disk space; with nine disks, parity accounts for only 11.1% of the total disk space. The disk drive parameters modeled by the simulator are shown in Table 1. We modeled the same IBM 0661 Model 370 disk drive as Holland and Gibson [Holl92].

There are two types of workloads presented to the simulator. The first is a synthetically generated stream of disk requests. These requests are uniformly distributed across the disk space, since one class of transaction processing applications exhibits little or no locality in their I/O streams. Typical request sizes range from 1KB to 6KB [Olso89, Holl92, Meno92]. We have chosen 4KB to match Holland and Gibson. These requests are issued to the disk subsystem in bursts. Therefore, the workload interarrival rate, or time between successive disk requests, is generated using an exponential distribution. To model a typical transaction processing environment, 75% of the disk requests are assumed to be reads [Meno92].

The average I/O rate for these synthetic traces is varied to create different workloads for the disk subsystem. Three I/O rates were used for the synthetic experiments—75, 100 and 125 I/Os per second—in order to simulate light, medium and heavy workloads. Average response time is used to compare the performance of the disk array under these workloads.

The second type of workload evaluated in this paper exhibits some data locality. This workload will be represented by two traces of disk requests. They will be analyzed in section 7.

The simulator assumes the disk array controller is responsible for rebuilding the failed disk. Thus rebuild time depends on how much rebuild data can be stored in the storage controller. Each disk drive writes rebuild data to its own buffer in the storage controller. The *rebuild unit* [Hou93a], or amount of rebuild data read each time from a surviving disk, is one track. The storage controller waits until a track of rebuild data is read from each surviving disk, regenerates the missing data, and then writes the data to the spare disk.

5 Impact of Track Piggybacking on Rebuild Time

We assume that the most important rebuild criterion is to minimize the impact of the rebuild process on the performance of a failed array. This requires that rebuild requests be serviced by a disk only when there are no application requests pending for that disk [Meno92]. As a result, the application sees marginally increased response times in Rebuild Mode compared to Degraded Mode.

The rebuild time depends on the rebuild algorithm. It also depends on how many data blocks can be buffered in the storage controller waiting for the spare disk to be available. A surviving disk can only read more rebuild data if there is room in the controller to store it. A larger buffer allows the surviving disks to read large amounts of rebuild data into the controller during idle times even though the spare disk may be busy. It also allows individual surviving disks to read rebuild data whenever they are idle, enabling them to slip with respect to each other.

Figure 3 compares rebuild times for the different algorithms when buffer space is limited to ten cylinders of data per disk. The minimum time required to rebuild the failed disk, in which case the application requests are completely ignored during rebuild, is labeled Rebuild Only and is shown for comparison purposes. Block Piggybacking does not significantly improve on Redirection of Reads. Since Block Piggybacking is a more complex algorithm than Redirection of Reads, it is probably not worth implementing. This confirms observations made by Holland and Gibson [Holl92] and Hou and Patt [Hou93b].

Track Piggybacking, on the other hand, provides superior rebuild times because many full tracks are rebuilt to the spare disk using piggybacked requests during the process of servicing application requests. Thus they do not have to be rebuilt using Minimal Update. Track Piggybacking reduces rebuild time by 6% and 25% for light and heavy workloads, respectively, when compared to Redirection of Reads. As

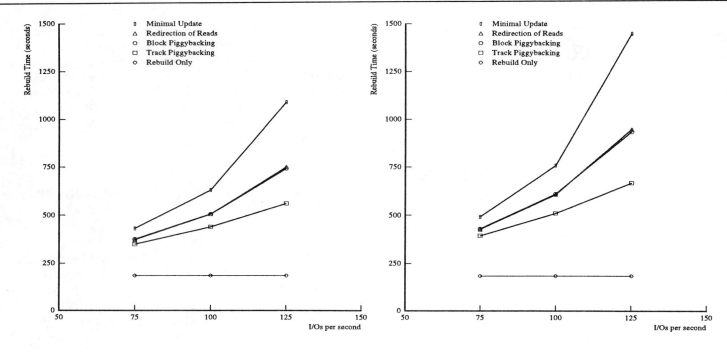

Figure 3: Rebuild Time as a function of I/O rate with ten cylinders of buffer space

Figure 4: Rebuild Time as a function of I/O rate with one cylinder of buffer space

can be seen, using Track Piggybacking and assigning a lower priority to rebuild requests increases rebuild times over Rebuild Only by 89% for light workloads and 204% for heavy workloads.

We found that rebuild time for each algorithm does not improve significantly when buffer space is increased beyond ten cylinders. Thus ten cylinders provide essentially the same rebuild times as an unlimited amount of buffer space.

Figure 4 compares rebuild times when there is one cylinder of buffer space per disk in the controller. Rebuild times increase as the surviving disks cannot slip as much with respect to each other during the rebuild process. Track Piggybacking improves rebuild time by 8% and 29% over Redirection of Reads for light and heavy workloads, respectively. If we compare figures 3 and 4, rebuild times for Track Piggybacking increase by 13% and 19% for light and heavy workloads, respectively, when buffer space is reduced to one cylinder per disk.

6 Impact of Track Piggybacking on Response Time

As discussed in the beginning of section 5, an important rebuild criterion is that the application sees marginally increased response times in Rebuild Mode compared to Degraded Mode. In essence, the rebuild requests should be nearly invisible outside the disk array.

Figure 5 shows the average response time for a RAID5 array using one cylinder of buffer space in the disk array controller. Normal Mode and Degraded Mode response times are shown for comparison purposes. Minimal Update has the highest average response time as every read request to the failed disk must be reconstructed by reading all surviving disks. The response time decreases for the more sophisticated rebuild algorithms since many read requests can be serviced by the spare disk.

It is interesting to note that for a heavy workload, the response times for the more sophisticated rebuild algorithms are lower than Degraded Mode response times. The reason is that any requests that can be redirected to the spare disk will reduce the heavy load on the surviving disks. Track Piggybacking does not reduce response times as much as Redirection of

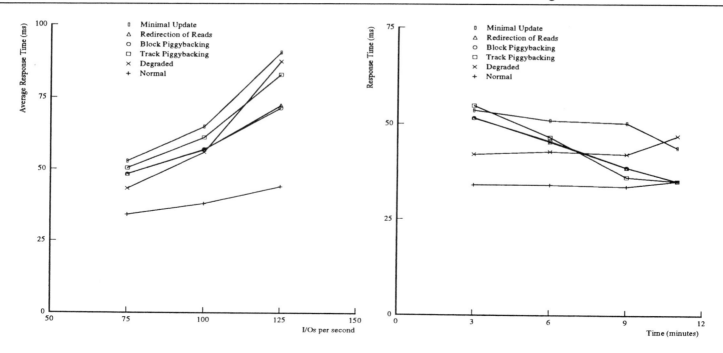

Figure 5: Average response time as a function of I/O rate to the disk array

Figure 6: Average response time during the rebuild period for a light workload

Reads and Block Piggybacking. For Track Piggybacking, each application request to the failed disk potentially reads an entire track of data from all surviving disks and can therefore delay any requests that follow. As a result, response times are increased by 4% and 15% over Redirection of Reads for light and heavy workloads, respectively. Redirection of Reads and Block Piggybacking have essentially the same response times.

Results for ten cylinders of buffer space as well as for an unlimited amount of buffer space in the disk array controller are similar and are not shown here. Response times for Track Piggybacking increase by up to 5% when buffer space increases from one cylinder to unlimited buffer space. Response times for Redirection of Reads increase by up to 3.3%.

Average overall response times, however, do not provide a complete picture of performance. Menon and Mattson [Meno92] suggest evaluating response times over a period of time, and we provide results here for the different rebuild algorithms. Figures 6 and 7 show the average response times measured in three-minute intervals during the length of the rebuild process. We limit buffer space in the disk array controller to one cylinder per disk, although similar results have been found for ten cylinders. Figure 6 shows results for

a light workload of 75 I/Os per second. All rebuild algorithms provide higher response times than Degraded Mode when the rebuild process starts, but their response times slowly drop below Degraded Mode and eventually match Normal Mode response times at the end of the rebuild process. The response time for Minimal Update does not return to Normal Mode because Minimal Update just finishes the rebuild process at time 11 minutes.

Block Piggybacking provides the same response time as Redirection of Reads. Track Piggybacking, on the other hand, begins with the highest response times and slowly drops below Redirection of Reads as Track Piggybacking finishes the rebuild process a little earlier. The overall difference between the two, however, is small. The disk array is underutilized for light workloads and has plenty of idle time to service rebuild requests. Thus optimizing the rebuild process is not very beneficial for light workloads, although there is some improvement between Minimal Update and the other more sophisticated rebuild algorithms.

It is interesting to note that the response time for Minimal Update also gradually decreases. The reason is that write requests that update rebuilt data on the spare disk are changed into *read-modify-write* requests. Read-modify-write requests require four disk accesses

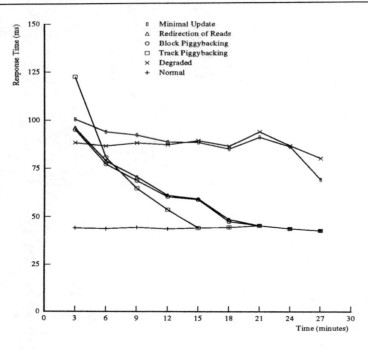

Figure 7: Average response time during rebuild for a heavy workload

while write requests to missing data require reading all surviving disks and writing the spare disk for a total of nine disk accesses.

Figure 7 shows average response times measured in three-minute intervals for a heavy workload of 125 I/Os per second. Track Piggybacking begins with much higher response times. It completes the rebuild process faster, however. This can be seen by observing that response times for Track Piggybacking merge with Normal Mode response times 15 minutes after beginning the rebuild process while response times for Redirection of Reads merge with Normal Mode response times 21 minutes after rebuild starts. In addition, Track Piggybacking provides superior response times compared to Redirection of Reads during half of its rebuild process. Thus Track Piggybacking only gives higher response times during half its short rebuild time while Redirection of Reads provides a lower overall response time during a longer rebuild time.

7   Impact of Track Piggybacking on Other Workloads

This section examines two traces of disk requests that are different from the synthetically generated

workload evaluated thus far in this paper because they exhibit some locality. The first trace, called *Order*, contains a transaction processing workload in which orders are generated for parts being sold by a machine parts distribution company. The second trace, called *Report*, has a batch workload in which reports are produced based on these orders generated during the day [Rama92]. These traces come from a commercial $VAX cluster^{TM}$ system running the VMS^{TM} operating system around 1988-89.

Ramakrishnan et al. [Rama92] reports that 2.1% of the files in the Order trace account for 98.6% of all read operations. Less than 2% of the files account for 95% of all write operations. The Report trace is similar to the Order trace in that a few of the files account for most of the disk accesses. The majority of disk requests in both traces are less than 5 KB in size.

The Order and Report traces were evaluated in order to find time periods during which the traces showed heavy disk traffic. This section will present results for these time periods. The Minimal Update, Redirection of Reads and Track Piggybacking rebuild algorithms are evaluated for rebuild times and response times, with response times for Normal and Degraded Modes provided for comparison purposes.

For the experiments in this section, data is striped across all disks in track units. With 24 KB of data per track and 512-byte data blocks, this gives a stripe unit of 48 blocks of data. By striping the data, the chances of developing data hot spots is reduced since consecutive tracks of data are placed on different disks. Striping data in track units allows most small disk requests to be serviced by a single disk, permitting many disk requests to be serviced concurrently by the disk array. It also allows many disks to cooperate in servicing a single large request.

A track rebuild unit is used to ensure low rebuild times and low response times. In addition, ten cylinders of buffer space in the disk array controller have been allocated.

The Order trace is evaluated first. Figure 8 shows response times as a function of time. Response times for Minimal Update, Redirection of Reads and Degraded Mode are initially very high, with Minimal Update and Degraded Mode starting at over 1000 ms while Redirection of Reads is closer to 500 ms. There are many requests being serviced at this time and operating with a failed disk simply increases response times. Redirection of Reads improves response times as the spare disk services some requests.

Track Piggybacking, on the other hand, has re-

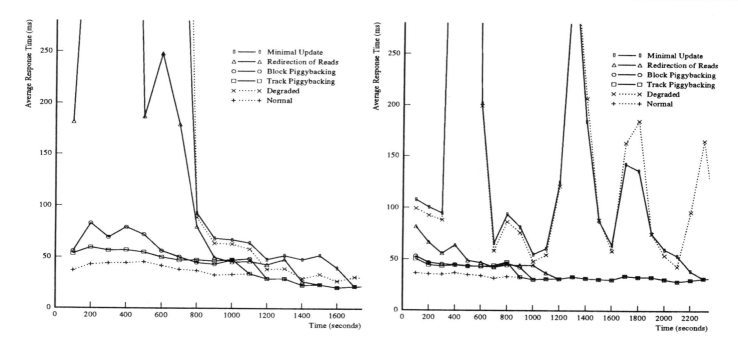

Figure 8: Response time and rebuild time for Order workload using RAID5

Figure 9: Response time and rebuild time for Report workload using RAID5

sponse times that are only 45% higher than Normal Mode and much lower than Redirection of Reads. There is a subset of data tracks that is frequently accessed. The first time these tracks are accessed, Track Piggybacking rebuilds the data to the spare disk. Thereafter, the spare disk can service all requests to this data, dramatically reducing the response time.

Low response times for Track Piggybacking also help reduce rebuild times. Track Piggybacking finishes rebuilding the failed disk at 989 seconds, followed by Block Piggybacking at 1088 seconds and Redirection of Reads at 1318 seconds. These times can be approximated from figure 8 by observing when the response times for the various rebuild algorithms merge with Normal Mode response times.

Block Piggybacking also has low response times, although they are up to 39% higher than Track Piggybacking response times. Block Piggybacking rebuilds frequently accessed data to the spare disk, which allows the spare disk to quickly service requests to this data. Block Piggybacking has higher response times than Track Piggybacking because it does not help the rebuild process as much as Track Piggybacking does. Therefore it cannot reduce the workload of the surviving disks as well as Track Piggybacking. In addition, Block Piggybacking only helps if there is temporal lo-

cality in data access, since data is accessed multiple times. Spatial locality, in which accessing one block of data means another nearby block of data is also likely to be accessed, is not exploited. Since Block Piggybacking is more difficult to implement than Track Piggybacking, it is probably not worth pursuing.

The Report trace is evaluated next in Figure 9. Similar to figure 8, Track Piggybacking provides the best response times, about 39% higher than Normal Mode at the beginning of the trace. It is followed closely by Block Piggybacking, which has up to 6% higher response times than Track Piggybacking. This time, however, Redirection of Reads also gives low response times since enough of the spare disk is rebuilt by the time the disk array is saturated at 400 seconds that many of the requests to the failed disk can be redirected to the spare disk. Response times for Redirection of Reads are just over twice Normal Mode response times but are quickly lowered until they are about 40% higher. Track Piggybacking response times are 38% to 93% lower than Redirection of Reads response times. The rebuild time for Track Piggybacking is 802 seconds, followed by Block Piggybacking at 868 seconds, Redirection of Reads at 1037 seconds and Minimal Update at 2120 seconds.

8 Conclusions and Future Work

We have evaluated rebuild times and response times for several rebuild algorithms and have proposed a new scheme, Track Piggybacking, for reducing the rebuild time for a RAID5 disk array while minimizing increased response times. When a block from the failed disk is accessed and that block has not been rebuilt to the spare disk, Track Piggybacking reads the corresponding tracks of data from each of the surviving disks in order to reconstruct the track on the failed disk that contains this block, and writes this track to the spare disk. Track Piggybacking consistently provides better rebuild times compared to other rebuild algorithms. Rebuild time is reduced by up to 29% over Redirection of Reads for a workload of uniformly distributed requests.

For heavy workloads, Track Piggybacking provides both higher and lower response times during a shorter rebuild process compared to other rebuild algorithms. The drawback is a 4% to 15% increase in the overall response time during the short rebuild time. We have shown that Track Piggybacking is a technique for efficiently using the spare disk during the rebuild process. It provides a good tradeoff between reduced rebuild times and increased application response times.

This paper also evaluated two disk traces that exhibited some data locality. The first time a request to data on the failed disk is serviced, that data is reconstructed to the spare disk via piggybacking. Subsequent accesses to this data and data on the same track can now be redirected to the spare disk. Thus frequently accessed data is automatically reconstructed to the spare disk under Track Piggybacking and, except for the first time that data is accessed, will always have good response times. Track Piggybacking provides good response times overall, generally less than 50% higher than Normal Mode and 38% to 93% lower than Redirection of Reads. More importantly, it provides good response times for frequently accessed data. At the same time, Track Piggybacking still provides low rebuild times, 23% and 25% lower than Redirection of Reads. For workloads that have some locality, Track Piggybacking provides both lower rebuild times and better response times than Redirection of Reads.

Track Piggybacking is a generic rebuild algorithm that can be applied to other disk arrays. In particular, mirrored arrays [Bitt88] and declustered parity arrays [Holl92] can benefit from the reduced rebuild times offered by Track Piggybacking. In addition, these two architectures have inherent mechanisms that can reduce response times, providing a balance for those cases where Track Piggybacking increases response time.

Track Piggybacking needs further investigation if it is to be used for all workloads. For instance, the increase in average overall response time needs to be addressed. One method for reducing response times is invoking Track Piggybacking only when the disk array can support the extra work. Track Piggybacking will then provide response times similar to Redirection of Reads. Of course, if the disk array is saturated, then piggybacking will not be invoked and Track Piggybacking will provide the same rebuild times as Redirection of Reads. Another possibility is rebuilding a partial track, such as half the track, instead of a full track each time. This may give some of the rebuild time advantages of Track Piggybacking while maintaining response times closer to Redirection of Reads.

Acknowledgements

This work is part of a larger research project in I/O being carried out at the University of Michigan. We thank Jim Pike of AT&T/GIS for providing the initial funding for our research. We thank Richie Lary of Digital Equipment for graciously providing disk traces for our experiments. We also wish to thank Mark Campbell, Roy Clark, Jim Gray, David Jaffe, Richard Mattson, Jai Menon, Spencer Ng, John Wilkes, and the members of our research group at Michigan, particularly Greg Ganger and Bruce Worthington, for all the technical discussions we have had on the various I/O issues.

Finally, our research group is very fortunate to have the financial and technical support of several industrial partners. We are pleased to acknowledge them. They include AT&T/GIS, Intel, Motorola, Scientific and Engineering Software, HaL, Hewlett-Packard and DEC.

References

[Bate85] K.H. Bates, M. TeGrotenhuis, "Shadowing Boosts System Reliability", *Computer Design*, April 1985, pp. 129-137.

[Bitt88] D. Bitton, J. Gray, "Disk Shadowing", *Proceedings of the Very Large Databases Conference*, September 1988, pp. 331-338.

[Gray90] J. Gray, B. Horst, M. Walker, "Parity Striping of Disk Arrays: Low-Cost Reliable Storage with Acceptable Throughput", *Proceedings of the 16th VLDB Conference*, August 1990, pp. 148-161.

[Gray91] J. Gray, D. Siewiorek, "High-Availability Computer Systems", *IEEE Computer*, September 91, pp. 39-48.

[Holl92] M. Holland, G. Gibson, "Parity Declustering for Continuous Operation in Redundant Disk Arrays", *Architectural Support for Programming Languages and Operating Systems*, 1992, pp. 23-35.

[Hou93a] R. Hou, Y. Patt, "Trading Disk Capacity for Performance", *Proceedings of the Second International Symposium on High-Performance Distributed Computing*, 1993, pp. 263-270.

[Hou93b] R. Hou, Y. Patt, "Comparing Rebuild Algorithms for Mirrored and RAID5 Disk Arrays", *ACM SIGMOD*, 1993, pp. 317-326.

[Kim86] M. Kim, "Synchronized Disk Interleaving", *IEEE Transactions on Computers*, November 1986, pp. 978-988.

[Livn87] M. Livny, S. Khoshafian, H. Boral, "Multi-Disk Management Algorithms", *SIGMETRICS*, 1987, pp. 69-77.

[Meno92] J. Menon, D. Mattson, "Performance of Disk Arrays in Transaction Processing Environments", *12th International Conference on Distributed Computing Systems*, 1992, pp. 302-309.

[Munt90] R. Muntz, J. Lui, "Performance Analysis of Disk Arrays Under Failure", *Proceedings of the Very Large Databases Conference*, 1990, pp. 162-173.

[Ng88] S. Ng, D. Lang, R. Selinger, "Trade-offs Between Devices and Paths In Achieving Disk Interleaving", *Proceedings of the 15th International Symposium on Computer Architecture*, 1988, pp. 196-201.

[Olso89] T. Olson, "Disk Array Performance in a Random IO Environment", *ACM Computer Architecture News*, September 1989, pp. 71-77.

[Patt88] D. Patterson, G. Gibson, R. Katz, "A Case for Redundant Arrays of Inexpensive Disks (RAID)", *ACM SIGMOD*, May 1988, pp. 109-116.

[Rama92] K.K. Ramakrishnan, P. Biswas, R. Karedla, "Analysis of File I/O Traces in Commercial Computing Environments", *ACM SIGMETRICS*, pp. 78-90.

[Redd89] A.L.N. Reddy, P. Banerjee, "An Evaluation of Multiple-Disk I/O Systems", *IEEE Transactions on Computers*, December 1989, pp. 1680-1690.

[Tera85] Teradata Corporation. "DBC/1012 Database Computer System Manual Release 2.0", Document No. C10-0001-02, November, 1985.

Dual-Crosshatch Disk Array:
A highly reliable hybrid-RAID architecture

Sunil K. Mishra, Sudheer K. Vemulapalli and Prasant Mohapatra
Department of Electrical and Computer Engineering
Iowa State University, Ames, Iowa 50011
Email: prasant@iastate.edu

Abstract —*In this paper, we propose a highly reliable RAID architecture called a Dual-Crosshatch Disk Array. It uses the proposed interleaved 2d-parity scheme, a low overhead triple-erasure correcting parity organization. It is a hybrid approach of RAID-4 and RAID-5 with one dedicated parity group and another parity group using block interleaved data and parity. The results obtained from simulations indicate that this architecture possesses extremely high reliability with low overheads, good degraded performance, and acceptable normal-mode performance.*

1 Introduction

To keep pace with the rapidly increasing processing power, disk array architectures have been proposed that can achieve high I/O capacity and performance. Disk array storage systems, such as Redundant Array of Inexpensive Disks (RAID) level-5 data organization [1], provide fault tolerance against disk drive failures, and are cost-effective and have good run-time performance. However, a storage subsystem consists of more than just disk drives. There are controllers for interfacing with the disk drives, cabling to provide data/control paths, power supplies, etc. For the disk array to be fault tolerant it must be able to tolerate failures in any of these components.

There are several known architectures for building disk array subsystems that are tolerant to failures in support hardware components and the disk drives [2]. These architectures employ parity protection, as in RAID-5, with block interleaved data and distributed parity, and are single-erasure tolerant. Such arrays have acceptable mean time to data loss (MTTDL) when the number of disks in the subsystem is small. However, the average number of disks in an installation is growing because of decreasing form factors and increase in the new forms of data needing massive storage capacity like audio and video for multimedia applications. It is projected that by year 2000, average commercial installation will need about 10 TB or more storage capacity [5]. To meet such large storage demand, average installations will need about 5,000 to 50,000 disks. Traditional arrays which can protect from concurrent failure of no more than one disk per parity group will have inadequate reliability for such large storage requirement.

In this paper we present a highly reliable and robust disk array architecture, the *Dual-Crosshatch Disk Array* (DCDA), that is capable of tolerating any three disk failures with minimum number of redundant disks, and any five controller failures. The DCDA uses a novel and efficient parity scheme, the *interleaved 2d-parity*, a variant of the 2d-parity scheme [3]. It is a

hybrid approach of RAID-4 and RAID-5 in the sense that one of the parity groups uses block interleaved data and stripped parity while the other uses dedicated parity disks, and hence the name *hybrid-RAID* architecture. The DCDA architecture has extremely high reliability with low check disk overhead, faster data recovery, good degraded-mode performance, and acceptable normal-mode performance making it an attractive solution for designing large disk arrays. The simulation results indicate that the DCDA architecture is order of magnitude more reliable than the crosshatch disk array which is shown to have higher availability than all the other existing disk array organizations. Furthermore, the MTTDL of DCDA is about 10^4 times more than that of the crosshatch disk array. The difference will be even more for the typical values of failure and repair rates.

The rest of the paper is organized as follows. In section 2, we review the advantages and disadvantages of the existing disk array architectures/organizations. In section 3 we describe the DCDA architecture. In section 4, we present the results followed by the conclusions in section 5.

2 Existing Disk Array Organizations

2.1 Single-Erasure Tolerant Architectures

A single-fault tolerant disk array, such as RAID-4 and RAID-5 architectures, provide data accessibility in the presence of any single-failure within the system. In these disk arrays data redundancy is obtained using the parity information for a group of disks, and parity is maintained on a check disk. When a disk fails, the data can be reconstructed by XOR-ing the data on other functional disks from that parity group. The RAID-4 organization uses block interleaved data and dedicated parity. In RAID-5 the parity information is also stripped among all the disks in a parity group along with the data. Stripping of parity results in better write performance [1].

Along with the disk drives, a disk array subsystem must also be tolerant to failures in the support components. Among the existing architectures that tolerate the failure of support hardware, the most prominent ones are Single Path Horizontal Array, Dual Path Vertical array, Dual Path Horizontal Array and Crosshatch Disk Array [2].

In the crosshatch disk array, the disk drives are dual ported and each disk drive is a member of two strings - a horizontal string and a vertical string, and the combination is unique for each drive. Crosshatch refers to the pathing topology of the array architecture. This architecture is tolerant to a single disk failure per parity group. As each drive can be accessed through two independent controllers, a single controller failure will not affect the data availability from any disk. Any double controller failure can render at most one disk inaccessible. Therefore, after the controllers are repaired, at most one disk needs to be rebuilt. Hence the repair time is less resulting in greater MTTDL.

2.2 Multiple-Erasure Correcting Codes

In order to maintain data integrity under more than one disk failure, more that one redundant disks are required. According to the coding theory, C concurrent disk failures can be tolerated using at least C redundant check disks [3]. Gibson et al. have presented the 1d-parity, 2d-parity, additive-3, and in general, the multidimensional parity scheme [3]. The 2d-parity, a double-erasure correcting code, can tolerate all sets of 3-erasures except the *bad 3-erasures*, and additive-3 code can correct all sets of 4-erasures except *bad 4-erasures* [3]. EVENODD encoding scheme, proposed by Blaum et al. [4], uses a special encoding scheme to store parity information, and can tolerate any 2 disk failures. Burkhard et al. have proposed maximum distance separable (MDS) codes capable of tolerating two or three concurrent disk failures using as many check disks [5].

3 Dual-Crosshatch Disk Array

Most of the existing architectures can tolerate only one disk drive failure per parity group. With the increase in the average number of disks in an installation, these traditional disk arrays may prove to be unreliable. In this section we describe the *Dual-Crosshatch Disk Array* architecture which can tolerate more concurrent disk drive and controller failures than any of the existing architectures.

3.1 The Interleaved 2d-Parity

In case of 2d-parity, parity is computed both along the rows and columns, and is stored in a check disk at the end of each row and column as shown in Fig. 1.a. In the proposed *interleaved 2d-parity* organization, the horizontal parity groups use block interleaved data and parity, unlike in the 2d-parity. The vertical parity groups use dedicated parity disks as in 2d-parity. As a result of parity stripping in the horizontal parity groups an extra check disk is required, compared to 2d-parity, for storing the vertical parity information. Fig. 1.b shows the interleaved 2d-parity scheme. A novelty of this parity scheme is that it can tolerate more number of concurrent failures than the number of redundant check disks required per parity group.

3.1.1 Failure Recovery – Under a single disk failure, data recovery is possible using either horizontal or the vertical parity group. For any double disk failure in any single parity group data can be recovered using the two orthogonal parity groups associated with the failed disks. However, when three disks fail, say disk 5, *P1*, and 9 as in Fig. 1.b, the data for disk 5 cannot be recovered using any single parity group. Data pertaining to disk *P1* and 9 can be reconstructed from the associated vertical and horizontal parity groups respectively; then the data on disk 5 can be reconstructed. If each parity group contains G data disks

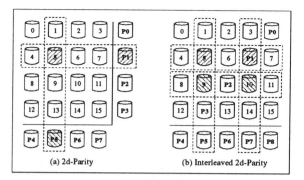

Figure 1: 2d-parity and Interleaved 2d-parity schemes

and a check disk, a total of $3G - 1$ disk accesses is required to construct the data on the three failed disks. Hence, the interleaved 2d-parity scheme is 3-erasure correcting.

The interleaved 2d-parity scheme can also tolerate all sets of 4-erasures except the *bad 4-erasure* as shown by shaded disks in Fig.1.b. The bad 4-erasures are unrecoverable as two disks fail in each of the four parity groups. With this parity organization, there is no chance of data loss for any number of disk failures until a bad 4-erasure occurs. For an NxN disk array using interleaved 2d-parity, a maximum of $2N-1$ disks can fail without the occurrence of a bad 4-erasure. The probability of occurrence of a bad 4-erasure is very low, and hence this scheme provides very large MTTDL.

3.1.2 Overhead – The *check disk overhead* is defined as the ratio of the number of check disks to information disks. For c check bits, the interleaved 2d-parity code can have a maximum of $(c-1)^2/4$ information bits as compared to the $c^2/4$ information bits for the 2d-parity code. In a disk array having G information disks per parity group and n such parity groups (we assume $n \leq G$), total of $n + G + 1$ check disks are required for the interleaved 2d-parity, whereas the 2d-parity needs $n+G$ check disks. When n is equal to G, the number of check disks required is $2G + 1$. The check disk overhead for this scheme is $(2G + 1)/G^2$ as compared to $2/G$ in case of 2d-parity. As G is more than 1, less than 3 redundant check disks are required (per data group) for this 3-erasure correcting code. The proposed parity organization is thus optimal in terms of check disk overhead.

3.1.3 Update Penalty – The *update penalty* [3] for a disk array is the number of check disk updates required for any write to an information disk. The minimum update penalty for any t-erasure correcting code is t [3]. For the proposed parity organization, any write to a disk needs updating the parity information in both the parity disks (in case of horizontal parity group, this is the disk containing parity for that block of data) and the dedicated parity disk corresponding to the vertical parity group. With reference to Fig.1.b, a write to disk 5 involves updating parity information in disk *P1*, *P5* and *P7*. Thus the proposed scheme has the update penalty of 3, and this is optimal for a triple-erasure-correcting code. Several schemes have been proposed to improve the write performance of RAID-5 [6], and can also be employed in this scheme to improve the performance. The performance issues are discussed in Section 3.3.

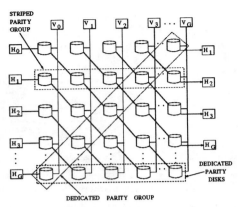

Figure 2: Dual-Crosshatch Disk Array.

3.2 The DCDA Architecture

Dual-Crosshatch Disk Array can be formally defined as an array in which the disk drives are dual ported and each disk is a member of two controller strings and two parity groups, and the combination is unique for each disk drive as shown in Fig. 2. The two controller strings are – (a) horizontal controller string and (b) vertical controller string. The two parity groups are – (a) horizontal-stripped parity group and (b) diagonal-dedicated parity group. The DCDA architecture uses the interleaved 2d-parity scheme, but instead of a vertical parity group we chose to use a diagonal parity to insure better performance. The DCDA can tolerate any three disk failures and all sets of 4-disks failures except the bad 4-erasures. As stated before, *crosshatch* refers to the pathing topology of the architecture. This architecture is referred to as *dual-crosshatch* because of the presence of additional diagonal-dedicated parity groups.

A single controller failure does not affect data availability from any disk. Under a vertical controller failure, access to at most one disk in each of the horizontal parity groups is affected. When a horizontal controller fails, disk access to at most two disks in a diagonal parity groups is affected. However, if we use a vertical parity group instead of diagonal parity group, in the case of vertical controller failure, access to all disks in that parity group is affected. All the parity disks of the diagonal parity groups are located in the last row of the disk array. This has the advantage that each of these disks are connected to two independent controllers. Since, parity information needs to be updated in these disks for a write to any data disk, two sets of controllers provide some degree of redundancy in the controller path. Any two controller failures renders at most one disk inaccessible in both the DCDA and the crosshatch diagonal disk array. Hence, after the controllers are repaired, only one disk needs to be rebuilt.

The DCDA architecture can tolerate any five controller failures as compared to triple controller failure tolerance of the crosshatch diagonal disk array. It can tolerate most of the six controller failures except for a few cases which cause bad 4-erasures. Hence, this architecture essentially eliminates the controller failure

as a contributor to the unavailability of the storage subsystem.

Assuming that G^2 information disks are organized in a square array of side G, then the DCDA organization needs $2G + 1$ redundant disks as compared to $2G$ redundant disks for EVENODD and the 2-erasure correcting MDS codes. For example, for a total of 1024 disks organized in a 32 x 32 array, 63 check disks would be required; so the check disk overhead is about 6.5%. The check disk overhead for the DCDA is nearly same as that of the EVENODD encoding scheme for large disk arrays, but the former can tolerate triple-erasures as compared to the double-erasure tolerance of the EVENODD parity scheme.

Considering a single disk failure with all the controllers functional – one set of controllers can access the functional disks of the horizontal parity group and the other set of controllers can access the functional disks of the vertical parity group. By using both the parity groups to reconstruct different blocks of data in parallel, data reconstruction time can be reduced. A shorter repair time decreases the probability of second disk failures and increases the MTTDL of the system.

The major advantage of the DCDA is that it is tolerant to more concurrent controller and disk drive failures than the other prior-art architectures. For example DCDA can tolerate any three disks and any one controller failures, or any disk and any five controller failures. Hence the DCDA is a better fault tolerant system as a whole than the existing architectures.

3.3 Performance Issues of DCDA

The performance of DCDA is comparable to that of crosshatch disk array except for the cases of small write and small read-modify-writes (RMW). Under normal-working-mode, a read request for a large chunk of data distributed over two or more parity groups can be handled in parallel using the two set of controller strings. Similarly, two different read requests for data spread over different parity groups can be handled in parallel. So, the read performance of the DCDA is nearly twice that of the RAID organizations.

Since DCDA uses the interleaved 2d-parity scheme, the update penalty for a write operation is 3. A small write needs four RMW accesses to the disk containing data and the three related parity disks. If both the controller strings are free, they can be used to write to the horizontal parity group and the dedicated parity disks in parallel. For example, a write to disk 5 in Fig. 1.b needs updating parity information on disk P1, P5, and P7. In the DCDA architecture, the horizontal controller string can be used to write to disk 5 and P1, while the other set of controllers can write to disks P5 and P7. For large write requests use of both the controller string ensures nearly same performance as RAID-5. So this high update penalty is partially nullified by the extra set of controllers. But when a controller string is busy, or under a few controller failures, the above assumption does not hold good, and small write and RMW performance may not be acceptable. We suggest the use of data cache in the controllers to improve the write and RMW performance. A number of schemes have been proposed for performance improvement using cache memory and other ways [6]. This architecture also has better degraded-mode performance characteristics as data on the failed disks can be reconstructed faster and made available. Hence this architecture has better throughput, in degraded mode, as compared to the other existing architectures.

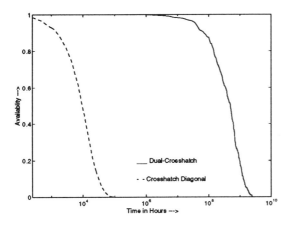

Figure 3: Availability vs Time for a 1024 disk array.

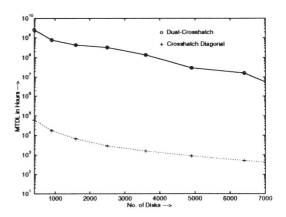

Figure 4: MTTDL vs Storage Capacity.

4 Results

The results presented in this section are based on event driven simulations of the different disk array architectures. We assume that the mean time to failure of the individual disks and controllers are independent and exponentially distributed. Typical value for MTTF of a disk drive is in the range of 50,000 to 400,000 hours and the corresponding MTTR with a hot spare is 1 to 2 hours. The corresponding values for the controllers (including the cable link) are 100,000 hours and 20 hours. With the typical values, the MTTDL of the DCDA is extremely high making it difficult to run the simulations for such large values of time. So we have considered the following values – MTTF and MTTR for disks to be 50,000 hours and 5 hours respectively, and those values for the controllers are 50,000 hours and 20 hours respectively. As we are comparing the availability of the two disk array architectures using these values gives an idea of the relative improvements.

In Fig. 3, we show the availability of the system with respect to time. A disk array consisting of 1024 disks arranged in a 32 x 32 array is considered here. As can be seen from the plot the availability of the DCDA architecture is very high compared to the crosshatch diagonal disk array. This is because the crosshatch disk array uses a single-erasure correcting code whereas the DCDA uses a triple-erasure correcting code. Furthermore, the occurance of the bad 4-erasure cases that result in data loss is very rare.

Fig. 4 shows the variation of MTTDL with storage capacity of both crosshatch diagonal and DCDA architectures. All the failure and repair rates used are same as those for Fig. 3. As can be seen from the plot, the reliability of both architectures decreases as the disk arrays are scaled up in dimension. But the MTTDL of the DCDA is more than that of the crosshatch diagonal disk array by a factor of around 45,000 for a disk array consisting of 1024 disks with the used parameters. This factor is even larger for the typical values of failure and repair rates as mentioned earlier in this section.

5 Conclusions

In this paper, the *Dual-Crosshatch Disk Array*, a new architectural alternative for configuring a redundant disk array is presented. Compared to the prior known array architectures, the DCDA provides higher fault tolerance, higher availability, better degraded-mode performance, and faster reconstruction. Since the DCDA uses the interleaved 2d-parity encoding scheme, the check disk overhead is almost same as that of the EVENODD and other double-erasure correcting schemes, but the DCDA can tolerate any triple-erasure. The array is fault tolerant to any five controller failures. Hence the DCDA is a robust and highly reliable storage subsystem.

References

[1] D.A. Patterson, G.A. Gibson, and R.H. Katz, "A Case for Redundant Arrays of Inexpensive Disks (RAID)," *ACM SIGMOD Intl. Conf. on Data Management*, pp. 109-116, 1988.

[2] S.W. Ng, "Crosshatch Disk Array for Improved Reliability and Performance," *Intl. Symposium on Computer Architecture*, pp. 255-264, 1994.

[3] Gibson et al., "Failure Correction Techniques for Large Disk Arrays," *Third Intl. Conf. on Architectural Support for Programming Language and Operating Systems*, pp. 123-132, April 1989.

[4] M. Blaum et al., "EVENODD: An Optimal Scheme for Tolerating Double Disk Failures in RAID Architectures," *Annual Intl. Symposium on Computer Architecture*, pp. 245-254, 1994.

[5] W.A. Burkhard, and J. Menon, "Disk Array Storage System Reliability," *Annual Intl. Symposium on Fault Tolerant Computing*, pp. 432-441, 1993.

[6] P.M. Chen et al., "RAID: High-Performance, Reliable Secondary Storage," *ACM Computing Surveys*, Vol. 26, No. 2, pp. 145-185, June 1994.

THE APPLICATION OF SKEWED-ASSOCIATIVE MEMORIES TO CACHE ONLY MEMORY ARCHITECTURES[†]

Henk L. Muller[‡]   Paul W.A. Stallard   David H.D. Warren

Department of Computer Science, University of Bristol, UK.

Working at: PACT, 10 Priory Road, Bristol. BS8 1TU, UK.

email: {henkm,paul,warren}@pact.srf.ac.uk

Abstract— *Skewed-associative caches use several hash functions to reduce collisions in caches without increasing the associativity. This technique can increase the hit ratio of a cache without significantly increasing the cost. In this paper we apply skewing to solve a problem particular to virtual shared memory architectures that are designed using only associative memories. When using ordinary set-associative memories the replication of data amongst many nodes leads to a reduced storage capacity in a specific set. Using skewing can alleviate this problem. Our results show improvement in the majority of cases.*

1 INTRODUCTION

Associative memories can contain arbitrary sets of data because the address of each datum is stored along with the data itself. These structures are particularly useful when a memory needs to be able to store a selection of data from a much larger address space, for example in a cache. Unfortunately, the flexibility of associative memory requires that each lookup searches the entire memory. Large associative memories are impractical for reasons of both speed and cost, leading to the design of set-associative memories. Such a memory comprises a number of *banks*, each consisting of an array of *slots*. When looking up a data item, a slot is selected within each bank by applying some hash function to the address of the data. From this collection of slots, known as a *set*, an associative selection is made. Each data item can reside in only one of the sets, but may reside in any slot within that set. Because the associativity is limited, a set-associative memory is simpler and cheaper to build.

The limited associativity means that set-associative memories do not perform as well as fully associative memories. When a set is full, no more items can be stored in that set, regardless of whether the rest of the memory is full or not. This problem manifests itself in multiprocessor virtual shared memory architectures, such as the DDM [4], KSR-1 [5] and COMA-F [6], that use (set-)associative memories as the sole storage of data. The designs are also known as

Cache Only Memory Architectures, or COMAs. If one data item needs to be replicated, it means that there is less room for other data in that set. Our initial solution to this problem was to use different hash functions on each processor. Each memory was therefore still set-associative, but the problem caused by the replication of data was removed.

The limited associativity also affects the caches of single processor architectures because there are restrictions on which working sets can be cached. The result is a low hit ratio for caches with low associativity. Seznec has proposed an alternative to set-associative caches known as *skewed-associative caches*, which use several hash functions to index the memory [1]. For only a small increase in complexity, *skewing* provides a closer approximation to fully associative memory than plain set-associative memory. The number of collisions between pieces of data that are frequently used is reduced and the hit ratio is increased. In this article we propose to use Seznec's skewing in multiprocessors that solely use (set-)associative memories. In addition to the benefits shown by Seznec, skewing also overcomes the the data replication restriction in COMAs. This means that we can run larger programs on the same machine before the memory is exhausted.

In Section 2 we give a description of skewing and the context in which we apply it. In Sections 3 and Section 4 we evaluate skewing, first under a worst case load, and then with real application programs.

2 SKEWING

Suppose that there are three data items, X, Y and Z, that are very frequently used and that hash onto the same set of a two-way set-associative cache. Only two can be stored in the cache; the third will be loaded on demand, evicting one of the other two. A skewed cache [1–3], uses a different hash function for each bank of the cache as depicted in Figure 1. The three data items may then clash in one bank but not in the other bank. Hence, all three data items can be stored in the cache at the same time: X in one bank, Y and Z in the other. This gives more freedom to how the data can be stored, and can consequently achieve a better hit ratio than a conventional cache. This is comparable to

---

[†]Work supported by ESPRIT OMI/HORN P7249.

[‡]Authors are ordered alphabetically.

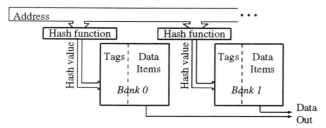

Figure 1: A skewed-associative memory.

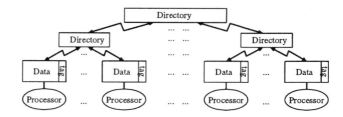

Figure 2: Architectures with only associative memories.

increasing the associativity, but at a lower cost.

The increased hit ratio (from the efficient storage of more varied working sets on each node) is one of the reasons why it is interesting to use skewing in an architecture that completely relies on (set-)associative memories. In addition, skewing also overcomes the data replication restriction that is apparent in these machines. Before discussing this loss of capacity and the solution in more detail we first discuss the class of architectures that is considered.

Cache Only Memory Architectures

As COMAs are built using only associative memories, they have a global address space but no fixed relation between the address of a data item and the node where it is stored. A data item can reside on any node, as long as the node has room to store it. The KSR-1 and DDM have a hierarchy of associative directories (see Figure 2) that store information about the data held in the branches below them. The COMA-F has a single directory and no hierarchy. The COMA-F and DDM implement the directories in a distributed fashion, in order to reduce contention. In all three architectures data can be shared and reside in the memories of each reading processor. The directories are used to locate a copy of an item, or to invalidate all other copies of an item (for writing).

When a data item is to be placed in a memory that is full, room must be made by evacuating one of the data items. In our implementation of the DDM, the item to be replaced is *inject*ed up into the hierarchy. As soon as an inject finds that another copy of the data exists, the inject is annihilated. If the inject carries the only copy in existence it will probe nodes until it finds a home. We have chosen a strategy that makes a breadth first search for an empty or shared item to be replaced.

The associative memories

The use of *set*-associative memories restricts the way data can be stored and shared in a way that a fully associative memory does not. Consider an example of a machine with a number of nodes, each with a 4-way plain set-associative memory. As there is only one hash function in the whole machine, each datum can be stored anywhere within exactly one set on each node. We call this collection of eligible sets across the machine the *global set*. Now assume one particular global set is 75% full with unreplicated items. If the application now wishes to share one of these data items, then this global set becomes full and no further items of that set can be shared. This is the case even if the rest of the memory is empty because any particular item is restricted to reside only in one particular global set.

It is not uncommon for an application to replicate data in a large number of nodes. As an example, synchronisation primitives such as locks and barriers are often implemented in software in such a way that the waiting processor spins on the lock location. The technique works on the assumption that the lock will be cached locally so the spinning action does not cause excessive network traffic. If the machine cannot keep a copy of the lock locally, the spinning processor has to fetch a copy from a remote node. The lock will then in turn displace some other item, which will have to find a home somewhere else.

Skewing can solve this problem. When the machine uses two (carefully chosen) hash functions, there is effectively one universal pool of memory, instead of s separate pools (where s is the number of sets in each associative memory). The choice of the hash functions is important, in that they must be *disjoint*. If the hash functions are H and G, and H hashes two addresses X and Y to the same value, then G must hash X and Y to different values. This can always be achieved provided that the number of addresses is less than the square of the number of hash values. If the hash values are for example 18 bits (with 64-byte items and 2-way associativity this gives 32 Mbytes memory per node), the item addresses must be restricted to 36 bits (4 T*bytes*).

Skewing allows specific items to be replicated, although the machine might need some time to settle while all the clashing items find a home. The number of hash functions required varies from one (for plain set-associativity), to ap (where a is the associativity and p is the number of processors) where each bank on each processor has its own hash function. Intermediate schemes can also be devised. For example, we could use the same two hash functions for the two way set-associative memories of all the odd nodes and another two functions for all the even nodes. In our evaluations we have compared three skewing schemes (with 2, 4, and $2p$ hash functions) with 2- and 4-way plain set-associative memories under worst case loads, and 2-way skewing versus 2-way plain for real applications.

The hash functions

Seznec gives criteria for a good set of hash functions for skewing [1]. These are: equitability (there are the same number of items competing for each slot); inter-bank dispersion (if two items clash in one bank, they do not clash in another); local dispersion in a single bank (because of spatial locality, a hash function should not map two near-neighbour items to the same location); and simple hardware implementation. Along with these criteria, Seznec also gives four functions that have the desired properties. For our experiments we needed a larger and extensible family of hash functions. We have chosen to use Tausworthe generators [7], functions normally used in cryptographic applications or random number generators. These functions satisfy all the above operational criteria but are not necessarily suitable for hardware implementation. We will see later which of these functions can be implemented cheaply.

A Tausworthe generator generates a series of bits. A new bit is added to the series by taking the exclusive-or of two or more bits from the last n bits of the series. When the bits are selected carefully, the generator can be applied $2^n - 1$ times before it repeats itself. Below we denote these Tausworthe functions with $T_n(x)$. T_n takes an n-bit string x as its parameter and returns an n-bit string consisting of $n - 1$ bits of x and one newly generated bit.

Different functions are obtained by repeatedly applying T_n (denoted $T_n^2, T_n^3, ..., T_n^k$). After taking a modulus with the number of sets, to give a set selection, one has a suitable family of hash functions: $h_{n,s,k} = T_n^k \bmod s, 0 \leq k < 2^n - 1$. Because T_n shifts only a single bit, two hash functions $h_{n,s,k}$ and $h_{n,s,k+1}$ do not show very good inter-bank dispersion [1]. This dispersion is improved when T_n is applied at least $\log_2 s$ times in a row. We therefore only use the functions $h_{n,s,0}, h_{n,s,\log_2 s}, h_{n,s,2\log_2 s}, \ldots$.

As said, not all of these functions can be implemented in hardware easily. If we use q to denote the number of the bit to be used in the exclusive-or, where $\frac{n}{2} \leq q < n$ so that $x^n + x^q + 1$ is a primitive polynomial [7], then only the functions $T_n^k, k < q$ can be implemented with one level of exclusive-or gates. In general one needs $\frac{k}{q}$ levels of exclusive-or gates. Although it is clearly not feasible to use the functions with high values of k in a hardware implementation, they are useful to study whether using many different hash functions is advantageous.

Evaluation strategy

The evaluation is performed in two stages. First we use a simple simulator that assumes random data distribution to determine worst case behaviour. Secondly we use an implementation of skewing in a detailed emulator of the DDM to show how real programs behave on an architecture with skewed-associative memories.

3 EVALUATING THE WORST CASE

The worst case is evaluated with a simulator, which uses a very simple model of a 16-node machine. Only the memory and the directory of each node are modelled. The network, processor and application program are abstracted away. To create a reasonable worst case, we assume that data is initially distributed without any replication. A fraction of the memory is filled with unreplicated data items and the rest is empty. The simulator then attempts to replicate an item over all nodes in the system. Each operation will, in general, cause other items to be evicted and these will have to find new homes. As the network is not modelled properly, the simulator selects a random node to be the new home. As the machine becomes full, finding homes for data becomes difficult and the machine will require more time to reach a stable condition. Eventually, the machine will be unable to find a home for certain data and the simulator will report a failure. If a steady state is found, each node tries to access its original data set to simulate the working sets being restored. This operation will either succeed, in which case the simulator reports how many items have been evicted during this process, or a stable situation (with the new data item inserted and all the original data restored to the node on which it was initially located) will not exist in which case the simulator reports a failure.

Five different hashing strategies are compared: 2- and 4-way plain set associativity, and three skewed schemes with 2, 4, and 32 hash functions. We create a worst case situation by only inserting an item in the machine if the hash value of its address clashes with the hash value of the item that is going to be replicated. In this way, we create a situation with as many collisions as possible. This requires a larger address space than the total amount of physical memory (to allow sufficient clashing items). In Figure 3, the number of evictions needed to reach the stable state is shown as a function of the fraction of the memory that has been filled with unreplicated items.

When using a skewed schemes, up to 60% of the memory can be filled with distinct items. There is an advantage in using four different hash functions instead of two, but the step from four hash functions to 32 does not show much improvement. The plain schemes give up quickly. With a-way set-associative memories, no more than $16(a - 1)$ distinct items (or $p(a - 1)$ for a machine with p processors) can be stored while still being able to replicate one item over the entire machine. This is a negligible fraction of the total memory. For a machine with skewed-associative memories almost the entire machine can be filled with items that might collide with one of the hash functions.

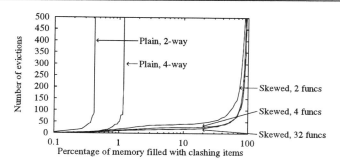

Figure 3: Worst case working set, 16 processors.

4 EVALUATING REAL PROGRAMS

The simulator is very simple and demonstrates the basic properties of a machine with skewed-associative memories and illustrates the effects of using multiple hash functions. For a more accurate evaluation, we have also implemented skewing in an emulator of the DDM [8]. This emulator is a complete DDM implementation, in software, on a parallel T800 transputer platform. For both plain and skewing we use two-way associative memories. Because of the small differences between the skewing schemes found in our simulation studies, only a simple skewing scheme with two hash functions per node ($h_{23,s,0}, h_{23,s,\log_2 s}$) has been tested on the emulator. These hashing functions are feasible in hardware.

Our set of benchmarks consists of four programs from the SPLASH-I suite (Water, Barnes, Mp3d and Cholesky); two parallel Prolog systems (Aurora and Andorra-I); three other numeric benchmarks, Wave, Wang and Nucleic; and one artificial program that is written to thrash the memory. This last program allocates a number of items that would clash in a plain set-associative memory and then starts accessing this data on various nodes to model the effect of replication. In a plain set-associative memory, this replication will always cause data to be replaced. With skewed-associative memories, the system can stabilise with all the working sets restored.

All programs are run with 64-byte items. The size of the DDM memory is restricted to the amount allocated by the application, rounded up to a power of two (our memory size must be a power of 2). Restricting the memory size to the amount allocated is a debatable choice as it may leave very little freedom for replication. It would be better to give each application what it allocates plus what it requires for sharing, but the sharing requirements are typically very dynamic and cannot be easily determined. We could, at any moment in time, measure how much data is replicated, but this would not tell us how much is usefully replicated, and how much is no longer required. The decision to give only sufficient space for the allocated data will result in limited free space and will help to demonstrate the effect of skewed-associativity. It should be noted that, with plain set-associativity, only one of these programs runs significantly slower in this amount of memory than with more memory.

Table 1: Thousands of messages sent.

Program and parameters	Memory, Kbytes		K Messages		Reduc-
	Needed	Present	Plain	Skew	tion
wave.240	1820	2048	271	855	-216%
wang.100000	4690	8192	725	721	0%
nucleic.p	251	256	705	698	1%
andorra.flypan	9968	16384	666	622	7%
mp3d.30000	1536	2048	11920	11091	7%
water.343	232	256	24052	21799	9%
aurora.psite	2290	4096	249	223	10%
cholesky.bcsstk14	1536	2048	7330	6393	13%
barnes.1000	728	1024	4799	4056	15%
thrashmemory.1	15361	16384	5356	73	99%

The results of running our benchmark programs with and without skewing are summarised in Table 1. For each application, the table shows the number of Kbytes allocated by the program, the total memory available on the machine, the number of messages sent with and without skewing and the reduction. The results were obtained on an 8-node configuration with a single level hierarchy, so the 8 directories at the leaf nodes communicate directly with the top nodes.

The number of messages generated during the execution of an application was chosen as a measure of the effectiveness of the memory system. If skewed-associative memory is better than plain set-associative memory, there will be more freedom in data distribution, the memory will be better utilised, and less messages need to be sent to fetch items on demand. Besides the number of messages, one could measure the execution time, or the amount of memory used.

We have explained earlier that it is hard to measure memory usage. The execution time is hardly affected in most programs for two reasons. Firstly, many of the messages are saved when the program is performing a synchronisation and would therefore be waiting anyway. Secondly, we have a multi-threaded execution model with four threads running on each node so latencies of extra messages can be hidden. Still, the reduction of the number of messages is very important because, on a multi-tasking machine, the network resources could be utilised by another application.

With a single exception, all programs show the expected improvement. The program that was written to thrash the memory benefits greatly from skewing with orders of magnitude fewer messages generated. Real applications also need fewer messages with skewing, by up to 15% for Barnes. In all these cases less of the machine's resources are being used. Less memory usage means that larger data sets could be run on the same machine and less network traffic makes the machine more efficient under a multi-tasking load. The number of messages is an average over the whole run. We expect that there is actually a much larger reduction in traffic during certain phases of the computation, and less reduction during others. In all cases we expect that we can run larger programs on the same machine before either the memory or the network bandwidth is exhausted.

There is one program that does not show improvement, and that actually behaves much worse. Wave, with 2 Mbytes of memory (256K per node) sends 3 times more messages with skewing than with plain set-associativity. The reason for this is that Wave needs more replication than there is free memory. Whatever strategy is chosen, the machine will thrash data. This effect is clearly visible when Wave is run with more memory. In this case the number of messages drops to 70000. Plain hashing will take care that thrashing is contained in a small area of the memory. When an item is requested that is not present at the moment, an item will be evicted, and injected to search for a home. In the meantime, the requested item will be brought to the node. The request for the data will overtake the injected item and fetch the data from a remote node. The injected item will thus certainly find a home at that node, because with plain hashing the item that has just been read must be in the same set. When using skewing the eviction process is identical, but because of the skewing, the item being read and the item being evicted, which clash on the node where the eviction starts, need not necessarily clash on the node where the data is obtained from. This means that the inject is not certain to find a home.

5 CONCLUSIONS

We have addressed a problem that is particular to virtual shared memory architectures in which all the memory is arranged set-associatively, such as the DDM, KSR-1 and COMA-F. Because of the sets, there is no single universal pool of memory, but a large number of small pools. Set-associativity causes each data item to be confined to a pool. Consequently, replication of data is limited within each pool; even though the rest of the memory may be unused.

This problem can be resolved by using multiple disjoint hash functions. This improves the capacity because items that clash in one location of the memory will not clash in another location. Different hash functions can be applied on different nodes, or within the same node. This latter option additionally reduces the number of collisions in each of the local memories; a property previously studied by Seznec.

We have shown that the use of two distinct hash functions on every node improves the performance of the memories in the DDM (and similar machines). In all but one of the cases studied, skewing is able to provide an improved sharing capacity over conventional set-associativity at minimal extra cost. This means that for given programs the memory is utilised more effectively, and we require less traffic.

The one exception to this has shown that skewing is not always a good thing. For applications that access and replicate data in particularly regular ways and when the memory is almost full, the misses with a conventional set-associative memory are predictable and confined to certain sets. In a skewed system the misses can affect a larger area of the memory resulting in a higher miss ratio. It should be noted however that this application was severely hampered by the restriction of the memory size. The plain set-associative case already issued three times more messages than the same program running with more memory.

In addition to local skewing, where each node uses the same two hash functions, the use of different hash functions on different nodes gives a further improvement. However, the simulator results presented suggest that this improvement is limited and of far less significance than the initial gain moving from plain set-associativity to local skewing.

Work on another implication of skewing, on the inclusion of multi-level directories without an excessive increase in associativity, is as yet incomplete but early investigations suggest that this is an interesting area for future research.

REFERENCES

[1] A. Seznec. A case for two-way skewed-associative caches. In *Proc. of the 20th ISCA*, pp 169–178, San Diego, California, May 1993. IEEE Computer Society Press.

[2] A. Seznec and F. Bodin. Skewed-associative Caches. In A. Bode, M. Reeve, and G. Wolf, editors, *PARLE '93*, pp 305–316, München, Germany, July 1993. Springer Verlag.

[3] N. Drach, A. Gefflaut, P. Joubert, and A. Seznec. About cache associativity in low-cost shared memory multi-processors. Technical Report 760, IRISA, Oct. 1993.

[4] D. H. D. Warren and S. Haridi. The Data Diffusion Machine—A Scalable Shared Virtual Memory Multiprocessor. In *Proc. of the Int. Conf. on FGCS*, pp 943–952, Tokyo, Japan, Dec. 1988.

[5] KSR. *KSR Technical Summary*. Kendall Square Research, Waltham, MA, 1992.

[6] P. Stenström, T. Joe, and A. Gupta. Comparative Performance Evaluation of Cache-Coherent NUMA and COMA Architectures. In *Proc. of the 19th ISCA*, pp 80–91, Gold Coast, Australia, May 1992. ACM Press.

[7] R. Jain. *The Art of Computer Systems Performance Analysis*. John Wiley & Sons Inc, New York, 1991.

[8] H. L. Muller, P. W. A. Stallard, D. H. D. Warren, and S. Raina. Parallel Evaluation of a Parallel Architecture by means of Calibrated Emulation. In *Proc. of the 8th IPPS*, pp 260–267, Cancun, Mexico, Apr. 1994. IEEE Computer Society Press.

MULTIPLE-BASE STORAGE SCHEME THEORY - APPLICATIONS TO 2D AND 3D PARALLEL ARRAYS

J. Jorda, A. Mzoughi, D. Litaize

IRIT / UPS

118 Route de Narbonne

31062 Toulouse cedex (France)

email: [jorda,mzoughi,litaize]@irit.fr

Abstract – *As the cycle time of CPUs or pipeline operators decreases faster than that of memories, a large number of memory banks must be used to increase the memory bandwidth. Consequently, special data mappings must be used to reduce bank conflicts. Such storage schemes allow conflict-free access to commonly used structures of parallel arrays. In this paper, we present a new class of storage scheme: multiple-base mappings[1]. After discussing properties of this general class, we introduce two specific schemes. They allow conflict-free access to common structures of 2D arrays as well as 3D arrays, and are characterized both by an efficient use of the number of memory banks, and simplicity of implementation.*

INTRODUCTION

Although processing units are getting faster, memory technology does not allow a reduction in their cycle time to the same extent. In order to be able to ensure appropriate feeding of processors, and in view of the considerable difference between processors' cycle times and those of memories, an increased number of physically independant banks (commonly several hundreds) must be used to enhance the memory bandwidth. Thus, the following problem comes to light: how can data be stored in different memory banks in such a way that it can be accessed without conflicts? This is particularly a problem when parallel arrays are to be stored.

Since 1970, a considerable amount of work has been published on the subject, and a renewal of interest by researchers in the last five years can be noted, because of the increasing gap between processor and memory cycle times. In fact, all these papers only deal with 2D array storage. However, high performance applications often require 3D arrays (flow modelling, Navier Stokes 3D, FFT 3D, etc...). So, classical mappings cannot be applied without restricting the ways in which common structures can be accessed without conflict.

We must now define the kind of structure that we would like to access without conflict. In a study made by FUJITSU in 1985 [7], it appears that 80% of accesses to a matrix are made by row, 10% with odd strides distinct from 1 and the rest in a rather random way. However, as this study dates from 1985, its relevance today could be somewhat doubtful. Actually, since then, standard mathematical libraries such as BLAS have appeared, which define call protocols and typical implementations for the most commonly used routines in scientific computing. It appears that a data arrangement penalizing access to columns (and therefore to submatrices) would not give satisfactory performance with algorithms of this type. We therefore base our argument on the most frequently used accesses in these routines which are: rows, columns, the forward diagonal and square blocks. The latter are very often used when algorithms are blocked [3]. In fact, blocked algorithms make it possible to recast the algorithms in terms of matrix-vector or matrix-matrix operations, which permits a better reuse of data than vector-based algorithms. Moreover, square blocks are also used on raster-graphics memory systems [2]. As mentioned above, current works only concern 2D arrays; we should then wonder which structures are to be accessed without conflict for 3D arrays. Since all 3D algorithms are made up of 2D ones, we consider that the structures to be accessed without conflict in 3D arrays are those mentioned above (in the case of 2D arrays) *in each one of the 2D sub-arrays: vertical, horizontal and lateral* (see figure 1).

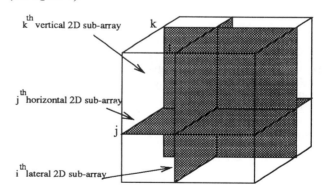

Figure 1: Example of lateral, vertical and horizontal 2D sub-arrays

In this paper, we present a new data mapping able to store both 2D and 3D arrays. First, we introduce the multiple-base skewing scheme class. Some general results concerning these mappings are demonstrated. Then, we present the 2D array storage mapping. It ensures conflict-free access to rows, columns, the forward diagonal and square blocks with an efficient use of the number of memory banks. In the third section, we extend this scheme in order to store 3D arrays with conflict-free access to the structures discussed earlier.

[1]This work is supported by the CNRS and the regional council of Midi-Pyrénées

DEFINITIONS AND GENERAL RESULTS

Let N be the number of memory banks of a computer. We have the following definitions:

Definition 1 *Let $D = [0, N-1]$. A linear skewing scheme with parameters (δ_0, δ_1) is a function μ from $[0, N-1]$ in $[0, N-1]$ defined by:*

$$\mu : \begin{cases} D \times D \to D \\ (x,y) \to \mu(x,y) = \delta_0 \cdot x + \delta_1 \cdot y \pmod{N} \end{cases}$$

The linear skewing schemes have been extensively studied by Lawrie [6]. This definition, given for 2D matrices, can be easily generalized to kD matrices and to matrices larger than N. He gave an example allowing the storage of $N \times N$ matrices in $2N$ memory banks, N being an even power of two.

We now introduce a generalization of linear skewing schemes, called multiple-base storage schemes. In fact, we have:

Definition 2 *Let $D = [0, N-1]$ and $(D_i)_{0 \le i \le p-1}$ a partition of a subset of D such that $D_i = [a_i, a_{i+1} - 1]$ with $a_i \in \mathbb{N}$, $a_0 \ge 0$, $a_p \le N$ and $\forall a_i \in [0, p-1]$ $a_i < a_{i+1}$. A multiple-base storage scheme with parameters $(\lambda, \delta_0, \cdots, \delta_{p-1})$ is a function μ defined by:*

$$\mu : \begin{cases} D \to D \\ x \to \mu(x) = \lambda \cdot x + b(x) \pmod{N} \end{cases} \quad with$$

$$b : \begin{cases} D \to D \\ x \to b(x) = \delta_{d(x)} \end{cases} \quad where \; d : \begin{cases} D \to [0, p-1] \\ x \to d(x) = i \end{cases} \quad such$$

that $x \in D_i$. The set $(\delta_i)_{0 \le i \le p-1}$ is called the set of storage bases.

This definition, given for a 1D vector, can also be generalized to matrices of higher dimensions and to larger matrices. For example [5], we can store a 2D matrix of size $(N-1) \times (N-1)$ in N memory banks (with N being an odd power of two, i.e. $N = 2^{2n+1}$) with the formula:

$$\mu_{i,j} = (\sqrt{N} + 1) \cdot i + 2 \cdot j + \lfloor \frac{j}{N} \rfloor \cdot (\sqrt{N} + 3)$$

One should note that this scheme is identical to that of Lawrie, except that we add a base for right sub-matrix items (see figure 2).

Let $D = [0, N-1]$, $(D_i)_{0 \le i \le p-1}$ be such a partition of D, and μ a multiple-base storage scheme having parameters $(\lambda, \delta_0, \cdots, \delta_{p-1})$.

Definition 3 *Let $\alpha = \gcd(\lambda, N)$. We use frame for an equivalence class modulo α. α is used for the period of the frame.*

Having, for all $i \in [0, p-1]$ $l_i = a_{i+1} - a_i + 1$ (l_i represents the length of the interval D_i). We have the following theorem:

Theorem 1 *A mapping μ, multiple-base storage scheme having parameters $(\lambda, \delta_0, \cdots, \delta_{p-1})$ is injective if and only if:*

- $\forall i \in [0, p-1]$ $l_i \le \frac{N}{\alpha}$ *and*

- $\forall i, j \in [0, p-1], i < j$ *we have $\delta_i \not\equiv_\alpha \delta_j$ or $\mu(a_i) \notin \{\lambda \cdot (a_j - k) + \delta_j \pmod{N}\}_{0 < k < l_i}$*

APPLICATION TO THE STORAGE OF 2D PARALLEL ARRAYS

The Parameters of the Storage Mapping

Let us consider a computer having N memory banks, where N is an odd power of two ($N = 2^{2n+1}$) greater than 8. We are going to store two dimensional parallel arrays of maximal size $(N-1) \times (N-1)$ in a computer of this kind. The structures we would like to access without conflict in these arrays are: rows, columns, the forward diagonal and the $2^n \times 2^n$ square blocks. Since the arrays we wish to store are two dimensional, the storage mapping formula we should use is:

$$\mu(x, y) = \lambda_i \cdot x + \lambda_j \cdot y + b_i(x) + b_j(y) \pmod{N}$$

In fact, $\lambda_i \cdot x + \lambda_j \cdot y \pmod{N}$ is the formula of a linear skewing scheme. Therefore, $b_i(x) + b_j(y)$ makes it possible to change the base of the mapping. This formula is therefore the general form of a two dimensional multiple-base storage scheme. We should now take an interest in $b_i(x)$ and $b_j(y)$.

Given a column number x and a row number y, the address unit must be able to compute the bank number very quickly. Therefore, $b_i(w)$ and $b_j(w)$ should change when w is a power of two. This implies a very simple logic which can therefore be implemented in a very efficient way and at low cost.

Numerous simulations gave the following parameters:

	i	j	
partition	$[0, N-2]$	$[0, \frac{N}{2} - 1]$	$[\frac{N}{2}, N-2]$
λ	1	$2^{2n} - (2^n - 1)$	
(δ_i)	(0)	$(0, 2^{2n} - (2^n - 1))$	

Thus, since $N = 2^{2n+1}$, the mapping formula is:

$$\begin{aligned} \mu(x,y) = {} & x + (2^{2n} - (2^n - 1)) \cdot y + \\ & \lfloor \frac{y}{2^{2n}} \rfloor \cdot (2^{2n} - (2^n - 1)) \pmod{N} \quad (1) \end{aligned}$$

Given a column number x and a row number y, we have seen how to compute the bank number of the corresponding item. The computation of the local address is very simple: the matrix element $a_{x,y}$ is stored in the bank number $\mu(x,y)$ at the address x. It should be noted that using an address computation of this kind implies a waste of memory: there are some holes. In fact, when an $(N-1) \times (N-1)$ matrix is stored in N memory banks with this scheme, the percentage of wasted memory is only:

$$\eta = \frac{N-1}{N \cdot (N-1)} = \frac{1}{N}$$

Conflict Free Access to Common Structures

Assume a computer having N memory banks (where $N = 2^{2n+1}$), and an $(N-1) \times (N-1)$ two dimensional array A stored by the equation 1. We have the following theorems:

Theorem 2 *Accesses to rows in A are conflict-free.*

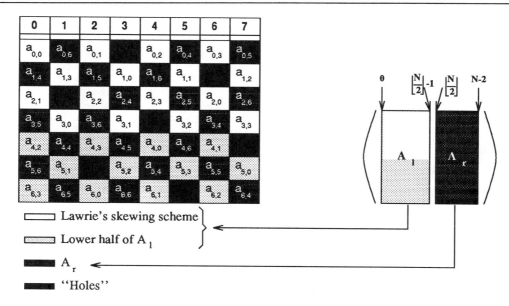

Figure 2: Example of two-base scheme storing 7×7 matrices in 8 memory banks

Theorem 3 *Accesses to columns in A are conflict-free.*

Theorem 4 *Accesses to the forward diagonal in A are conflict-free.*

Theorem 5 *Accesses to the $2^n \times 2^n$ square blocks in A are conflict-free.*

We have proved that the mapping:

$$\mu(x,y) = x + (2^{2n} - (2^n - 1)) \cdot y + \lfloor \frac{y}{2^{2n}} \rfloor \cdot (2^{2n} - (2^n - 1)) \pmod{N}$$

ensures conflict-free accesses to rows, columns, the forward diagonal and $2^n \times 2^n$ square blocks of a $(N-1) \times (N-1)$ maximal size 2D array.

We may now wonder if this mapping could be extended in order to store 3 dimensional arrays. That is to say: how many $(N-1) \times (N-1)$ arrays can be stored with the mapping presented above, ensuring in addition conflict-free access to common structures in each of the 2D sub-arrays? We will now discuss this question.

APPLICATION TO THE STORAGE OF 3D PARALLEL ARRAYS

The Parameters of the Storage Mapping

We still consider an N memory bank system, where N is an odd power of two (e.g. $N = 2^{2n+1}$). We have seen how to store, in a computer of this kind, $(N-1) \times (N-1)$ maximal size 2D arrays. We would like to extend this mapping to the storage of $(N-1) \times (N-1) \times (N-1)$ maximal size 3D arrays. The structures we would like to access without conflict in these arrays are the rows, the columns, the depths, the forward diagonal in each of the horizontal, vertical and lateral 2D sub-arrays and the $2^n \times 2^n$ square blocks located in the sub-arrays above.

Since the arrays we wish to store are three dimensional, the general formula of the storage mapping is:

$$\mu(x,y,z) = \lambda_i \cdot x + \lambda_j \cdot y + b_i(x) + b_j(y) \pmod{N}$$

(for the same reasons as in the previous section). We have shown in the previous section that the formula

$$\mu_{2D}(x,y) = x + (2^{2n} - (2^n - 1)) \cdot y + \lfloor \frac{y}{2^{2n}} \rfloor \cdot (2^{2n} - (2^n - 1)) \pmod{N}$$

ensures the storage of $(N-1) \times (N-1)$ arrays. We have then tried to store 3D arrays by storing the consecutive vertical 2D sub-arrays, each of the latter *with the same formula as that used for the case of 2D arrays*. In order to ensure conflict-free access to the vectors of the lateral and horizontal 2D sub-arrays, the storage of the vertical sub-arrays should be done by adding a base. Therefore, the general form of a mapping of this kind is:

$$\mu_{3D}(x,y,z) = x + (2^{2n} - (2^n - 1)) \cdot y + \lfloor \frac{y}{2^{2n}} \rfloor \cdot (2^{2n} - (2^n - 1)) + \lambda_k \cdot z + b_k(z) \pmod{N}$$

Numerous simulations have given the following parameters:

partition	k	
	$[0, \frac{N}{2} - 1]$	$[\frac{N}{2}, N-2]$
λ	$2^{2n} + 2^n - 2$	
(δ_i)	$(0, 2^{2n} + 2^n - 1)$	

Then, the mapping formula is:

$$\mu(x,y,z) = x + (2^{2n} - (2^n - 1)) \cdot y + (2^{2n} + 2^n - 2) \cdot z + \lfloor \frac{y}{2^{2n}} \rfloor \cdot (2^{2n} - (2^n - 1)) + \lfloor \frac{z}{2^{2n}} \rfloor \cdot (2^{2n} + 2^n - 1) \pmod{N} \quad (2)$$

The local address is simple to compute: it is proportional to the line number for each vertical 2D sub-array (as in the previous section); the difference is that it is necessary to add $N - 1$ when storing the next vertical 2D sub-array.

We can, now, prove that access to the vectors mentioned earlier is conflict-free. All proofs will not be specified here (for more details, see [4]).

Conflict Free Access to Common Structures

Assume a computer having N memory banks (where $N = 2^{2n+1}$), and an $(N - 1) \times (N - 1) \times (N - 1)$ three dimensional array A stored by the equation 2. We have the following theorems:

Theorem 6 *Accesses to the rows of A are conflict-free.*

Theorem 7 *Accesses to the columns of A are conflict-free.*

Theorem 8 *Accesses to the depths of A are conflict-free.*

Theorem 9 *Accesses to the forward diagonal in the vertical 2D sub-arrays of A are conflict-free.*

Theorem 10 *Accesses to the forward diagonal in the horizontal 2D sub-arrays of A are conflict-free.*

Theorem 11 *Accesses to the forward diagonal in the lateral 2D sub-arrays of A are conflict-free.*

Theorem 12 *Accesses to the vertical $2^n \times 2^n$ square blocks of array A are conflict-free.*

Theorem 13 *Accesses to the lateral $2^n \times 2^n$ square blocks of array A are conflict-free.*

Theorem 14 *Accesses to the horizontal $2^n \times 2^n$ square blocks of array A are conflict-free.*

CONCLUSION

Since there is a big gap between processor and memory cycle times, numerous data mappings have seen the light of day since 1970. Now, since the use of new (such as superpipeline or superscalar) techniques has increased this gap, there has been a renewal of interest in storage schemes during the last five years. However, existing mappings only deal with 2D array storage, thus neglecting 3D array based applications (Navier Stokes, Flow Modelling, etc...).

In this paper, we have introduced a new class of data mappings: multiple-base schemes. Multiple-base skewing schemes are very close to linear ones, but they are more powerful, which we have demonstrated using two examples. In the first we stored a 2D array with a mapping of this kind which presents numerous advantages:

- good use of memory bandwidth (since the maximal size of stored arrays is $(N - 1) \times (N - 1)$);

- conflict-free access to rows, columns, forward diagonal and $2^n \times 2^n$ square blocks;

- an easy-to-compute local address and memory bank number since it involves only modulo 2 arithmetic.

In the second example, we extended the mapping above to the storage of 3D arrays. Thus, $(N-1) \times (N-1) \times (N-1)$ arrays can be stored with this storage scheme; accesses to common structures are conflict-free, and the computation of addresses is simple. It is, then, very well suited to the computation of a large amount of data.

It should be noted that only two-base skewing schemes have been presented. We are now extending this work by studying multiple-base data mappings. Thus, synchronized accesses for multiprocessors are currently being studied, as well as the storage of higher dimensional arrays. Multiple-base schemes are both of theoretical and practical interest, and should then be studied extensively.

REFERENCES

[1] P. Budnik and D. J. Kuck, "The Organization and Use of Parallel Memories", *IEEE Trans. Comput.*, vol. C-20, pp. 1566–1569, Dec. 1971.

[2] B.Chor, C.E. Leiserson, R.L. Rivest and J.B. Shearer, "An application of number theory to the organization of raster-graphics memory", *Journal of the ACM*, Jan. 1986.

[3] J. J. Dongarra, I. S. Duff, D. C. Sorensen and H. A. Van Der Vorst, "Solving Linear Systems on Vector and Shared Memory Computers", *SIAM*, 1993 (second printing).

[4] J. Jorda, A. Mzoughi and D. Litaize, "Storing 2D and 3D Arrays Using Multiple-base Storage Schemes", Technical Research Report, IRIT-UPS (Toulouse, France), IRIT/95-11-R (March 1995).

[5] J. Jorda, A. Mzoughi and D. Litaize, "Semi-Linear and Bi-Base Storage Schemes Classes: General Overview and Case Study", *9th International Conference on Supercomputing*, July 1995.

[6] D. H. Lawrie, "Access and Alignment of Data in an Array Processor", *IEEE Trans. Comput.*, vol. C-24, pp. 1145-1155, Dec. 1975.

[7] H. Tamura, Y. Shinkai, F. Isobe "The Supercomputer FACOM VP system", *Fujitsu Sc. Tech. J.*, March 85

UNSCHEDULED TRACES AND
SHARED-MEMORY MULTIPROCESSOR SIMULATION

Patricia J. Teller
New Mexico State University
P. O. Box 30001/CS
Las Cruces, NM 88003-8001, USA
teller@cs.nmsu.edu

Abstract -- *Trace-driven simulation of shared-memory multiprocessor (MP) architectures can be driven by scheduled or unscheduled traces. A scheduled trace is a set of traces generated by running a parallel program on a real or simulated MP. An unscheduled trace is an augmented trace generated by running a parallel program on a uniprocessor. Scheduled traces can be used to drive simulations of only MP systems with the same number of processors as the MP that generated the trace (MP_{trace}) and are limited to the same static task-to-processor scheduling as MP_{trace}. In contrast, unscheduled traces offer much more flexibility: they can be used to simulate dynamic task-to-processor scheduling, MP systems of different sizes, and multiprogramming. In addition, an unscheduled trace allows the task granularity of traced programs to be varied from simulation to simulation and operating system references to be inserted dynamically during simulation.*

INTRODUCTION

The majority of trace-driven simulation studies of shared-memory multiprocessors (MPs) use scheduled traces (*e.g.*, [1] and [2]). A *scheduled trace* is a set of address traces, which is called a *trace-set*, that was generated by a real or simulated MP while executing or interpretively executing a parallel program. For such a trace-set, there is a one-to-one mapping of the processors (PEs) of the MP that generated the trace (MP_{trace}) to the individual trace files of the trace-set.

Thus, a scheduled trace can be used to simulate only a MP (MP_{sim}) with the same number of PEs as MP_{trace}. Moreover, the task-to-processor scheduling of MP_{trace} is enforced on MP_{sim}, and this is not necessarily the scheduling that would occur if the associated workload was executed on MP_{sim}. Since architectural differences between MP_{trace} and MP_{sim}, *e.g.*, different cache organizations or management, can yield different task-to-processor schedulings [3], simulations using scheduled traces can be inaccurate. Inaccuracy of a simulation driven by a scheduled trace also can be the result of other timing dependencies that are inherent in the traced program [4] and/or perturbations due to trace generation techniques [5].

As noted in [5], abstraction is the key to accurate trace-driven simulation. Abstraction decouples the environment of MP_{trace} from the environment of MP_{sim}. It is the key to techniques that eliminate timing dependencies associated with lock and barrier synchronization, and it is the key to eliminating timing dependencies associated with dynamic task scheduling of tasks with deterministic traces, where work is assigned to idle PEs on a first-come, first-served basis. Goldschmidt and Hennessy [5] note that they do not know of any simulators that have accomplished this.

Our simulation methodology accomplishes this and more by using unscheduled traces to drive shared-memory MP simulations. An *unscheduled trace* is generated by executing an instrumented parallel program on a uniprocessor. The use of such traces to drive MP simulations offers considerable simulation flexibility:

* one unscheduled trace can be used to simulate MP systems with different numbers of PEs than MP_{trace},

* an unscheduled trace allows dynamic task-to-processor scheduling,

* an unscheduled trace can be used to drive simulations with variable task granularities,

* multiple unscheduled traces can be used to drive simulations of multiprogrammed MPs, and

* reference sequences associated with such events as timer interrupts, context-switching, and page-fault handling can be inserted dynamically during simulation.

This paper briefly describes how unscheduled traces permit this flexibility in MP simulation by (1) defining the class of programs that we address, (2) describing the nature of an unscheduled trace, and (3) explaining our simulation methodology. Finally, we discuss our related research in this area. For more details about this research refer to [6].

PARALLEL PROGRAM CLASS

Although we believe that MP trace-driven simulation using unscheduled traces is applicable to a larger class of programs, currently we restrict our attention to shared-memory SPMD programs [7] that are comprised of the task groups and have the characteristics described below.

Task Group Definitions

Any parallel program can be described by its *process flow graph* (*pfg*), an acyclic, directed graph in which nodes represent task groups and edges represent execution (precedence) constraints. A *task group* is a set of tasks, not necessarily maximal, sharing the same precedence relationship. Multiple tasks intended for concurrent execution are in the same task group. A task group that contains no tasks serves solely as a synchronization point in the program. Each task of a task group at the source of an edge of the *pfg* must be completed before any task in the task group at its sink can be executed.

In the case of SPMD programs, each task group may have associated import tasks, which often update private variables. A process must execute all import tasks encountered in its execution of the program, and the import tasks associated with a *pfg* node must be executed by each process prior to executing any other tasks belonging to that node. Therefore, such a node could be expanded to two nodes joined by an edge, where one node only contains the import tasks, and the other, the successor of the first, contains all the other tasks in the task group.

The class of SPMD programs to which we currently restrict our attention are those that are comprised of homogeneous task groups of the following types:

serial-wait: A task group comprised of one task that is to be executed by only one process. An edge emanates from its associated *pfg* node.

parallel-wait: A task group comprised of tasks which are intended to be executed concurrently. An edge emanates from the *pfg* node of such a task group.

replicate: A task group with p copies of a replicate task, an import task which is to be executed by all of the p processes that are cooperatively executing the program. As an import task, it carries the inherent precedence constraints described above.

barrier: A task group comprised of one, possibly null, task, that is to be executed by all processes cooperatively executing the program. A barrier can be represented in the *pfg* as either a task group with no tasks that serves solely as a synchronization point, or an edge in the *pfg*.

Program Characteristics

Timing dependencies can lead to dynamic definition of process control flow and the private and/or shared variables referenced by a process, dynamic task scheduling and barrier synchronization, and unsynchronized read-write access to shared data. We restrict our attention to dynamic task scheduling and programs that have inherent timing dependencies that can lead to dynamic

barrier synchronization. The programs to which we restrict our attention have the following characteristics.

1. Process control flow, other than that associated with the distribution of tasks that can be executed concurrently or tasks that are to be executed by any one process, is not determined by conditional branches that have outcomes directly or indirectly dependent on shared variables.

2. Individual tasks do not have inherent timing dependencies. Each is statically defined, *i.e.*, the control flow and referenced private and shared variables of a task except those defined in 3 below, are deterministic. Thus, the trace of each task is deterministic.

3. Shared variables that can be read and written concurrently, such as iteration variables, accumulators, and counters associated with barriers, are accessed using atomic fetch-and-φ type primitives [8].

UNSCHEDULED TRACES

An unscheduled trace is generated by running an instrumented SPMD program on a real or simulated uniprocessor. The generated trace contains information regarding each memory reference and markers that identify different types of tasks (*i.e.*, basic scheduling units), precedence relationships among tasks, and user-specified program regions (*e.g.*, the initialization or termination section of the program). Given such an augmented address trace as input, a preprocessor builds a task queue in which this information is embedded. A MP simulator, like our simulator (SIM), given such task queues as input, controls the scheduling of a job's tasks to PEs of MP_{sim}, consistent with their precedence relationships. Marker-based techniques such as this also have been used by So, *et al.* [9] to implement SPAN, which analyzes a process flow graph and provides information to assist the user in attaining maximal speedup, and by Agarwal and Cherian [10] to study the performance of shared-data caches in large-scale MPs.

The generation of an unscheduled trace necessarily includes the insertion of markers. These markers perturb the trace in the sense that they change the addresses associated with instructions recorded in the trace. Perturbations of this type can be eliminated by using techniques such as those proposed by Borg, Kessler and Wall [11].

The unscheduled traces that we currently are working with were generated using a modified version of PSIMUL [12]. Originally, PSIMUL was designed to interpretively execute a parallel program and generate a scheduled trace of virtual addresses referenced during that execution of the program. We modified PSIMUL, with the cooperation of researchers at IBM's T. J. Watson Research Center, to allow the generation of an unscheduled trace of the execution of the parallel program on a uniprocessor. Note that when an unscheduled trace is used to simulate an

MP system, MP_{sim} is assumed to have a global (*i.e.*, system-wide) virtual address space. The private address spaces unique to processes are distinguishable from one another as long as one process is associated with each PE. The shared address space is referenced identically by all processes. Conceptually one page table is shared by all processes. Of course, process identifiers could be affixed to private address spaces to distinguish them.

SIMULATION METHODOLOGY

Our simulation platform (Figure 1) uses explicit simulation and analytical modeling of system components and data structures to simulate a MP. Input parameters to SIM define the number of PEs, memory modules, and I/O devices in MP_{sim}; the size and organization of caches, TLBs, and memories; characteristics of the PEs, memories, and the network, such as instruction execution time, memory access time, and network switch time; and characteristics of the operating system, such as time quantum, context-switch time, and scheduling policy. It is assumed that a process is associated with each simulated PE. SIM is comprised of a scheduler and a modeler. As we briefly describe below, the scheduler schedules tasks to PEs and the modeler assesses processing delays to tasks. We have used SIM to study MP caches, TLBs, memory organizations, and demand paging (*e.g.*, [13], [14], and [15]).

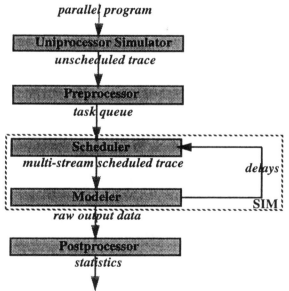

Figure 1. Simulation Platform.

Dynamic Task-to-Processor Scheduling

Since knowledge of precedence constraints among tasks and identification of synchronization point is embedded in the task queue that is input SIM, dynamic task-to-processor scheduling can be simulated. Thus,

SIM is able to simulate MP systems of any size and can dynamically schedule tasks to PEs of MP_{sim}, either on a task completion or time-slice basis.

SIM schedules tasks to PEs and dynamically creates a scheduled multi-stream virtual address trace. This is done by updating each simulated PE's state at each simulation step, and generating a memory reference whenever applicable. Simulated PEs do not necessarily generate a memory reference at each simulation step due to events that carry a time delay, such as cache or TLB misses and context switches. When a PE encounters such an event during the execution of a task, a time delay is assessed on the PE. The length, in terms of time, of a simulation step is dependent on the minium of the delays associated with the simulated PEs. In this way, although an unscheduled trace is statically defined, the information that it contains is sufficient to allow many different scheduled multi-stream virtual-address traces to be created dynamically from it.

Variable Task Granularity

An unscheduled trace contains sufficient information to allow the aggregation of multiple tasks in a task group, thus, it allows task granularity to be varied from simulation to simulation. When an unscheduled trace is used to build a task queue, each task group is encapsulated and contains a prologue. A task group's prologue indicates, among other things, the number of tasks in the group; and the tasks in a task group are delineated by markers. Given this information and an input parameter that defines how the tasks in a task group should be aggregated, SIM dynamically creates "new" tasks (which are each an aggregate of multiple tasks of a group) and schedules them to PEs. In this way, task granularity can be varied from simulation to simulation. The number of tasks in a task group denotes its maximum exploitable parallelism. This can be traded off for larger task granularity to achieve better load balancing and a reduction in the overhead associated with context switches.

Multiprogramming

To simulate a time-shared multiprogrammed MP system, SIM is given multiple unscheduled traces as input. Each trace is considered a job; a selected algorithm, *e.g.*, a round-robin algorithm, is used to schedule the tasks of multiple jobs to PEs of MP_{sim}.

The virtual address spaces as defined by the traces are used to form a single global virtual address space. This is accomplished by associating a job information file with each trace, which indicates the smallest and largest virtual memory addresses referenced in the trace. At the beginning of a multiprogram simulation, the data in the input traces' job information files are used to compute and record base and offset page numbers for each trace. A trace's offset is the page number of the smallest referenced address (smallest referenced address / page size). A trace's base is the next available page in the global address space. During simulation, the global virtual page number of a memory reference of job_i ($trace_i$) is computed as:

$$referenced\_address >> (page\_size - offset_i) + base_i$$

If demand paging is not simulated, then the size of physical memory is considered to be equal to that of the global virtual address space. Otherwise, the sizes can be different.

Canned Trace Segments

Most traces do not include operating system references. As noted by Smith [16], it is difficult to include in the trace of operating system references, the effects of input/output, and the effects of context switches and interrupts. Thus, far we have captured the effect of input/output (with respect to demand paging) and context switches with respect to task scheduling. Currently we are working on a way of dynamically inserting trace segments associated with these and other operating system events into the scheduled multi-stream traces that SIM's scheduler generates. Canned trace segments associated with operating system events will be representative of a Unix-like operating system and, depending upon the associated event, will be inserted either at specific time intervals or at the points of occurrence of the event.

Using a similar method, it is not difficult to simulate synchronization points such as barriers. All that is needed is a canned trace segment that includes the address of the synchronization variable associated with the barrier. Given this, SIM can, for example, simulate the barrier by having each process spin on that variable until all processes have entered the barrier. Currently an implicit or explicit barrier is simulated by only locking the task queue until all processes have encountered the barrier. In the near future, we will be implementing the above-described method.

RELATED RESEARCH

Currently, our Performance Research Group, in collaboration with the Parallel Architecture Research Lab. at New Mexico State University's Department of Electrical and Computer Engineering, is building two platforms to generate uniprocessor and unscheduled MP traces. Once these platforms are in place, we will further investigate the applicability of unscheduled traces. Traces generated via these platforms will be added to a data base of traces being established at New Mexico State University for the use of researchers and educators worldwide (anonymous ftp to tracebase.nmsu.edu).

ACKNOWLEDGEMENTS

Many thanks to my colleagues, Richard Oliver and Qidong Xu, for their careful reading of this paper.

REFERENCES

[1]   R. Thekkath and S. Eggers, "Impact of sharing-based thread placement on multithreaded architectures," in *Proc. of the 21st Int. Symp. on Computer Architecture* (ISCA), pp. 176-186, 1994.

[2]   A. Cox and R. Fowler, "Adaptive cache coherency for detecting migratory shared data," in *Proc. of the 20th Annual ISCA*, pp. 98-108, 1993.

[3]   M. Dubois, J. Skeppstedt, L. Ricciulli, K. Ramamurthy, and P. Stenstrom, "The detection and elimination of useless misses in multiprocessors," in *Proc. of the 20th Annual ISCA*, pp. 88-97, 1993.

[4]   M. Holliday and C. S. Ellis, "Accuracy of memory reference traces of parallel computations in trace-driven simulation," *IEEE Trans. on Parallel and Distributed Systems*, 3 (1), Jan. 1992.

[5]   S. Goldschmidt and J. Hennessy, "The accuracy of trace-driven simulations of multiprocessors," Computer Systems Lab., Stanford Univ., 1992.

[6]   P. Teller, "Unscheduled traces and shared-memory multiprocessor simulation," New Mexico State Univ., 1994.

[7]   F. Darema-Rogers, D. George, V. Norton, G. Pfister, "A single-program-multiple-data computational model for EPEX/FORTAN," IBM RC11552, 1986.

[8]   A. Gottlieb and C. Kruskal, "Coordinating parallel processors: a partial unification," Ultracomputer Note No. 34, New York Univ., 1981.

[9]   K. So., A. Bolmarcich, F. Darema, and V. Norton, "A speedup analyzer for parallel programs," in *Proc. of the 1987 Int. Conf. on Parallel Processing*, pp. 653-662, Aug. 1987.

[10]   A. Agarwal and M. Cherian, "Adaptive backoff synchronization techniques," in *Proc. of the 16th ISCA*, pp. 396-406, 1989.

[11]   A. Borg, R. Kessler, and D. Wall, "Generation and analysis of very long address traces," in *Proc. of the 16th Annual ISCA*, pp. 270-281, 1990.

[12]   K. So, F. Darema-Rogers, D. George, V. Norton, and G. Pfister, "PSIMUL -- a system for parallel simulation of the execution of parallel programs," IBM RC11674, 1986.

[13]   P. Teller and Q. Xu, "Memory-based multiprocessor translation-lookaside buffers: multiple paging arenas vs. large size TLBs," in *Proc. of the 1994 IEEE 13th Annual Int. Phoenix Conf. on Computers and Communications*, pp. 47-53, Apr. 1994.

[14]   P. Teller and A. Gottlieb, "Locating multiprocessor TLBs at memory," in *Proc. of the 27th Hawaii Int. Conf. on System Sciences,* pp. 554-563, Jan. 1994.

[15]   Q. Xu and P. Teller, "Unified vs. split TLBs and caches in shared-memory MP systems", to appear in *Proc. of the 9th Int. Parallel Processing Symp.*, Apr. 1995.

[16]   A. Smith, "Cache memories," ACM Computing Surveys, 14, pp. 473-530, Sept. 1982.

A Framework of Performance Prediction of Parallel Computing on Nondedicated Heterogeneous NOW *

Xiaodong Zhang   Yong Yan

High Performance Computing and Software Laboratory
The University of Texas at San Antonio
San Antonio, Texas 78249

Abstract

We study performance predictions for parallel computing on nondedicated heterogeneous networks of workstations (NOW). Our approach is based on a two-level model. On the top level a semi-deterministic task graph is used to capture the parallel execution behavior including the variances of communications and synchronizations. On the bottom level, a discrete time model is used to quantify effects from nondedicated heterogeneous network systems. An iterative process is used to determine the interactive effects between network contention and task execution. We validated the prediction model using experiments on a nondedicated heterogeneous network of workstations.

1   Introduction

Effective scheduling of application programs on a nondedicated heterogeneous network of workstations (NOW) is crucial to obtaining good parallel computing performance. The heterogeneity and time-sharing effects make the scheduling more complex and difficult. It is necessary to use predicted performance results to assist the scheduler in finding an efficient mapping between an application program and the system. In this paper, we focus on developing a practical performance prediction methodology for NOW systems.

A two-level prediction framework has been widely used in existing prediction methods [1] with the objective of separating the performance factors arising from the application level and from the system level. These prediction methods differ in the way they model the non-deterministic nature of system effects of inter-process communication events, contention for shared resources, and program structures. These methods fall into two classes: stochastic

methods and deterministic methods. In stochastic models, non-determinism is reflected by taking into account the variance or distribution of execution times of the synchronizing processes in the program. However, as surveyed and shown in [1], stochastic models proposed so far have not been evaluated with regard to efficiency and accuracy for actual applications because these models require complex solution techniques as well as the assumption of exponential task times for analytical tractability. In deterministic models, the variance of execution times is assumed to be negligible and these execution times are represented as deterministic quantities. Intuition and conceptual simplification are major advantages of deterministic models which make it possible to predict a real application. In [1], [2] and [5], the authors have demonstrated the deterministic assumption to be practical. A major limitation of deterministic models is that they fail to reflect the effect of network contention, which is an important factor affecting the executions of many applications in network architectures. This motivates us to develop a semi-deterministic two-level model to predict the performance of heterogeneous NOW systems. Our model is aimed at keeping the advantages of deterministic models while taking network contention into consideration through a reasonable amount of experiments.

In a performance prediction methodology, a high-level abstraction of an application plays a critical role in deciding modeling complexity and applicability. Based on the distributed programming paradigm used in PVM [4] and other programming systems, a task graph, called the program execution graph (PEG), is used to describe the execution as well as the communication and synchronization relations of a parallel computation. To separately quantify effects of the program structure and the system, the communication and synchronization points are distinguished in the PEG of an application. The CPU requirement of each distributed task is approximated by a deterministic function of input parameters. At the system level, a workload model is developed to reflect the effects of a nondedicated environment. To capture the effects of network contention, communication delay is estimated by combining applica-

*This work is supported in part by the National Science Foundation under research grants CCR-9102854 and CCR-9400719, and under instrumentation grant DUE-9250265, by the U.S. Air Force under research agreement FD-204092-64157, and by the Air Force Office of Scientific Research under grant AFOSR-95-1-0215.

tion communication patterns with an empirical delay function. The heterogeneity of the underlying network system is quantified by the computing power of each workstation. The interaction between performance factors of the two levels is modeled through an iterative process.

We predict the execution time, speedup and efficiency of a parallel computation using the performance metrics proposed in [6]. The performance bottlenecks and overhead latency distribution are also predicted. To validate our prediction method, one numerical program is implemented and measured in a nondedicated heterogeneous NOW.

2 Heterogeneous computing models

2.1 Heterogeneous network model

A heterogeneous network (HN) can be abstracted as a connected graph $HN(M, L)$, where $M = \{M_1, M_2, ..., M_m\}$ is set of workstations of different types (m is the number of machines) and $L \subseteq M \times M$ is a set of bi-direction links.

In practice, a network system is usually nondedicated. Each workstation has a certain owner workload distribution which significantly affects the execution of a parallel task on it. In a nondedicated environment, only the idle CPU cycles are used to execute tasks of a parallel computation. We use a discrete time model to capture the effects of the owner workload. For each workstation $M_i (1 \leq i \leq m)$ in the system, the owner workload distribution on M_i is described by the following two parameters:

- O_i: the average execution time of the owner task on M_i.

- P_i: the average probability of the owner task arriving on M_i during a given time step. Moreover, we assume that the owner think time on M_i satisfies a geometric distribution with mean $\frac{1}{P_i}$.

The owner processes are assumed to have priority over parallel processes and can preemptively schedule a parallel computation only when a time slice is expired or parallel tasks wait for communications. In addition, a owner task can start its execution at least one time slice later after the completion of the preceding owner task.

2.2 Heterogeneous programming model

An application program can be abstracted by a two-level model. The top level of the model is represented by a standard program data-dependence graph, (PDG graph) [3], which is used to describe the parallelism in a program. The bottom level of the model is a task execution graph, which determines the order of a parallel computation on each node. In order to avoid discussing the task scheduling problem, we focus on the program execution graph in the two-level model assuming a given scheduling scheme.

In task execution graph, a parallel program, denoted by $A(I)$, where I represents a vector of input-parameters, is assumed to have m distributed tasks $A_1(I)$,

$A_2(I), ..., A_m(I)$, cooperating on a parallel computation through message passing. Task $A_i(I)$ ($1 \leq i \leq m$) is assigned and executed on machine M_i. The size of program $A(I)$ is denoted by $|A(I)|$, which can be defined as the number of required operations to solve $A(I)$ [6]. Thus, $|A(I)| = \sum_{i=1}^{m} |A_i(I)|$.

2.3 Quantify the heterogeneity

In heterogeneous network systems, workstations may have different memory systems, different CPU speed and different memory and I/O capacity. The relative computing power among a set of heterogeneous machines varies with applications. In order to quantify the relative computing power among the workstations in a network system, for each machine M_i, a computing power weight $W_i(A)$ with respect to application A is defined as follows:

$$W_i(A) = \frac{S_i(A)}{\max_{j=1}^{m}\{S_j(A)\}} \quad i = 1, ..., m \quad (1)$$

where $S_i(A)$ is the speed of M_i to solve A on a dedicated system.

Formula (1) indicates that the power weight of a machine refers to its computing speed relative to the fastest machine in a system. The value of the power weight is less or equal to 1. Since the power weight is a relative ratio, it can also be represented by measured execution time. If we define the speed of the fastest machine in the system as one basic operation per time unit, the power weight of each machine denotes a relative speed. If $T(A, M_i)$ gives the execution time for computing program A on machine M_i, the power weight can be calculated by the measured execution times as follows:

$$W_i(A) = \frac{\min_{j=1}^{m}\{T(A, M_j)\}}{T(A, M_i)} \quad (2)$$

By formula (2), we know that if a basic operation executes t_u cycles on the fastest machine, it executes t_u/W_i cycles on machine M_i.

Due to limitations of memory size, cache size and other hardware components, the power weights usually change with the problem size of an application. The work reported in [6] observes that the computing power weights of workstations in a network system are constant if the data size of an application is scaled within the memory bound of each workstation. In this paper, we only discuss the cases with constant computing power weight.

3 Methodology

Figure 1 gives the framework of the prediction method. The PEG graph is a major abstracting mechanism in the high-level model. The low-level model is responsible for abstracting system characteristics, such as owner workload, heterogeneity and network contention. A predictor generates predicted performance results using the performance parameters provided by the two-level model.

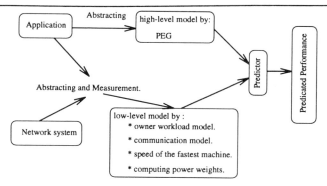

Figure 1: Performance prediction framework.

3.1 High-level model: abstracting application characteristics

To separate the non-deterministic effect of network contention and to analyze the execution of a task, the communication points involving a send or receive operation in a task are distinguished from other local computations in the task. Each communication operation has one of the following two types:

- *Nonblocking send*: it does not block the execution of a process sending a message.

- *Blocking receive*: it must wait for the arrival of the required message.

Another commonly used communication model, the blocking send/blocking receive model, can be implemented by the above model. Based on communication points, the local computations of a distributed task are abstracted as a number of segments, each segment comprising all the local computations between two successive communication points in a task.

To precisely represent the execution of a parallel computation, the PEG graph is formally defined as a 4-tuple $(C, \rightarrow, \Pi, \Theta)$ structure where

- C is the set of computational segments and communication points in distributed tasks of an application. A communication point or a computational segment is termed an operation unit, which is the basic timing unit. For simplicity of the discussion, c_{ij} is used to represent a basic operation unit in i-th task of an application.

- $\rightarrow$ is the precedence relation between operation units in C, which is defined as follows:

    - For two operation units c_{ij}, c_{ik} in the same distributed task, $c_{ij} \rightarrow c_{ik}$ if and only if c_{ij} executes immediately before c_{ik}.

    - For a pair of communication operations: c_{ij} in the i-th task and c_{kr} in the k-th task, $c_{ij} \rightarrow c_{kr}$ if and only if c_{ij} sends the k-th task a message which will be received by c_{kr}.

- Π is a message function. For each send operation c_{ij}, $\Pi(c_{ij})$ gives the total number of message bytes sent by c_{ij}.

- Θ is a complexity function which gives the number of basic operations in a computational segment with respect to the fastest machine in the system. A basic operation is one of the following three types of operation (paper [6] reports that these operations have major effects on the computing speed of a workstation):

    - I/O operation,

    - Memory read/write, and

    - Floating-point operation.

Generally, for segment c in an application, $\Theta(c)$ is a function of the input parameters of the application.

To be consistent with control flow structures of most network programs, we assume that the PEG graph comprises one main task which starts other tasks and collects their results. To get the PEG graph of an distributed program, the following steps are executed:

1. Inlining subroutines: which inlines all the subroutines into the main program to get a flat program structure.

2. Indexing the loop body: In order to distinguish operation units within different loop iterations, the loop index variable is used to index operation units in a loop iteration. (Here we assume that a loop is represented by *"for"* type.)

3. Normalizing the loop body: For each loop including communication operations, it is normalized to start with a communication operation.

3.2 Low-level Model: abstracting system effects

In the low-level model, important system overhead and latency factors, such as the communication startup time, the observed network bandwidth and the message delivery delay in the absence of contention are directly measured from a NOW system.

3.2.1 Measuring computing speed

To calculate the execution time of a task segment, the speed, denoted by S_f, of the fastest machine is measured based on the following sample data.

- $f_{i/o}, f_{r/w}, f_{fp}$: the frequency of occurrence of each type of operation: I/O, memory read/write and floating point operations respectively.

- $\bar{t}_{i/o}, \bar{t}_{r/w}, \bar{t}_{fp}$: average execution time of each type on the fastest machine in the system.

The speed is calculated as follows:

$$S_f = \frac{\frac{f_{i/o}}{\bar{t}_{i/o}} + \frac{f_{r/w}}{\bar{t}_{r/w}} + \frac{f_{fp}}{\bar{t}_{fp}}}{f_{i/o} + f_{r/w} + f_{fp}} \tag{3}$$

In practice, the frequency of occurrence of each type of operation in an application may not change with the problem size of the application. For this type of applications, the speed can be simply measured by a set of sample input parameters, and can be used to predict scaling performance. Specifically, if the complexity function Θ of an application is given, the speed S_f can be approximately calculated by $S_f = \frac{\Theta(c)}{t_c}$ for a computational segment c that has execution time of t_c on the fastest machine. Finally, with the speed S_f, the speed of one of the rest machines of the system, denoted by S_i, can be represented using the computing power weight as $S_i = S_f \times W_i$. (W_i is the computing power weight of M_i.)

So, the execution time of segment c on any dedicated machine $M_i (1 \leq i \leq m)$, denoted as $T_{cpu}(c, M_i)$, can be expressed as

$$T_{cpu}(c, M_i) = \frac{\Theta(c)}{S_i} = \frac{\Theta(c)}{S_f \times W_i}. \qquad (4)$$

Here, T_{cpu} is considered the required CPU time of a segment.

3.2.2 Modeling network contention

In the presence of network contention, the delay of a network communication is mainly determined by two factors: the number of processes simultaneously competing for the network bus and the order in which the competing processes will use the bus. If the execution order of competing processes using the bus can be predicted, then the communication delay for each message-passing operation could be determined by a measured network latency function. In order to simulate the random aspects of using the bus among a set of competing processes, a random function $Net\_random(k)$ is used to abstract the network to determine the next process that uses the bus from $Net\_Q(t)$, a queue of competing processes at time t. The determination of $Net\_Q(t)$ is done in the high-level model.

3.2.3 Modeling owner workload effect

For a computational segment c, the function $T_{cpu}(c, M_i)$ gives the amount of CPU time demanded in a dedicated environment. The real execution time of c in a nondedicated environment is affected by the execution of the owner workload. To model the effect of the owner workload, a discrete time model is used, which, as described in Section 2, assumes a geometric distribution with mean $\frac{1}{P_i}$ for the owner think time. When an owner process starts, an executing parallel task is suspended and the owner process is immediately started to execute for O_i time on machine M_i. Once the owner processes complete execution, the parallel task restarts execution and is guaranteed to execute for a time period of τ, the slice time for scheduling in the operating system (for example, the size of a time slice is one second in the Sun workstation system), before the owner may issue another process requesting the machine.

From the owner workload model, it is known that computational segment c must finish in a time period of $T_{cpu}(c, M_i) + O_i \times \frac{T_{cpu}(c, M_i)}{\tau}$ at machine M_i. The execution time of a computational segment at a single machine is thus the CPU demand time plus the time to complete any owner processes that occur during the segment tenure in the system:

$$T_r(c, M_i) = T_{cpu}(c, M_i) + n \times O_i,$$

where n is the number of owner process requests. The owner process can make a request after each scheduling slice time when the task uses the machine. Hence the number of owner requests is binomially distributed:

$$n = Bin(v, j, P_i) = \binom{v}{j} P_i^j (1 - P_i)^{v-j}, \qquad (5)$$

where $v = T_{cpu}(c, M_i)/\tau$. Thus, the expected execution time of segment c at machine M_i is

$$T_r(c, M_i) = T_{cpu}(c, M_i) + \sum_{j=0}^{v} O_i \times j \times Bin(v, j, P_i). \quad (6)$$

3.3 Performance prediction

The main function of the predictor is to calculate communication delay using the performance parameters from the two-level model. To predict the parallel execution performance of an application, the predictor has four input parameters: PEG graph, the execution time function of computational segments given by equation (6), communication startup time and communication transmission rate, and a random function that simulates bus arbitration. Then the predictor outputs the starting and ending times of the execution of each operation. The basic process flow of the prediction algorithm is given as follows, where two work queues, Q_w and Q_{net}, are used, respectively, to store executable operations and to store send operations competing for the network in a time period t.

Step 1 : Initially, Q_{net} is set to empty, and Q_w stores all the executable operations that have in-degree of 0 in the PEG graph.

Step 2 : For each computational segment and each receive operation in Q_w, do: (1) directly compute its ending time using the input parameters and its starting time, (2) update the starting time of its immediately-following operation according to its ending time, and (3) put its immediately-following operation into Q_w if the immediately-after operation become executable.

Step 3 : Put all the operations of Q_w which compete for the network at current time into Q_{net}, and determine the first operation to use the bus by the given random function. Based on communication transmission rate and startup time, calculate the ending time of the chosen operation.

Step 4 : If Q_w is not empty, goto step 2. Otherwise, if Q_{net} is not empty, goto step 3. The iteration process continues until both queues are empty.

From the above prediction algorithm, the parallel computing time of application A is calculated as

$$T_p = max\{T_e[i]|(1 \leq i \leq m)\}, \quad (7)$$

where $T_e[i]$ is the ending time of the last operation in the i-th task.

4 Prediction validation

To evaluate a parallel computation in a nondedicated heterogeneous NOW, *speedup* and *efficiency* are two useful metrics which have been defined in [6]. The speedup is

$$SP_A = \frac{min_{j=1}^{m}\{T(A, M_j)\}}{T_p}, \quad (8)$$

where $T(A, M_j)$ is the sequential execution time for application A on heterogeneous machine $M_j(j = 1, 2, \cdots, m)$. The efficiency is

$$E = \frac{\sum_{j=1}^{m}(W_j \times |A_j|/S_j)}{\sum_{j=1}^{m}(T_p - T_j^o)W_j}, \quad (9)$$

where $|A_j|/S_j$ represents the CPU busy time needed to execute A_j, the task allocated onto machine M_j, and T_j^o is the time of machine M_j executing the owner workload.

To validate our performance prediction method, a matrix multiplication program is abstracted by the two-level model proposed in Section 3, and implemented using PVM on a heterogeneous network system that consists of three SPARC 10-30 workstations, five SPARC 5-70 workstations and five SPARCclassic workstations, connected by an Ethernet. Within the memory bound of workstations, SPARC-5, SPARC-10 and SPARCclassic have power weights 1, 0.83 and 0.45 respectively for the matrix multiplication.

In the performance prediction and measurement of the program, the matrix size was fixed at 1000×1000, which was bounded by the memory limit of $16MB$ on SPARC-classic workstations. The system computing power was scaled either by increasing the number of workstations (physical scaling), or by upgrading the workstations' power (power scaling). The owner task size was set to 2 seconds, i.e., $O_i = 2$ $(1 \leq i \leq m)$. In our performance evaluation, two types of owner workload environments were used: a dedicated environment with 0% owner workload utilization and a nondedicated environment with 10% owner workload utilization. The slice time for scheduling is 1 second. The precisions of prediction results are evaluated as follows:

$$diff = \frac{|\text{measured result} - \text{predicted result}|}{\text{measured result}} \times 100\%.$$

Table 1 reports the difference for prediction results on execution time (T), speedup (SP) and efficiency (E). All the maximal differences are less than 10% and most of them are less than 5%.

physical scaling	u_o=0%			u_o=10%		
	T	SP	E	T	SP	E
	4%	4%	8%	6%	6%	9%
power scaling	u_o=0%			u_o=10%		
	T	SP	E	T	SP	E
	3%	3%	2%	3%	3%	4%

Table 1: Maximal difference between prediction and measurement

5 Conclusion

In this paper, we propose a practical performance prediction framework to provide useful results to guide the task allocator and scheduler in a heterogeneous network system. Based on this framework, we are conducting comprehensive analysis of the characteristics of parallel computing in nondedicated heterogeneous NOW systems.

Acknowledgement: We are grateful to N. Wagner for his careful reading of the paper and his constructive comments and suggestions.

References

[1] V. S. Adve, *Analyzing the behavior and performance of parallel programs*, Ph. D Thesis, University of Wisconsin-Madison, Dec. 1993.

[2] D. Culler, R. Karp, D. Patterson, A. Sahay, K. E. Schauser, E. Santos, R. Subramonian and T. Eicken, "LogP: Towards a realistic model of parallel computation", *PPOPP'93*, May 1993, pp. 1-12.

[3] J. Ferrante, K. J. Ottenstein, and J. D. Warren, "The program dependence graph and its use in optimization", *ACM TOPL*, Vol. 9, No. 3, 1987, pp. 319-349.

[4] V. S. Sunderam, "PVM: a framework for parallel distributed computing", *Concurrency: Practice and Experience*, 2(4) 1990, pp. 315-339.

[5] X. Zhang, Z. Xu and L. Sun, "Performance predictions on implicit communication systems", *SPDP'94*, Oct. 1994, pp. 560-568.

[6] X. Zhang, Y. Yan, and H. Yang, *Evaluating the performance and scalability of heterogeneous computing on networks of workstations*, Tech. Rep., High Performance Lab., Uni. of Taxas at San Antonio, Nov. 1994.

[7] X. Zhang, Y. Yan, and K, He, "Latency metric: an experimental method for measuring and evaluating program and architecture scalability", *JPDC* , Vol. 22, No. 3, 1994, pp. 392-410.

A COMPUTATIONAL MODEL FOR A CLASS OF SYNCHRONIZED DISTRIBUTED MEMORY MACHINES

James J. Carrig Jr. and Gerard G. L. Meyer
Department of Electrical and Computer Engineering
The Johns Hopkins University
Baltimore, MD 21218
jcarrig@mail.ece.jhu.edu, gmeyer@mail.ece.jhu.edu

Abstract – *This paper presents a model suitable for tuning the performance of algorithms on synchronized distributed memory machines, such as the CM-5. The model fits between uncomplicated models which are useful for high-level algorithm design, and very detailed models needed for machine design. Validation is provided by comparison to CM-5 measurements while varying problem, algorithm, implementation, and machine parameters. The model is applied to analyze CM-5 performance, provide guidelines for efficient implementation, predict scalability of algorithms, and to suggest system enhancements.*

INTRODUCTION

Compiling and running old Fortran source code on a new computer is often the first phase in porting an application. This first phase frequently does not produce efficient code, and so a second phase is required where the component algorithms are tuned for a more optimal implementation. Quite often the application, its algorithms, their implementation, and even the computer system are parameterized. When this flexibility exists, a model-based approach is advantageous.

The practicality of using computational models has led to the development and use of many different models for algorithm design. PRAM models have been used extensively to simulate shared memory machines [6, 9]. Many people have noted that the trend in parallel computing is to build distributed memory machines which emulate shared memory operation [1, 5]. Although these machines may look like shared memory machines to the user, Cypher and Sanz note that they do not perform like shared memory machines [4]. More appropriate models need to be developed for performance tuning.

We believe that the most useful models are built upon two distinct sets of parameters. The first set, the *architectural parameters*, depends only on the system being modeled. These parameters specify either the quantity of hardware or the speed of the system. The second set, the *execution parameters*, are derived by consideration of how the algorithm will be mapped to the architecture. Therefore the values of the execution parameters may depend on the application, algorithm, implementation, and architectural parameters. Once these relationships are known, the optimal algorithm and implementation parameters may be determined mathematically. If two systems were defined using the same execution parameters, porting code from one system to the other would still require the effort of mathematically determining the optimal parameters, but would not require the extraction of any new execution parameters.

Even if a model-based approach were shown to yield superb results, no one will use the model if each use requires the manual extraction of a large number of parameters. This concern immediately brings two questions to mind. How well can real systems be modeled with only a few parameters, and is it possible that a wide range of systems can be modeled with a common set of parameters? In this paper we address only the first question. We demonstrate that 8 parameters are sufficient to capture many of the intricacies of one of the most widely used distributed memory computers: Thinking Machine Corporation's CM-5.

SDM MODEL

The Synchronized Distributed Memory (SDM) model consists of a program manager (PM) fully interconnected to p ordered processing nodes, each capable of producing v simultaneous outputs. The main program is contained in memory attached to the PM, but user variables may also be stored in local memories attached to each node. Unlike a real CM-5 with vector units, the model has only one memory per node.

The program is always in exactly one of four states: the PM is busy performing arithmetic computation, the PM is broadcasting the step instructions to the nodes, the nodes are busy performing identical arithmetic computation, or there is some communication of user data. The total execution time t of a given algorithm on the SDM model is defined in (1) as the sum these four parts:

$$t = t_{PM} + t_B + t_A + t_C \qquad (1)$$

where t_{PM} is the sum of all time spent performing PM arithmetic, t_B is the sum of all time spent broadcasting steps, t_A is the sum of all time spent in node arithmetic, and t_C is the sum time of all communication.

In the first state, the PM is busy performing arithmetic while the nodes are idle. Execution time is simply modeled as the weighted sum of algebraic computations:

$$t_{PM} = \sum_a n_a^{PM} \tau_a^{PM} \qquad (2)$$

where n_a^{PM} represents the number of algebraic operations of type a, τ_a^{PM} is the time to compute an operation a, and $\sum_a$ is the sum over all types of arithmetic.

In the second state, the PM broadcasts instructions to the nodes. These instructions initiate computation or communication on the nodes, and assure correctness of the parallel algorithm. It is because of these frequent broadcasts that we call the model *synchronized*. As a consequence, algorithms are executed in steps where each step executes instructions contained in one broadcast. Otherwise, these broadcasts may be viewed as pure overhead since no user data is exchanged.

A broadcast contains computation or communication instructions, but not both. Define N_A as the number of arithmetic steps, and N_C as the number of communication steps. Assuming each broadcast requires time τ_S, we derive t_B as the product below:

$$t_B = (N_A + N_C)\tau_s. \tag{3}$$

In the third state, a parallel computation step, all nodes perform identical arithmetic operations in identical time, though some may act on dummy data. A node with n parallel multiplications to perform will perform them v at a time in the same time it takes it to perform $\lceil n/v \rceil$ serial multiplications. It is therefore appropriate to discuss the number of serial computations executed by the nodes in a computation step.

Let $n_a(i)$ be the number of serial computations it takes to complete the arithmetic a at step i. The execution time for a computation step is then the weighted sum of the number of arithmetic operations at each step by the time for each operation. The total time spent in node arithmetic, t_A, is found by summing over all operations at each of the N_A arithmetic steps:

$$t_A = \sum_{i=1}^{N_A} \sum_a n_a(i)\tau_a. \tag{4}$$

In the fourth state, a communication step, each node decodes the communication instructions, packs whatever data it must communicate, and transmits data. Communication is point-to-point. A node is free to send and receive at the same time since receiving is assumed to take no time or resources. Define $t_C^k(j)$ as the time for the kth node to communicate at step j. The communication time for a step is found by taking the maximum over all of the nodes. The total communication time, t_C, is the sum time of each of the N_C communication steps:

$$t_C = \sum_{j=1}^{N_C} \max_{k=1,\dots,p} t_C^k(j). \tag{5}$$

In order to calculate $t_C^k(j)$, we must consider the three phases of communication: decoding, packing, and transmission. τ_d is the time needed for the nodes to decode the communication instructions and determine what communication is required. τ_p relates the time needed to pack the data for transmission to the amount of data being packed. τ_t relates the time needed to transmit a vector to the number of vectors being transmitted.

Communication falls into two categories: a directly specified constant shift, and everything else. A constant shift requires each node to send the same amount of data to a node whose address differs by a constant amount. Directly specified means that indirect addressing is not used so the shift can be recognized by the compiler. An indicator function I_s, is set to 1 to flag a directly specified constant shift.

When $I_s = 1$ decoding is not required, and the data is packed very efficiently. If $n_w(j)$ is the number of *words per vector shifted*, the packing for this communication takes time $n_w(j)\tau_p$. Transmission is proportional to $n_t^k(j)$, the number of *vectors transmitted* from node k at step j. All communication other than a directly specified constant shift, takes time τ_d to decode and a packing time dependent on the number of different words to be transmitted. $n_p^k(i)$ is the number of *words packed* by the kth node at step i. For example if node 1 sends nodes 2-4 m vectors of length r elements (words) at step 5, the m vectors of length r are packed and transmitted to 3 nodes so $n_p^1(5) = mr$ and $n_t^1(5) = 3m$. Since this is not a shift, $I_s(5) = 0$ and $n_w(5)$ is not relevant.

The communication time for each node as described above, is defined in (6):

$$t_C^k(j) = \begin{cases} \tau_d + n_p^k(j)\tau_p + n_t^k(j)\tau_t : I_s = 0 \\ n_w^k(j)\tau_p + n_t^k(j)\tau_t : I_s = 1. \end{cases} \tag{6}$$

ALGORITHM SPECIFICATION

Algorithm performance on a real machine is closely related to program specification, consequently more information than is provided in a dependency graph is needed to obtain accurate performance estimates.

Data must be grouped into vectors. In the SDM model, a vector is any small grouping of variables always assigned to the same node and treated as a unit. This grouping is designed to account for the parallel and serial designation of the CM-5 array axes. CM-5 arrays are distributed across node memory by division along parallel axes. Elements with identical parallel indices are always stored in the same physical memory.

The vectors and arithmetic must be assigned to nodes, and ordered in slightly more detail than required by the dependency graph. In particular, the ordering requires that the nodes perform identical arithmetic at all times, each producing at most v simultaneous outputs. For now we call each operation occurring at the same time a mini-step. These mini-steps will be grouped together to form the steps needed for the model inputs.

Once the computation and data assignment have been specified, it is possible to determine the communication between nodes by analyzing the data mapping and the dependency graph. Often the data mapping is parameterized by p, the number of nodes, which is unknown at compilation. As p increases, say from 1 to N, communication generally increases as well. If communication is required for any value of p, then communication is always generated by the compiler. For those values of p for which the communication is unnecessary, we view the operation as a node sending data to itself.

The final part of the specification, needed to determine t_B, is dependent upon implementation decisions. A computation step consists of all consecutive mini-steps involving node computation unbroken by branching, and a communication step consists of all consecutive mini-steps which involve data communication unbroken by branching. This division requires some knowledge of the way in which the algorithm will be programmed.

Algorithms for large numerical problems will generally be programmed using loops. Since a loop contains branching statements, knowledge of which mini-steps are

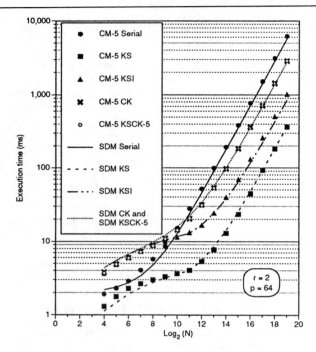

Figure 1: *Comparison of the CM-5 performance to the SDM model prediction for 5 communication patterns as the recursion length N varies by nearly 5 order of magnitude.*

in the interior of the loop is required to determine step boundaries. Similarly, some knowledge of the communication implementation is required. If communication step j consists only of a directly specified constant shift, then $I_s(j)$ is set to 1 for that step.

VALIDATION

Our validating examples are based on linear recursion algorithms because such algorithms arise in problems of interest to us and because many linear recursion algorithms have been proposed, providing an example of when a model based approach could be used to help in the selection of the optimal algorithm and implementation.

Linear recursion problems are characterized by two positive integers: the recursion length N, and the recursion dimension r.

Linear Recursion Problem: Given an $r \times r$ matrix F, $\{u_i \in E^r | i = 1, \ldots, N\}$ compute $\{z_i \in E^r | i = 1, \ldots, N\}$ such that

$$z_i = \begin{cases} u_1 & : \ i = 1 \\ Fz_{i-1} + u_i & : \ i = 2, \ldots, N. \end{cases} \quad (7)$$

Many linear recursion algorithms have been presented in the literature. We present results for a Serial algorithm, two implementations of an algorithm proposed by Kogge and Stone (KS and KSI), an algorithm proposed by Chen and Kuck (CK), one hybrid algorithm (KSCK-5), a partitioned version of the Kogge and Stone algorithm (PKS), and a partitioned version of the Chen and Kuck

algorithm (PCK) [7, 3]. PKS is presented in more detail by Kruskal, Rudolph, and Snir [8]. The partitioned algorithms possess a design parameter m, have low communication requirements, and approaches the minimum computational bound presented by Snir [10].

The u_i's and z_i's are treated as vectors in the implementation, meaning that they are considered indivisible during data distribution. Each algorithm is mapped to the SDM model with the u_i's distributed as evenly as possible, and an identical distribution is used for the z_i's.

For the partitioning algorithms, m specifies both the number of independent calculations in stages 1 and 3, and the size of the stage 2 recursion. In general, m should be large enough to keep all the nodes active and small enough to keep the stage 2 communications minimal. Since each node of the SDM model produces v outputs on identical operations, computations are mapped to nodes v at a time. When $m < pv$, there are not enough independent calculations to keep all the nodes busy.

For recursions with $N \gg pv$, $m = pv$ minimizes serial computation. We will demonstrate, however, that when step overhead dominates the execution time, it is advantageous to specify more simultaneous activities so performance is optimal for some $m > pv$. On the other hand, when communication dominates the execution time, not all the available nodes should be used, so $m < pv$ is optimal.

Since we are primarily interested in problems where N is very large, stages 1 and 3 consist of many steps and will, of course, be programmed using loops. Step boundaries for the first and third stages then depend on an implementation parameter, s, the level of loop unrolling.

The level of loop unrolling corresponds directly to the number of mini-steps computed at each iteration of the loop. Since each iteration of the loop involves branching, only the s mini-steps within the same loop iteration can be packed into each same step. If s is small, the overhead of the step broadcast τ_S may be large compared to the useful computation at each iteration.

The serial algorithm begins with the data mapped onto node memory, so that 2 communication steps are required. The first *gathers* the u_i from each node to the PM, the second *scatters* the z_i from the PM to each node.

The KS algorithm requires $log_2(N)$ alternating steps of computation and communication. KS may be programmed using directly specified shifts, so $I_s(j) = 1$. The KSI algorithm is simply another implementation of KS where indirect addressing has been implemented via an index array.

The CK and KSCK-5 algorithm computation steps are similar to KS, but the communication steps are quite different. CK requires non-overlapping *local broadcasts*.

Table I: *The 8 architectural parameters and corresponding values sufficient to simulate a CM-5.*

p	v	$\tau_{\times+}^{PM}$	$\tau_{\times+}$
32-1,024	32	1.4 μs	3.5 μs

τ_s	τ_d	τ_p	τ_t
48 μs	81 μs	76 μs	5.1 μs

That is, at step j of the CK algorithm, $N/2^j$ vectors are broadcast to 2^{j-1} destinations. The KSCK-5 algorithm communicates similarly to the KSI algorithm for the first 5 steps, and identically to CK thereafter.

The SDM architecture parameters values used to simulate a CM-5 are given in Table I. Of course p, the number of nodes in the SDM model, is chosen to be the number of nodes on the CM-5. Each CM-5 node has 4 vector units with 8 outputs so we take $v = 32$. The time for a step broadcast, τ_S, corresponds to the time to start a *PE code block* on the CM-5, which has been estimated by Thinking Machines Corporation to be 48 μs [11]. The remaining parameters were chosen to fit the data in Figure 1.

Figure 1 compares model prediction to CM-5 measurement as the problem size varies from 16 to 512K (2^{19}) by powers of 2, with $r = 2$. Symbols represent CM-5 measurements, whereas lines connect model predictions. Since predictions were computed only for N equal to powers of 2, the effect of the ceiling functions is not shown.

Let t_p represent one of the model predictions shown in Figure 1 and t_m be a corresponding measurement. For each measurement, we compute the maximum percent deviation as:

$$\epsilon = 100 \times \frac{|t_p - t_m|}{t_m}. \qquad (8)$$

The measurements shown in Figure 1 deviate from the model by less than 20%. Including all of the measurements in [2], the SDM model never deviates from CM-5 measurements by more than 30%.

PERFORMANCE TUNING

Each partitioning algorithm possesses a design parameter m, which controls the amount of parallelism in the algorithm. When $m = 1$ the algorithm is serial. When $m = N$ there is no partitioning at all. The selection $m = pv$ is just large enough to utilize all available nodes, minimizing computation when $N \gg pv$. With the data mapping in the previous section, m also parameterizes the implementation.

In stages 1 and 3, a large value of m packs more computation into each step, reducing the number of steps, and therefore t_B. However, increasing m also increases the arithmetic and communication times for stage 2. This trade-off is shown in Figure 2. We will show below that the growth of the term t_C during stage 2 ultimately limits the number of parallel nodes which can be effectively used.

There is a second means of reducing t_B, loop unrolling, which has no effect on stage 2. Consider a recursion with $r = 2$ and $m = 2,048$. When $s = 1$, t_B accounts for 69% of the execution time of stages 1 and 3. When $s = 16$, that is the loops are unrolled 16 levels, the influence of t_B is reduced to 18%.

Figure 2 shows the behavior of the PKS algorithm for both $s = 1$, and $s = 16$ as a function of m. Notice that the optimal value of m decreases as s increases.

We conclude by considering the speedup predicted by the model. For each value of p the optimal value of m is independently derived, and the speedup over one node is computed as:

$$S_p = t_1/t_p \qquad (9)$$

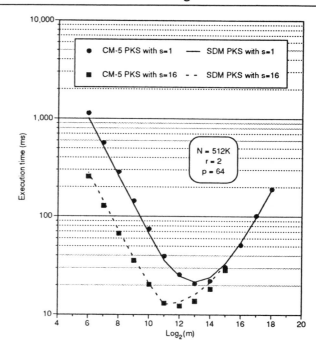

Figure 2: *The influence of loop unrolling and algorithm granularity are captured by model parameters s and m respectively. Although every point in the graph solves the same recursion, performance varies by a factor of 100 when s and m are varied. The SDM model may be applied to select the parameters which yield the optimum performance.*

where t_p is the execution time of the parallel algorithm on p nodes. We compare the ideal linear speedup ($S_p = p$), the algebraic speedup ($t_B = t_C = 0$), and the speedup predicted for the best CM-5 algorithm (PKS with $s = 16$ and the optimal m selected) to the speedup computed from actual CM-5 measurements. Because there is no such thing as a 1 node CM-5, we computed the speedup over 32 nodes and normalized that value to lie on the SDM model curve. Figure 3 presents the case with $N = 512K$ and $r = 2$. While the algebraic speedup continues to improve beyond 4,096 nodes, the speedup predicted by the SDM model levels off after only 256 nodes. The speedup never decreases since the optimal m may yield an algorithm which does not use all available nodes.

CONCLUSIONS

We presented a parameterized model for synchronized distributed memory machines and a selection of parameters for which the model predictions closely match CM-5 performance measurements. The SDM model is designed to assist in the porting and tuning of algorithms. As a consequence, it is somewhat more detailed than models used in algorithm design, and somewhat less detailed than models used for machine design.

The SDM model is neither an attempt to uncover the mapping from Fortran source code to execution time, nor an attempt to model every operation possible on a real computer system. On the contrary, it lumps together the

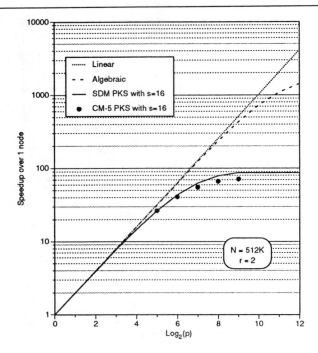

Figure 3: *Although the algebraic speedup of the PKS algorithm is nearly ideal, the SDM model provides a much more realistic prediction.*

appropriate system activities in order to present a view of the machine which is consistent and intuitive to a programmer. When possible, we model algorithm activities without reference to the Fortran code implied by these activities. To accomplish this goal, we have chosen to model many system activities as "software overheads," but this view is not exact, and there undoubtedly exists other views which give equally satisfying results.

We demonstrated that the model successfully predicts fairly low-level optimizations such as loop unrolling and alternative communication specifications. We saw that the notion of algorithm steps is crucial to obtaining high performance on SDM machines, and that the cost of communication can limit the scalability of even well designed algorithms to a point which is far below the algebraic scalability.

Using the SDM model, we have identified two software improvements. First, unnecessary communication is often generated because the layout is parameterized by the number of nodes, which is unknown at compilation. By specifying the number of nodes at compilation, the precise data layout can be determined, so unnecessary communication could be eliminated. Second, the data parallel programming model provides more synchronization than is necessary for sequential correctness of the algorithm. Applying both of these enhancements to the PCK algorithm on our typical problem ($N = 512K, r = 2$), we would expect performance on 64 nodes to improve by a factor of 3.7.

This paper presents evidence that 8 parameters are sufficient to model a real computer system to within 30% accuracy while widely varying the problem, algorithm, implementation, and hardware parameters. The model can then be used to select the best algorithm, and the corresponding optimal set of parameters. Future work is required to address questions regarding the systematic generation of models, the sensitivity of the model parameters, and to determine if a common set of parameters can be used as inputs to model a range of real computer systems.

ACKNOWLEDGMENTS

This work was supported in part by a grant of HPC time from the DoD HPC Shared Resource Center, Army High Performance Computing Research Center at the University of Minnesota CM-5, and also by the National Center for Supercomputing Applications under grant number ASC930006N.

REFERENCES

[1] Bell, G. "Ultracomputers a Teraflop Before Its Time," *Communications of the ACM* **33**, (August 1992), pp. 27–47.

[2] Carrig Jr., J. J., and Meyer, G. G. L. *A Model for Synchronized Distributed Memory Machines.* Electrical and Computer Engineering, The Johns Hopkins University, 95-05, (February 1995), 25 pp.

[3] Chen, S.-C., and Kuck, D. J. "Time and Parallel Processor Bounds for Linear Recurrence Systems," *IEEE Trans. Comput.* , (July 1975), pp. 701–717.

[4] Cypher, R., and Sanz, J. L. C. *The SIMD Model of Parallel Computation*, Springer-Verlag, 1994, 517 pp.

[5] Dongarra, J. J., Duff, I. S., Sorensen, D. C., and van der Vorst, H. A. *Solving Linear Systems on Vector and Shared Memory Computers*, Society for Industrial and Applied Mathematics, 1991, 256 pp.

[6] Hwang, K. *Advanced Computer Architecture: Parallelism, Scalability, Programmability*, McGraw-Hill, 1993, 771 pp.

[7] Kogge, P. M., and Stone, H. S. "A Parallel Algorithm for Efficient Solution of a General Class of Recurrence Equations," *IEEE Trans. Comput.* , (August 1973), pp. 786–792.

[8] Kruskal, C. P., Rudolph, L., and Snir, M. "The Power of Parallel Prefix," *IEEE Trans. Comput.* , (October 1985), pp. 965–968.

[9] Leighton, F. T. *Introduction to Parallel Algorithms and Architectures: Arrays · Trees · Hypercubes*, Morgan Kaufmann, 1992, 831 pp.

[10] Snir, M. "Depth-size Trade-offs for Parallel Prefix Computation," *J. Algorithms* , (June 1986), pp. 185–201.

[11] Thinking Machines Corporation. *CM-5 CM Fortran Performance Guide*, Cambridge, Massachusetts, 1994, 197 pp.

An Analytical Model of the Standard Coherent Interface 'SCI'

A.J. Field and P.G. Harrison
Department of Computing, Imperial College
180 Queen's Gate, London SW7 2BZ U.K.

Abstract

We present a new analytical performance model of the IEEE P1596 Standard Coherent Interface operating on the default unidirectional ring architecture. The performance metrics are derived from the equilibrium probability of a cache line being in a given state; these are found by solving a set of fixed point equations. From this we derive expressions for the message traffic emanating from each node and over the ring taking into account the relevant traffic priorities in SCI. Further analysis then yields the mean memory access time and processor utilisation. We demonstrate the application of the model by comparing the performance of two different node configurations.

1 Introduction

In this paper we present an analytical model of a ring-based shared-memory multiprocessor operating the IEEE P1596 Standard Coherent Interface (SCI) [3] and we assume an implementation of the protocol on the default unidirectional ring network. Full details of the protocol may be found in [2,3], for example, but we note the key point which is that, in the absence of a shared bus for broadcasting coherency information, the sharers of a cache line are represented explicitly in the form of a *sharing list*. We demonstrate the use of this model by comparing the performance of two internal node architectures when the protocol is being used as an interconnect medium for multiple bus-based multiprocessors.

To date, the only analytical model relevant to SCI that we are aware of is that of [5] which uses an M/G/1 model for analysing slotted ring networks, suitably adapted for a subset of the SCI traffic. The paper does not address SCI cache coherency, however, and so does not model the message traffic generated by each node to maintain a coherent global memory. Our approach is based on that of [4] for modelling bus-based coherency protocols; it differs in its characterisation of memory usage and in the way it models ring traffic

and queueing for the cache and memory.

1.1 Workload Assumptions

We assume that memory accesses are either to *private* or *shareable* memory and that shareable memory is further classified into *control variables* and *shared areas*. Control variables comprise synchronisation variables like semaphores and locks for controlling access to shared structures; these variables are typically relatively small in number but addressed with relatively high frequency. The shared areas contain larger data structures, typically with access priveliges governed by control variables, that are relatively infrequently accessed. It is assumed that the control variables are memory mapped in such a way that they may *all* reside in the SCI cache at the same time so that there are no conflicts in the cache between accesses to these variables. The cache thus contains disjoint sets of lines: those which may contain both control variables and shared area data (we call this "Region I" of the cahe) and those which may contain only shared area data ("Region II"). Within the control variables and shared areas, memory accesses are assumed to be uniformly distributed; within each region, the cache lines are then statistically identical.

2 A New Model for SCI

The approach we have taken uses the basic method in [4] to obtain the equilibrium cache line state probabilities and uses them to determine the message arrival rates for a separate model of the ring.

We assume the following parameters:

α – the probability of a read

β – the probability of a hit in the SCI cache

γ_1 – the probability that a memory access is to a control structure

γ_2 – the probability that a memory access is to the shared areas

γ_3 – the probability that a memory accesses is to private data

σ – the probability that an access to private data, which is mapped into the global address space, is held locally

K – The number of nodes

N – The total number of shareable memory blocks

m – The number of (shareable) lines containing control variables

n – The capacity (in blocks) of each SCI cache

τ – The rate at which a busy processor leaves the "think" state (note that 1τ is the mean think period)

Note that $\gamma_1 + \gamma_2 + \gamma_3 = 1$. We aim to determine the probability, π, that a processor is busy doing useful work and also the mean time to service a memory request.

Cache Line States

A memory block is either **Home** (uncached) or **Cached** and cached copies may either be *clean* or *dirty*. There are eleven basic line states although eight of them we will subdivide further in accordance with the type of data they contain (control variables or shared area data). The states are: **1 Private Clean**–The location contains a clean copy of private (non-shareable) data; **2 Private Dirty**; **3 Only Clean**–The location contains the only cached copy of a memory block and is clean; **4 Only Dirty**; **5 Head Clean**–As 3 but it is at the head of a list containing at least two members; **6 Head Dirty**; **7 Mid Clean**–The location is in the middle of the list (i.e. at neither the head nor tail) and the cached copy is clean; **8 Mid Dirty**; **9 Tail Clean**–As 5 but the location is at the tail of the sharing list; **10 Tail Dirty**; **11 Invalid**. For Region I of the cache we annotate states 3–10 with the subscript 'a' if the (shared) block contains control information and 'b' if it contains shared region data. We make the assumption that different types of data do not co-exist on the same cache line which is easily ensured by a suitable allocation of global memory addresses. The cache lines in Region I therefore have nineteen possible states, and those of Region II eleven.

2.1 Actions generated by a processor

A given processor emerges from the think state at rate τ, and according to the state of the cache location it is accessing will generate one of the following processor *actions* (the distinguishing subscripts 'a' and 'b' are omitted as the actions are the same for each): **Creation (CR)**–A read or write miss on a location in state 11 not cached by other processors; **Addition (AD)**–As above but already cached. **Deletion-Creation (DC)**–A read or write miss on a *home* block mapping to a cache location previously in state 3–10; **Deletion-Addition (DA)**–A read miss on an already *cached* block mapping to a location in state 3–10; **Deletion-Reduction (DR)**–A write miss on a *cached* block mapping to a location in state 3–10; **Take-Reduction (TR)**–A write hit on a block mapping to a location in state 7–10; **Head-Reduction (HR)**–A write hit on a location in the processor's cache but in state 3,4; **Invalid-Reduction (IR)**–A write miss on a *cached* block mapping to a location in state 11.

To add to a sharing list a processor sends a short message to the memory, which sends another short message to the current head, which in turn sends a long message containing the up-to-date copy of the block to the processor. Deleting from a list requires either one or two short messages depending on whether the cache line is at the head/tail or middle of its sharing list. Note that for each Deletion operation in state 2 then a *writeback* of the line will also be performed, requiring one additional long message to the memory.

2.2 The analytical model

We assume that the evolution of the state of a given location in a cache follows a Markov process, independent of the states of other locations. This process is irreducible, aperiodic and has a finite state space and thus has a steady-state. A separate Markov process is established for each cache region for the reasons already stated.

Let q_j be the steady state probability that a given cache location in Region I is in state j for $j = 1, 2, 3_a, 3_b, ..., 10_b, 11$. It is also the average proportion of cache locations in state j in equilibrium. Let q_j' be the same for Region II, $j = 1, ..., 11$.

State Transitions

The state transitions occur as a result of read and write operations to the cache. We imagine we are looking at a particular "observed" line of a cache and consider the various read/write operations which can change the state of that line. We omit much of the detail but note that the length of the mid portion of a sharing list is assumed to be geometrically distributed. With this assumption we can easily find the mean sharing list length (L_a, L_b for Region I and L' for Region II).

Transition Rates

The processors leave the think state at rate $\pi\tau$ and we consider those memory accesses which cause a change in line state. Note that the transition rates do not cover transitions from a state to itself, for example as a result of a read hit in states 1–10. Consider Region I first and define η_r to be the probability that, given a cache miss on a type r memory block, the block is cached elsewhere and ϵ_r to be the probability that a cached copy of a type r line is clean (r is either a or b). These are easily defined in terms of the q_j and q_j'.

We will not list every transition rate (the full list contains 30 rules), referring the interested reader instead to [1]. However, we consider a few specific cases to explain the basic principles:

- $s \rightarrow 2$, i.e. a transition from any state to the state "Private Dirty". This happens when private data (held in the global address space), whose address maps to the observed cache line, is written to. The associated transition rate is thus $\pi\tau\frac{(1-\alpha)\gamma_3}{n}$.

- $\{1, 2, 3a, ..., 10a\} \rightarrow 6b$. A cache line holding private data or control variables will transit to "head dirty" if a read to a dirty, and already cached, copy of a shared area variable maps to the observed line address. The transition rate is thus $\pi\tau\frac{\alpha\gamma_2\eta_b(1-\epsilon_b)}{n}$.

- $5a \rightarrow 7a, 6a \rightarrow 8a$. This happens on a remote read miss from on of the (average $K - L_a$) processors who do not currently have a cached copy of the control variable in the observed line. A new entry will be added to the sharing list making the observed line state transit from "head" to "mid". The associated rate is $\frac{(K-L_a)\alpha\gamma_1}{m}$.

The transitions for Region II are derived similarly except there are only eleven states and 14 rules since control variables map only to Region I.

The balance equations can be derived from the transition rates; these are not linear, but define a *fixed point* on the vector of steady-state probabilities for each region and so in practice are solved iteratively.

The probability that a processor emerging from a think state generates each action are easily determined in the form of a table δ indexed by state and action. For example $\Pr\{CR$ in state $11a\} = \delta_{11a,CR} = 1 - \eta_a$, and $\Pr\{$ DA in state $7a\} = \delta_{7a,DA} = \alpha\eta_a(1 - \beta)$.

Message Streams

Each action makes the processor send and receive a certain number of long or short messages through the ring. In order to calculate the time these messages take to be transmitted we need to calculate the rate at which they are produced. The next step requires us to define six new tables S, S', L, C, C' and M representing the number of short messages (Region I and Region II respectively), long messages, cache accesses (Region I and Region II respectively) and memory accesses, indexed by the line state and the action initiated.

For example, an AD action first experiences a cache miss and then takes a copy of the required block, adding it to the head of the current sharing list. It thus issues one short message to memory requesting the head of the sharing list (involving a memory access); this in turn sends a short message to the head of the list (involving a cache access), which in turn sends a long message containing the latest copy of the line to the originating processor (this must be written to the processor's cache and therefore involves one further cache access). Note that some messages may be destined for the local memory block of the processor and so will create no ring traffic. These are explicitly annotated with an * in the table. Thus, for the example above we have, $S_{11,AD} = 1^* + 1$, $S'_{11,AD} = 1^* + 1$, $L_{11,AD} = 1^*$, $C_{11,AD} = 3$, $C'_{11,AD} = 3$ and $M_{11,AD} = 1$. Similarly for the other actions/states.

2.3 Mean transmission time for a message

Given the above tables we can now determine the mean transmission time of a message around the ring. All messages issued by a transmitting node will perform one full circuit of the ring (i.e. through $K - 1$ ring buffers). The receiving node will extract the incoming packet and, at the same time, pass it on as an echo packet which we will assume is the same length as a short packet.

In SCI messages originating from the ring have priority over messages in the transmit queue originating from the node and so we appeal to Cobham's formulae to determine the mean transmission time of short and long messages around the ring, T_s and T_l respectively, on the assumption that the messages arriving at a processor are Poisson. This analysis assumes that the length of a long message is a fixed multiple M of the length of a short message. The various transmission times and utilisations can then be expressed in terms of the transmission time of a short message, t_{short} which is determined by the SCI link speed.

2.4   Cache/Memory Access Delay

We model the cache/memory controller as a first-come-first-served queue and appeal to a separate M/G/1 model of the queue in which the service time distribution is a probabilistic mixture of the (constant) cache and memory access times, t_{cache} and t_{mem} respectively. We will use the model to compare two alternative node architectures: the first is the conventional SCI model of Figure 1 and the second a similar design in which the memory is located on the system bus (as in a conventional bus-based multiprocessor). We first define the probabilities P_I and P_{II} that a memory reference addresses the cache in Regions I and II respectively:

$$P_I = \gamma_1 + (\gamma_2 + \gamma_3(1-\sigma))\frac{m}{n}$$

$$P_{II} = (\gamma_2 + \gamma_3(1-\sigma))\frac{n-m}{n}$$

Note that $P_I + P_{II} = 1 - \gamma_3\sigma$ which we denote by P_s and we write $\overline{P_I} = P_I/P_s, \overline{P_{II}} = P_{II}/P_s$.

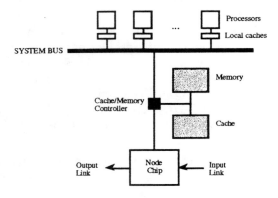

SYSTEM BUS

Figure 1: Standard Node Architecture

Case 1: Standard Memory Model

The mean number of cache and memory accesses (n_c and n_m respectively) for each action can be found from P_I and P_{II}, the equilibrium line state probabilities and the various tables defined earlier. From these we can find the rate at which cache (λ_c) and memory (λ_m) accesses are produced by a processor:

$$\lambda_c = \tau P_s + \lambda_t \frac{n_c}{n_c + n_m}$$

$$\lambda_m = \tau\gamma_3\sigma + \frac{\tau(\gamma_1 + \gamma_2)(1-\beta)}{K} + \lambda_t \frac{n_m}{n_c + n_m}$$

$\lambda_t = \lambda_s + \lambda_l$ is the rate at which messages are produce by a node. Short messages are produced at the rates:

$$\lambda_s = \tau P_I \sum_{s,a} q_s\, \delta_{s,a}\, S_{s,a}$$

$$+ \tau P_{II} \sum_{s',a} q'_{s'}\, \delta'_{s',a}\, S'_{s',a}$$

The equation for λ_l is identical except that L and L' replace S and S' respectively. Since the total arrival rate of cache and memory accesses is $\lambda_{cm} = \lambda_c + \lambda_m$ we can obtain the mean queuing time at the cache/memory controller from the Pollaczek-Khinchine formula:

$$Q_{cm} = \frac{\lambda_c t_{cache}^2 + \lambda_m t_{mem}^2}{2(1 - \rho_{cm})}$$

where ρ_{cm} is the cache/memory controller utilisation given by: $\rho_{cm} = \lambda_c t_{cache} + \lambda_m t_{mem}$ The mean time to service a memory request is then

$$
\begin{aligned}
T = {} & P_s\big[\frac{\lambda_s T_s + \lambda_l T_l}{\tau P_s} \\
& + n_c(Q_{cm} + t_{cache}) + n_m(Q_{cm} + t_{mem}) \\
& + p_{hit}\,(Q_{cm} + t_{cache})\big] \\
& + \sigma\gamma_3(Q_{cm} + t_{mem})
\end{aligned}
$$

where

$$p_{hit} = \overline{P_I}\left(1 - \sum_{s,a}\delta_{s,a}\right) + \overline{P_{II}}\left(1 - \sum_{s',a}\delta_{s',a}\right)$$

The processor utilisation is then:

$$\pi = \frac{\frac{1}{\tau}}{\frac{1}{\tau} + T} = \frac{1}{1 + \tau T}$$

Case 2: Revised Memory Model

This second design is similar to that of the original except that requests to the memory from the SCI ring may now have to compete with other requests on the local system bus; we model the additional (local) bus coherency traffic by an additional parameter τ' in the model. Equally, however, some accesses to memory, specifically private accesses which require no intervention from the SCI cache, can proceed independently of the SCI node controller. Thus the relative performance of the two designs is determined by the nature of the workload.

We model is essentially the same as the one above expect the quantities $\lambda_{cm}, Q_{cm}, \rho_{cm}$ are found to be

$$\lambda'_{cm} = \tau P_s + \lambda_t$$

$$Q'_{cm} = \frac{\lambda'_{cm} t_{cache}^2}{2(1 - \rho'_{cm})}$$

$$\rho'_{cm} = \lambda'_{cm} t_{cache}$$

by an analysis of the new cache/memory traffic.

3 Numerical Results

We now show some preliminary numerical results assuming the fixed parameters $\tau = 1.8 \times 10^6$ references/sec, $N = 10^6$ memory blocks, $n = 10^3$ lines per cache, $m = 10^2$ control variable lines $\alpha == 0.7$, $\gamma_2 = 0.2$, $\sigma = 0.95$ and $\beta = 0.98$. We further assume $t_{cache} = 240ns, t_{mem} = 480ns$ (note that reading and writing a line involves several memory cycles). The SCI ring is assumed to be fibre optic and is taken to operate at 1Gbit/sec so that 16 byte short messages $t_{short} = 64ns$ to transmit from one ring node to the next. A long message is taken to be 80 bytes on average, i.e. 5 times that of a short message. The additional coherency traffic on the system bus is taken to be 10% of the total memory access rate, that is $\tau' = 0.1\tau$.

Figure 2 shows the *elongation* for varying K for $\gamma_1 = 0.1$ for bothe original and revised node archiectures, and Figure 2 the same for $\gamma_1 = 0.3$. The elongation is a measure of the overhead, on a single memory access time, that is imposed by SCI on a read miss:

$$elongation = \frac{T}{\beta\, t_{cache} + (1 - \beta)\, t_{mem}}$$

An analysis of the memory (i.e. bus) queues reveals that the critical factor affecting the models is contention for access to the node memory. In the revised model memory accesses from remote processors have to contend with local system bus traffic, including local coherency protocols. Equally, however, remote accesses to the SCI cache can now be serviced at the same time as accesses to the local memory. This leads to a tradeoff between the two designs with the trends being hard to predict in the absence of a model. For this parameterisation the crossover occurs between $\gamma_1 = 0.1$ and $\gamma_1 = 0.3$ as shown.

Elongation

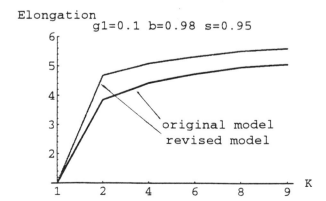

Figure 2: Elongation as a function of K ($\gamma_1 = 0.1$)

Elongation

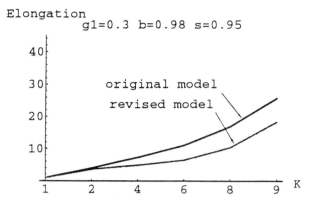

Figure 3: Elongation as a function of K ($\gamma_1 = 0.3$)

A more detailed discussion of the model and its sample output may be found in [1]. We are currently validating the model with respect to an execution-driven simulation of the SCI protocol running codes from the SLASH suite. SPLASH programs are executed in order to measure not only performance metrics like elongngation, but also the parameters used in the model. By running the model with these parameters we can then compare the model output with that of the original executing SPLASH code. Preliminary results show a promising match for MP3D at least but this exercise is, as yet, incomplete.

References

[1] A.J. Field and P.G. Harrison. "An Analytical Model of the Standard Coherent Interface 'SCI'", Internal Report, Department of Computing, Imperial College, 1995.

[2] S. Gjessing, D.B. Gustavson, J.R. Goodman, D.V. James and E.H Kristiansen. "The SCI Cache Coherence Protocol". In *Scalable Shared Memory Multiprocessors*, M. Dubois and S. Thakkar, eds., Kluwer academic Publishers, Norwell, Mass. 1992

[3] The IEEE. "IEEE P1596 Standard Specification". IEEE Publication, 1989.

[4] A.G. Greenberg and I.Mitrani "Analysis of Snooping Caches". *Proc. of Performance 87, 12th Int. Symp. on Computer Performance*, Brussels, December 1987.

[5] S.L. Scott, J.R. Goodman and M.K. Vernon. "Performance of the SCI ring". *In Proc. of the 19th Annual Int. Sym. on Computer Architecture.* May 1992.

HIDING MISS LATENCIES WITH MULTITHREADING ON THE DATA DIFFUSION MACHINE[†]

Henk L. Muller[‡]   Paul W.A. Stallard   David H.D. Warren

Department of Computer Science, University of Bristol, UK.

Working at: PACT, 10 Priory Road, Bristol. BS8 1TU, UK.

email: {henkm,paul,warren}@pact.srf.ac.uk

Abstract— *Large parallel computers require techniques to tolerate the potentially large latencies of accessing remote data. Multithreading is one such technique. We extend previous studies of multithreading by investigating its use on the Data Diffusion Machine (DDM), a virtual shared memory machine in which data migrates according to its use. We use a detailed emulator to study DDM's with up to 72 nodes, allowing the scalability of multithreading to be tested further than in other studies. The results are promising and show that the applications tested can all benefit from multithreading on the DDM. Most applications however reach the ceiling of their parallelism. We briefly discuss how the results may generalise to other architectures.*

1 INTRODUCTION

Virtual shared memory (VSM) is an increasingly common design goal for parallel computers, aiming to support a simple shared memory programming model on a scalable architecture. Scalability necessarily implies that memory must be distributed. The Data Diffusion Machine (DDM) [10] is one of a number of similar VSM designs which aim to reduce the costs of remote memory access by allowing data to freely migrate to the processors that need it. By using only (set-)associative memories to store the data, there is no connection between the address of a data item and the node where it is stored. Instead, data is moved automatically by the hardware to the nodes which need it, thereby increasing the proportion of local accesses. Making all memory set-associative can be seen as a logical progression of other VSM designs. When the cache of other machines is made large enough, the machine becomes virtually identical to a machine with only set-associative memory.

Although the DDM and similar designs reduce the frequency of remote memory accesses, they do not eliminate them entirely. As machines become larger, the cost of remote accesses becomes greater. This phenomenon is aggravated by the ever widening gap between processor speeds and memory speeds. Some means of tolerating the remote access latency is therefore essential. Several techniques exist to tolerate or hide latency including prefetching, weak consistency models and multithreading [1].

Local memories in the DDM, and caches in other VSM architectures, can be built with prefetch mechanisms that take advantage of address locality to reduce misses (or at least reduce miss penalties). When an item is accessed, a simple prefetching scheme will ensure the next item in the address space is also available—if it is not, a fetch will be issued on the assumption that the program will need this data soon. This can be enhanced by letting the compiler or programmer insert explicit prefetching instructions in the code. Prefetching works very well in practice but it is limited. To mask very large miss penalties would require prefetching many items ahead. The further ahead the prefetching must predict, the less likely it becomes that the prediction will be correct. Many of the prefetched items will therefore not be used, while the network will become congested with prefetch messages. When real demand fetches occur, they can find themselves queued behind speculative prefetches.

The DDM currently supports a strong consistency model (sequential consistency) for VSM. Weakening the consistency model allows the machine to relax the conditions under which it is forced to stall. Processor consistency [2] allows a write miss to proceed in parallel with subsequent read accesses, whereas even weaker models [3,4] allow all operations, except those explicitly for synchronisation, to run in parallel. Weakening the consistency affects the programming model as the programmer and/or compiler must ensure that the memory is consistent at the correct times. Weaker consistency will only help to reduce miss penalties if the algorithm only requires consistency at defined synchronisation points, or if the frequency at which the memory needs to be made consistent is low.

Multithreading is a conceptually simple idea in which the processor is able to switch threads and proceed with other useful work while the miss is being serviced. Essentially, multithreading relies on two things: a fast mechanism of thread switching that can take advantage of idle times of the order of microseconds; and the presence of some other

---

[†]Work supported by ESPRIT OMI/HORN P7249.

[‡]Authors are ordered alphabetically.

work to do. In the case of multiprogrammed machines, this other work may well be another application, but we will restrict ourselves to the case of running a single application with the goal of the shortest execution time possible. In this case, multithreading relies on parallelism in the application by running a number of threads on each processor, all working on the same problem. The overhead of thread switching may reduce the potential gain from multithreading. In addition, all the threads on the same node share the processor cache and will interfere with each other's cache performance. This interference can be constructive, although in general it will not be.

To implement multithreading, the processor must make a decision when to do something else—this decision, and the thread switch itself, will take time. The longer this time, the longer the potential latency must be before it makes switching threads worthwhile. Many processors now have the capability to keep several threads active at the same time allowing thread switches to happen much faster. With sufficiently fast switching, multithreading can be used to hide the latency of local memory accesses. Some processors switch threads every cycle so eliminating the decision overhead and making it possible to even hide the latency of processor cache hits. Prefetching and weak consistency do not help with this very fine grain latency, although compiler technology can help to reorder instructions in such a way that most of the cache latency can be hidden.

The latency hiding methods described all attempt to solve the same problem—how can the processor be kept busy given that it is connected to a relatively slow memory? However, implementing all the options together will not work as some of these techniques will interfere with each other.

Multithreading looks very promising, and it seems a suitable choice for the DDM. It can hide the latency not only of remote accesses, but also of local memory accesses and even of accesses to the processor's on-chip cache. However, multithreading is not without its drawbacks. In this article we concentrate on how much of this potential can actually be realised.

The precise model of multithreading adopted, the way it is incorporated in the DDM, and the implementation strategy are presented in Section 3. We have run a set of benchmarks on the DDM with between 1 and 72 processors and between 1 and 6 threads per processor. The results of each application are presented and analysed in Section 4. The results as a whole are discussed in Section 5. In that section we also discuss how the choice of multithreading model, the choice of architecture (the DDM) and the parameters of the architecture (e.g. memory and cache latency) affect the results. Changing the parameters will have an impact on the results, but we believe that similar trends will be visible on other architectures with similar parameters.

2 OVERVIEW AND RELATED WORK

Multithreading has been proposed for a number of purposes. It allows processors to use simpler and deeper pipelines, provides the user with a multithreaded execution model and can be used to tolerate the latency of slow (relative to the processor) memory operations. For the purposes of this paper we will concentrate on the latter. Of course, multithreading is not without its cost and the increased functionality and performance must justify the extra hardware to maintain a number of thread states simultaneously in the processor, the large register file and the increased scheduling complexity.

The concept of multithreading is not new. A number of machines have used a multithreading processor and various academic studies have been performed. To put the work of this paper in context, we present here a brief summary.

The earliest multithreaded processors worked on the switch-every-cycle principle. HEP [5] and MASA [6] were two such processors that maintained a number of contexts in hardware. The execution pipeline was simplified by the knowledge that the previous instruction from each instruction stream (context) would have completed execution before the next one commenced, and hence pipeline interlocking was not needed. This gave a pipelined execution unit without the complication of resolving instruction dependencies and with sufficient contexts, these processors could hide memory access latency very successfully. The major disadvantage of the approach was that with only one thread running, the maximum achievable performance is limited to processor's peak performance divided by the number of pipeline stages. In the case of HEP, with an 8 stage pipeline, the performance consequences were severe. To keep 8 stages busy requires 8 contexts that are ready to run: given that some contexts will be stalled on memory accesses, HEP required a large number of contexts.

Switching every clock cycle is probably unnecessary. Modern RISC processors can achieve very high processor utilisations with a single thread using various techniques of pipelining, renaming and out of order execution. There are still memory latencies that cannot be hidden, especially in multiprocessor machines, so multithreading can still play an important role in increasing performance. Switching less often than every cycle has a number of advantages: it requires less active threads and allows single thread performance to be unaffected. The processor automatically uses instruction level parallelism to hide some of the latency leaving more of the coarse grained parallelism available for multiple processors. The thread switches could occur on every memory access, on every shared memory access or on every cache miss.

Weber and Gupta [7] looked specifically at the use of multithreading in a directory based, sequentially consistent, cache coherent machine. Their study assumed a switch-on-miss strategy that switched threads if a read access hit an invalid entry or if a write access hit an invalid entry or a shared entry. They presented results for machines with 16 processors obtained using an off-line trace generator and a trace-driven simulator. The simulator modelled all the major components of the machine (caches, protocol, buses and network) and assumed a non-zero thread switch time. Software load imbalance was ignored by stopping the simulation as soon as one of the threads ran out of work. The network was based on a cross bar switch that enabled remote references to be satisfied in 26 cycles (compared with 18 for a local reference).

Their results showed substantial performance benefits from multithreading for the small machines investigated, provided the switching overhead was low. As one would expect, there is less to be gained going from two to four threads per node than from one to two. With high switch times (16 cycles) multithreading degrades performance. They also investigated the effect of cache interference and showed that it was always negative for their experiments.

Later work by the DASH group at Stanford investigated the effects of various methods of latency hiding, including multithreading [1]. These experiments used the DASH simulator with 16 processors fed by a closely coupled trace generator. This work demonstrates some benefits of multithreading but concludes that other forms of latency hiding are more effective for their machine. The benefit of multithreading varied according to the application from good to very poor. Much of the negative effects can be attributed to the interference between threads in the small (4 Kbytes) direct mapped coherent caches.

Boothe and Ranade [8] produced results for predicted performance for much larger machines, up to 1024 nodes. Their model was quite simple, assuming a fixed latency for the network and hence not modelling the network congestion expected with larger machines. The simulations for this work were based on a smaller number of processors and the results scaled accordingly. The scaling is not described in detail but seems to extrapolate the small machine results without compensating for the reduction in grain size or any limit to the parallelism available in the application.

Laudon *et al.* have recently proposed an approach to multithreading called interleaving [9]. An interleaved processor statically allocates threads to slots in the processor pipeline. Two threads, for example, can be interleaved by allocating the odd slots to one thread, and the even slots to the other thread. On a t-threaded processor, interleaving reduces the thread instruction rates by a factor of t, and hence reduces the gap between the processor and memory speeds by a factor of t. t can be chosen so that a single thread has a reasonable execution speed while it is sufficiently long to hide most latencies. For very large miss penalties this scheme might be less effective than a switch-on-miss ap-

Figure 1: An example DDM hierarchy.

proach because the latter is able to schedule a new thread instead of having to run at a degraded performance.

The multithreading results that we present here were obtained by implementing multithreading in one particular implementation of the Data Diffusion Machine. We have evaluated large configurations (up to 72 nodes with 432 threads) with a very detailed emulator. Although these results do not directly apply to other architectures or even DDM's with different timing characteristics, we believe that other machines will have similar behaviour when similar degrees of multithreading are used.

3 THE ARCHITECTURE

The Data Diffusion Machine (DDM) [10] is a scalable virtual shared memory architecture. In the DDM, data has no fixed home location but instead migrates around the machine according to where it is being used. This flexibility in data distribution coupled with the strong consistency model provided by the DDM protocol combines a simple programming model with good performance. We have built a detailed emulator [11] of the DDM that implements the machine totally in software and operates on a parallel machine. From the development of this emulator we have gained confidence in the feasibility of the DDM [12].

The DDM memory

Each node in the DDM has a set-associative memory. A hierarchy of directories is used to keep track of each data item, as shown in Figure 1. Each directory knows where the data items below it are stored. Conceptually the directories are organised in a tree structure. For performance reasons directories are split. As an example, the eight top level circles in Figure 1 are a distributed implementation of the *single* top level directory. A transaction driven protocol is used to copy data from one node to another or to invalidate all but one outstanding copies so that the last copy can be updated. The DDM protocol provides a strong consistency model (currently sequentially consistency is supported although processor consistency may offer performance benefits without compromising programmability).

The DDM that we evaluate in this paper uses point-to-point serial links for the interconnect. These links are simple and cheap to implement, but are relatively slow. If the data is not present in the local memory, the miss penalty

Table 1: Latencies in clock ticks

Instruction	2 clock ticks
Cache hit	5 clock ticks
Local memory hit	20 clock ticks
Local memory miss	800-2000 clock ticks

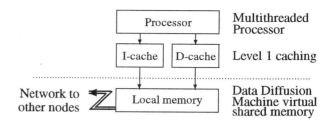

Figure 2: The architecture.

will be large, in the order of 1000 clock cycles. The timings are given in Table 1 in system clock ticks, which is the (arbitrary) time unit used by the DDM emulator. Accessing local data is relatively cheap, accessing data that is missed in the local memory has to come from a remote memory. These accesses are routed over the network and the miss penalty will be large. The exact timing depends on the distance to the remote memory and the traffic density. We believe that these latencies are realistic for state of the art systems. In future systems the gap between processor and memory will widen, and more levels of caching might be employed.

The multithreaded processor

We have assumed an aggressively multithreading processor in our evaluation. A thread switch occurs on every memory access. Instructions accessing the memory can be recognised early in the pipeline, and we assume that another thread can be scheduled without a performance penalty. The effects of different thread switching times have been studied by Weber [7]. We briefly discuss the implications of using a less aggressive multithreading model in the discussion at the end of this paper.

A schematic drawing of a single DDM node is given in Figure 2. We assume that the instructions are accessed from an instruction cache with a very high hit ratio. We ignore all instruction traffic from instruction cache to the memory, and also ignore the latency that an instruction miss might cause to the processor. The cache and local memory are shared between the various threads.

The emulator

The architecture described above has not yet been implemented in hardware. The performance figures presented in this paper were obtained using a calibrated emulator of the DDM. The emulator implements the Data Diffusion Machine, and therefore virtual shared memory, on a parallel T800 transputer system in software. The emulator is functionally correct and can execute any shared memory program in parallel. In order to obtain sensible performance figures from the emulator, the various components of the system are calibrated so that their timing characteristics reflect the timing model of our target architecture. This is true for all the components, whether they be mod-

elled in software or mapped on to the components of the emulation platform. The transport of messages over the network topology is emulated using the timing of the network components, resulting in reliable values for miss penalties, incorporating delays due to routing and congestion. The emulator is fully distributed and as such is fast enough to allow us to run full applications with realistic data sets, even on very large architectures.

The emulator used in this paper is an enhanced version of the one published earlier [11]. Three major improvements have been made: the emulator supports an arbitrary fanout (not limited to the fanout of T800 transputers); splitting of the DDM hierarchy is completely implemented; and multithreading is supported. The functional implementation of multithreading is relatively simple as the transputer supports this naturally. Threads are modelled using transputer processes. To implement the timing of the threads, the emulator maintains one clock per node to determine whether the processor is being used, plus one clock per thread to determine when the thread can be scheduled. The transputer hardware is used to maintain the scheduling list.

The transputer platform on which the emulator is run has limited memory (4 Mbytes per node) which contains the application code, the stacks and data segments of each thread, the modelled processor cache, the DDM set-associative memory and the code that emulates the directory controller. The restricted memory forces us to limit the size of the DDM memory to 2 Mbytes per node. The 16 Kbytes data cache is 4-way set-associative with a write-through policy, while the 2 Mbytes of local memory are 2-way set-associative, with an invalidate on write policy enforcing strong consistency. Despite the restrictions on memory size, we are still able to run quite large data sets which in most cases represent realistic problems.

The size of a data item (the unit of transport and the unit of sharing) is 64 bytes, and the prefetch strategy is set to prefetching only the next item on a read miss. A deeper level of prefetching had an adverse influence on the performance. The topologies that we used were single-level hierarchies with a split of 4 for up to 8 nodes. Two-level hierarchies were used for up to 48 nodes, with a split of 2 in the first level and a split of up to 8 in the top level. The 72 node topology has also two levels, but it is only split in

Table 2: The benchmark set.

Application & Parameter		Sync	10^6 Instructions
Wave	240 3	Local	133
Wang	100000	Global	35
Aurora	Proteins200	Local	286
Water	343 2	Global	1055
Barnes	4000 3	Global	660
Cholesky	tk23.O	Local	1450

the top level due to the limited number of processors in our emulation platform.

The application programs

The programs that we have used as a benchmark are listed in Table 2. Wave is a program that simulates part of the North Sea and simulates the change in water height at each point on a grid over time. Wang is a numerical application that solves a tridiagonal matrix using Wang's algorithm. Aurora is an or-parallel Prolog system on which we are running the Proteins application. Protein sequences consist of chains of amino acids and this application finds all occurrences of a set of functionally significant sections in a database of 200 proteins. The remaining benchmarks; Water, Barnes and Cholesky are programs from the SPLASH-I suite [13] and are well documented elsewhere. Most of these benchmarks have a great deal of parallelism that can be exploited by a multithreaded architecture. Aurora/Proteins200 and Cholesky/tk23.O however, have limited parallelism due mainly to the particular data sets used. The means by which algorithms synchronise is particularly important for multithreaded architectures. Global synchronisation, between each step of a simulation for example, makes it very hard for multithreading to realise its full potential. Load imbalance becomes very damaging and the latency of synchronisation cannot be hidden by other threads as they are all involved in the synchronisation. Barnes, Water and Wang all rely heavily on global synchronisation. Algorithms that use only local synchronisation would be expected to perform better under multithreading. Program data sets were chosen to be as large as feasible while still fitting in our memory and not taking too long to run.

4 EXPERIMENTS AND RESULTS

Each application has been run on a number of machines with between 1 and 72 processors with either 1, 2, 4 or 6 threads per node. The results of each application are presented in graphs of relative speed against the number of leaf nodes. Each graph has two Y-axes; the left hand one shows the speedup relative to the single processor time with 6 threads, and the right hand one shows the speedup

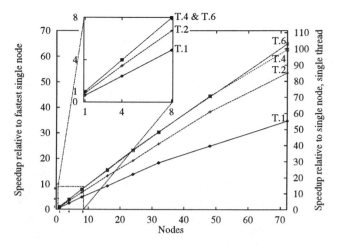

Figure 3: Speedup graph for Wave 240

relative to the single threaded single processor point[a]. The speedup lines for 1, 2, 4 and 6 threads are labelled T.1, T.2, T.4 and T.6 respectively. Due to memory limitations on our experimentation platform not all configurations could be tested. Aurora and Cholesky could not be run with 6 threads per node or with 4 threads per node on a single processor.

The performance of Wave is shown in Figure 3. Wave clearly benefits from multithreading. On a single node the execution time with six threads is some 35% faster than with a single thread (this is hardly visible, so we have magnified this portion of the graph). This single node speedup is achieved by hiding the latency of processor cache misses and results in very high processor utilisation. With multiple processors, multithreading is also able to hide the latency of remote references. A single thread achieves a self-referential speedup of 54 on 72 nodes. With six threads, the self-referential speedup is 65 on 72 nodes. The ability of Wave to utilise the potential of multithreading is shown by the total speedup from a single thread on a single node to six threads on 72 nodes, which is just over 102. Each additional thread is able to hide a portion of the latencies incurred by the other threads. As expected therefore, the benefits of adding extra threads decreases, as can be seen by the greater improvement moving from 1 to 2 threads versus the improvement moving from 2 to 4 or from 4 to 6.

In Figure 4, Wang shows a slightly different behaviour. With only a single thread, Wang has a speedup of 58 on 72 processors. A second thread on each processor clearly improves performance. Two more threads help at 48 nodes, but do not add performance at 72 nodes, whereas six threads

---

(a) This is not exactly true: the precise points used as reference are the execution time of the configuration with the highest efficiency multiplied by the number of nodes of that configuration; and the execution time of the single thread configuration with the highest efficiency multiplied by the number of nodes. This prevents seemingly super linear speedups.

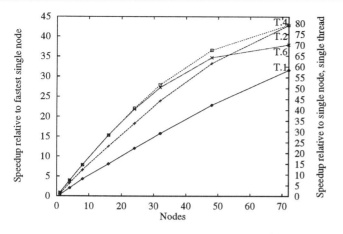

Figure 4: Speedup graph for Wang 100000

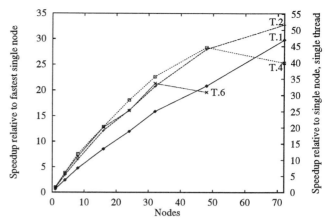

Figure 6: Speedup graph for Water 343

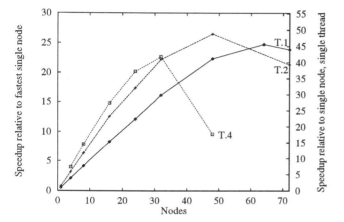

Figure 5: Speedup graph for Aurora/Proteins

done, the performance deteriorates with additional processors. This behaviour is shown clearly in Figure 5 (note that the line for 6 threads is missing due to memory limitations). The single threaded performance increases up to 64 nodes, with two threads the optimal performance is at 48 nodes (96 threads in total) and with four threads the optimal performance is at 32 nodes (128 threads in total). Notice that with Aurora the optimal speed is reached with more threads in total when there are more threads per node. The dynamic creation of work by Aurora and increased locality cause more parallelism to be usefully exploited. The overall optimal performance is (for the topologies we tested) with 48 nodes and two threads per processors.

In Figure 6, Water shows that for 72 processors, 2 threads run faster than 1 or 4 threads in that order (6 threads does not run on 72 processors as there are too few tasks). The bad performance for four threads per node is caused by load imbalance. With 4 threads per node on 72 nodes there are 288 threads. Dividing 343 molecules over 288 threads results in 55 threads with 2 molecules and 233 threads with one molecule. The default thread allocation policy allocates the first 4 threads to node 0, the next 4 to node 1 etc., in order to maximise locality. For Water this results in 13 nodes with 4 threads having two molecules each (8 molecules per node), one node with 3 double and 1 single molecule threads (7 molecules per node) and 58 nodes with 4 single molecule threads each (4 molecules per node).

To test this hypothesis we have done two simple tests with Water. If we change the run time support to distribute the work more evenly (breadth first) the speedup figures for 4 threads on 72 nodes improve from 40 to 48. A side effect is that the miss ratios increase by a factor of 1.5 as some locality has been lost. If we modify the source of Water instead, we can distribute work evenly while retaining locality. In this configuration, the miss ratios are unaffected but the speedup remains at 48. This change had no effect because the multithreading was hiding the higher miss ra-

result in lower performance than four. This is caused by a sequential bottleneck in Wang's algorithm. Wang splits the matrix in as many parts as there are threads, clears them in parallel, sequentially performs an operation on all border elements, and then in parallel goes through all submatrices again. More threads means that the parallel phases become shorter, while the sequential phase becomes longer (there are more borders). With 72 nodes and 6 threads per node there are 432 threads, which means that the parallel phase solves matrices of size $100000 / 432 = 231$ while the sequential phase works on a matrix of size 432.

The Aurora system uses a dynamic scheduler to split the available work amongst the available workers. Both the partitioning and the distribution of work are dynamic. Each worker behaves very much like a sequential Prolog engine and the only major overhead is incurred when a worker switches its context to a different part of the search tree. With larger numbers of threads, the granularity of work becomes finer and the frequency with which the scheduling overhead is incurred increases. When the granularity is so fine grained that the overhead dominates the useful work

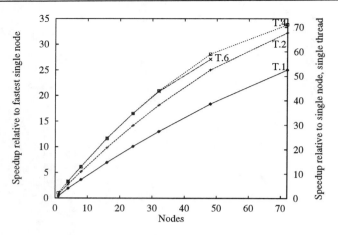

Figure 7: Speedup graph for Barnes 4000

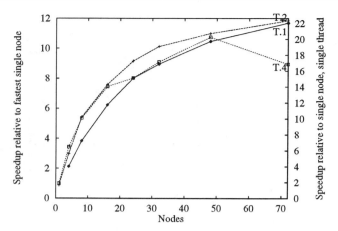

Figure 8: Speedup graph for Cholesky tk23.O

tios of the previous experiment. These results are partly in accordance with an extensive study on multithreading and locality by Thekkath and Eggers [14]. They have shown that load imbalance is the major factor affecting execution time under multithreading. However, in contrast with that study, we observe a clear deviation in miss ratios. We expect that this is due to machine and associative memory size, as their miss ratios are reported for small machines. Our experiments support their conclusion that the execution time on large machines is hardly affected by data allocation policies providing improved locality.

The results of Barnes are shown in Figure 7. The self-relative speedup of four threads on 72 nodes is only half the ideal speedup. The reason for this seems to be that Barnes uses alternate phases of computation and global communication. It is only during the computation phase that the program can benefit from multithreading. During the global communication phase, multithreading cannot improve performance because all threads are waiting for data and the processor will therefore be underutilised. During this phase multithreading can have a negative impact on the performance because the network might be overloaded.

The results of Cholesky are shown in Figure 8. Cholesky partitions the work statically, but the work is farmed out dynamically (as opposed to, for example, Water where the work is statically partitioned *and* statically allocated to a thread). If we ignore the line with 4 threads for the time being, one can see that there is a limit in speed that is approached by both 1 and 2 threads per node. The difference in speed between 1 and 2 threads per node is a factor of 1.5 for a small number of nodes but the difference is marginal on 72 nodes. Four threads per node does not perform very well and this might have to do with the way the work is distributed. Threads of Cholesky spin if there is no work and this can lead to deteriorated performance. A similar effect would also be visible when running with 2 threads if a sufficiently large number of processors were employed.

5 DISCUSSION AND CONCLUSIONS

The results from these experiments demonstrate that all the applications tested can get a real benefit from multithreading on the DDM. Our single threaded results are broadly comparable to the (limited) results available from other architectures and these have been significantly improved with the use of multithreading. Earlier studies had already shown that multithreading could be useful on small machines where there was likely to be excess parallelism available. We have now shown how much larger machines can benefit with many hundreds of threads. Although other architectures may well benefit from multithreading in a similar way, it is clear that the DDM is able to exploit this technique and produce close to ideal speedups for many applications until the limit of parallelism is reached.

It should be noted that all programs, except Wave, run out of parallelism on the chosen data sets when running them with 6 threads on 72 nodes, but for differing reasons. The parallelism exploited by Aurora when running Proteins becomes too fine-grained, the sequential phase of Wang becomes too long in comparison with the parallel phases, Water suffers from load imbalance because the input set is too small for so many tasks, Barnes suffers from a global communication phase and Cholesky has a clearly visible lack of work. In all cases this causes 6 threads on 72 nodes to have lower performance than 4 threads on 72 nodes, and sometimes 4 threads per node to have a lower performance than 2 threads per node.

Multithreading uses excess parallelism, but these programs with these data sets do not have any excess parallelism remaining at 72 nodes with 6 threads on each node. So, can we do more with this limited parallelism, by changing the parameters of our machine, or by adopting another multithreading model?

The ratios of processor speed to cache speed, memory speed and network speed directly affect the performance of multithreading, as does the ratio of these to the time required for the processor to switch thread. As multithreading is a technique to hide latency, the slower the memory system is compared to the processor, the better multithreading works but the more parallelism is used up to mask the latency of on-node cache and memory. If the thread switching time is significant compared to any of the levels of the memory system then it becomes pointless to switch threads for accesses to that memory level [7].

We have chosen a multithreading model that switches on every memory operation. Even more aggressive models, switching on every cycle or interleaving, would give better performance overall, but lower speedups. If we were only to switch on cache misses, or even only on local memory misses, the performance of the whole machine would be lower, but the speedup would be better. Less parallelism would be used up to mask local memory accesses leaving more available to execute instructions in parallel. Eventually, this would lead to almost the same level of performance, but would require more nodes since the processors are spending more time idle (waiting on a cache hit or a local memory hit). One would have to balance the cost of more nodes with the cost of implementing a more aggressive multithreading model.

To improve the ultimate performance, one can envisage that the compiler reorders the memory operations so that (part of) the memory latency is overlapped with the instruction stream. This means that the processor will be fully utilised with fewer threads per node, allowing more nodes to be used, resulting in shorter execution times.

Finally, one can employ other latency tolerating techniques at the same time [1]. A weak consistency model makes programming harder but will leave more parallelism to run the program in parallel. Prefetching is more complex. We have restricted the prefetch to a single item on each read miss because genuine requests can become blocked behind prefetches in the network. More prefetching was clearly harmful. A protocol could be designed to allow prefetches to be overtaken by ordinary requests, but care must be taken to preserve consistency.

For other types of VSM architecture, with ordinary memories and coherent caches instead of set-associative memories, we would also expect to see benefits from multithreading. However, the DDM's set-associative memories are typically larger than the coherent caches of other architectures, reducing the negative interference between the threads and allowing positive interference to emerge. Further experiments are needed to show if this effect is noticeable.

Although similar experiments on other machines will yield different results, the trends that have appeared seem to be inherent in the parallel application. With different ratios of processor speed to network, or with different levels of

prefetching or less aggressive multithreading, the features may appear earlier or later but they will appear.

REFERENCES

[1] A. Gupta, J. Hennessy, K. Gharachorloo, T. Mowry, and W.-D. Weber. Comparative Evaluation of Latency Reducing and Tolerating Techniques. In *Proc. of the 18th ISCA*, pp 309–318, Toronto, Canada, May 1991. ACM Press.

[2] J. R. Goodman. Cache Consistency and Sequential Consistency. Technical Report WISCN-1006, Dept. of Comp. Science, University of Wisconsin, Madison, Mar. 1989.

[3] S. V. Adve and M. D. Hill. Weak Ordering - A New Definition. In *Proc. of the 17th ISCA*, pp 2–14. Institution of Electrical and Electronic Engineers, June 1990.

[4] K. Gharachorloo, D. Lenoski, J. Laudon, P. Gibbons, A. Gupta, and J. Hennessy. Memory Consistency and Event Ordering in Scalable Shared-Memory Multiprocessors. In *Proc. of the 17th ISCA*, pp 15–26. Institution of Electrical and Electronic Engineers, June 1990.

[5] J. S. Kowalik, editor. *Parallel MIMD Computation: the HEP supercomputer and its applications*. MIT Press, 1985.

[6] R. H. Halstead, Jr and T. Fujita. MASA: A Multithreaded Processor Architecture for Parallel Symbolic Processing. In *Proc. of the 15th ISCA*, pp 443–451, June 1988.

[7] W.-D. Weber and A. Gupta. Exploring the Benefits of Multiple Hardware Contexts in a Multiprocessor Architecture: Preliminary Results. In *Proc. of the 16th ISCA*, pp 273–280, June 1989.

[8] B. Boothe and A. Ranada. Improved Multithreading Techniques for Hiding Communication Latency in Multiprocessors. In *Proc. of the 19th ISCA*, pp 214–223, Gold Coast, Australia, May 1992. IEEE Computer Society Press.

[9] J. Laudon, A. Gupta, and M. Horowitz. Interleaving: A Multithreading Technique Targeting Multiprocessors and Workstations. In *Proc. of the Sixth ASPLOS-VI*, pp 308–318, San José, CA, Oct. 1994. ACM.

[10] D. H. D. Warren and S. Haridi. The Data Diffusion Machine—A Scalable Shared Virtual Memory Multiprocessor. In *Proc. of the Int. Conf. on FGCS*, pp 943–952, Tokyo, Japan, Dec. 1988.

[11] H. L. Muller, P. W. A. Stallard, D. H. D. Warren, and S. Raina. Parallel Evaluation of a Parallel Architecture by means of Calibrated Emulation. In *Proc. of the 8th IPPS*, pp 260–267, Cancun, Mexico, Apr. 1994. IEEE Computer Society Press.

[12] H. L. Muller, P. W. A. Stallard, and D. H. D. Warren. An Evaluation Study of a Link-Based Data Diffusion Machine. In *Proceedings of the International Workshop on Support for Large Scale Shared Memory Architectures*, pp 115–128, Cancun, Mexico, Apr. 1994.

[13] J. P. Singh, W.-D. Weber, and A. Gupta. SPLASH: Stanford Parallel Applications for Shared-Memory. Technical report, Computer Systems Laboratory, Stanford University, 1991.

[14] R. Thekkath and S. J. Eggers. Impact of Sharing-Based Thread Placement on Multithreaded Architectures. In *Proc. of the 21st ISCA*, pp 176–186, Chicago, Illinois, Apr. 1994. IEEE Computer Society Press.

Hierarchical bit-map directory schemes on the RDT interconnection network for a massively parallel processor JUMP-1

Tomohiro Kudoh†   Hideharu Amano‡   Takashi Matsumoto*   Kei Hiraki*

Yulu Yang ‡   Katsunobu Nishimura‡   Koichi Yoshimura†   Yasuhito Fukushima†

†Tokyo Engineering University
‡Keio University
*The University of Tokyo

Abstract

JUMP-1 is currently under development by seven Japanese universities to establish techniques of an efficient distributed shared memory on a massively parallel processor. It provides a memory coherency control scheme called the *hierarchical bit-map directory* to achieve cost effective and high performance management of the cache memory. Messages for maintaining cache coherency are transferred through a fat tree on the RDT(Recursive Diagonal Torus) interconnection network. In this report, we discuss on the scheme and examine its performance. The configuration of the RDT router chip is also discussed.

1 Introduction

JUMP-1 is a massively parallel processor prototype developed by collaboration between 7 Japanese universities[4]. The major goal of this project is to establish techniques for building an efficient distributed shared memory on a massively parallel processor. For this purpose, a sophisticated methodology called Strategic Memory System (SMS) is proposed [11][4].

The cache directory and management scheme is one of the most important issues in the SMS. In traditional distributed memory systems, a directory entry is associated with every cache line. However, this requires a large amount of memory for a massively parallel processor which provides both a large number of processors and a large amount of memory space. In the SMS, each node processor shares a global virtual address space with two-stage TLB implementation, and the directory is attached not to every cache line but to every page, while the data transfer is performed by a cache line.

For the directory entry associated with each page, a hierarchical bit-map directory scheme was introduced[10][11]. In addition, to reduce the size of directory, a directory reduction scheme has been proposed. However, the original scheme [1] is built based on a simple n-ary tree network, and some extensions are required for the use in JUMP-1.

---

[1] The original scheme[10] is called the pseudo-fullmap directory scheme consisting of the hierarchical multicast(directory) scheme described here, local, and 1-to-1 communications.

In this paper, we extend the original hierarchical bit-map directory scheme to apply on an interconnection network RDT(Recursive Diagonal Torus) which is proposed for JUMP-1. In addition to the original directory reduction scheme, two other schemes are also proposed. In Section 2, JUMP-1 and its interconnection network RDT are briefly introduced. In Section 3, the hierarchical bit-map directory scheme and the directory reduction schemes are introduced. In Section 4, the implementation of the hierarchical bit-map directory schemes on the RDT is described. These schemes are compared with other directory schemes and the number of redundant packets are evaluated in Section 5. Finally, the implementation of the RDT router chip which enables hierarchical bit-map directory schemes are described.

2 JUMP-1 and the RDT

2.1 Structure of JUMP-1

As shown in Figure 1[4], JUMP-1 consists of clusters connected with an interconnection network RDT[14]. Each cluster also provides a high speed point to point I/O network connected with disks and high-definition video devices.

Each cluster is a bus-connected multiprocessor (Figure 2[4]) including 4 coarse-grained processors(CPU), 2 fine-grained processors (Memory Based Processor or MBP) each of which is directly connected to a main

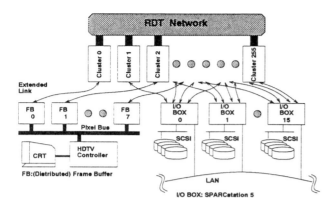

Figure 1: Structure of JUMP-1

memory and the RDT router chip. A CPU is an off the shelf RISC processor (SUN Super-Sparc+) which performs the main calculation of the program. The MBP, the heart of JUMP-1, is a custom designed fine-grained processor which manages the distributed shared memory, synchronization, and packet handling. The first prototype of JUMP-1 which is scheduled to be available in this winter provides 256 clusters, thus, 1024 processors.

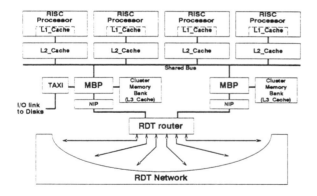

Figure 2: Structure of a JUMP-1 cluster

2.2 Interconnection network RDT

The RDT is a network consists of recursively formed two-dimensional square diagonal toruses. In order to reduce the diameter, preparing bypass links for the diagonal direction is the best way for the torus network. Assume that four links are added between a node (x, y) and nodes $(x \pm n, y \pm n)$. Here, n is called the *cardinal number*. Then, the additional links form a new torus-like network. The direction of the new torus-like network is at an angle of 45 degrees to the original torus, and the grid size is $\sqrt{2}n$ times of the original torus. Here, we call the torus-like network the rank-1 torus. On the rank-1 torus, we can make another torus-like network (rank-2 torus) by providing four links in the same manner. Figure 3 shows rank-1 and rank-2 toruses

when n is set to be 2. The RDT consists of such recursively formed toruses.

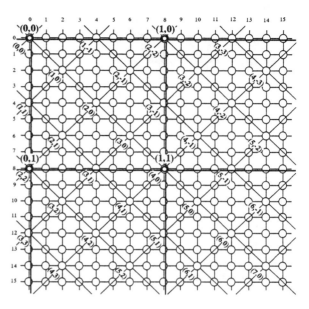

Figure 3: Upper rank toruses

Recursive Diagonal Torus RDT(n,R,m) can be defined as a class of networks in which each node has links to form base (rank-0) torus and m upper toruses (the maximum rank is R) with the cardinal number n. Note that, each node can select different rank of upper toruses from others.

The RDT in which every node has links to form all possible upper toruses is called the perfect RDT (PRDT(n,R)) where n is the cardinal number (usually, 2) and R is the maximum rank. Although PRDT is unrealistic because of its large degree (4(R+1)), it is important as a basis for establishing routing algorithm, broadcasting/multicasting, and other message transfer algorithms.

The JUMP1 must be scalable to the system with ten thousand nodes. In this case, m is set to be 1 (degree = 8). For this number of nodes, the maximum rank of upper toruses is 4. Thus, the RDT(2,4,1) is treated here.

In the RDT, each node can select different rank toruses from others. Thus, the structure of the RDT(2,4,1) also varies with the rank of toruses which are assigned to each node. This assignment is called the *torus assignment*. Various torus assignment strategies can be selected considering the traffic of the network. If the local traffic is large, the number of nodes which have low ranks should be increased. However, complicated torus assignment introduces difficulty to the message routing algorithm and implementation. For the JUMP-1, we selected a relatively simple torus assignment shown in Figure 4.

In this assignment, a node has eight links, four for the base (rank-0) torus and four for rank (1-4) torus (Most of links for upper rank toruses are omitted in Figure 4). Note that a node with a rank torus has neighboring

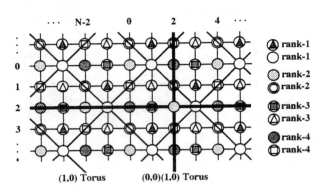

Figure 4: Torus assignment used in the JUMP-1

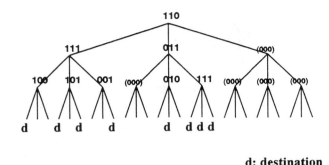

d: destination

Figure 5: Hierarchical bit-map directory scheme

nodes with toruses of other three ranks. Therefore, any rank torus can be used with a single message transfer between neighboring nodes. This property reduces the diameter and average distance between nodes.

The RDT provides various advantages as an interconnection network of the JUMP-1. It includes two dimensional mesh, and simple near-optimal routing algorithm enables smaller diameter than that of the hypercube (11 for 2^{16} nodes) with smaller degree (8 links per node). Moreover, it includes the fat-tree of mesh structure. Using this structure, hierarchical multicast can be efficiently supported. This feature is important to implement the hierarchical bit-map directory scheme treated here.

3 Hierarchical bit-map directory scheme

Most of conventional non-bus based shared memory multiprocessors equip a cache directory whose entries are associated with cache lines. However, in a massively parallel machine both with a large number of processors and a large address space, a large amount of memory required for the directory will be unacceptable. In JUMP-1, directory entries are associated with pages while the data are transferred by a cache line[11] in order to reduce the required amount of memory for the directory. Using this strategy, both the expansion of the directory memory and the congestion of the network caused by the large size message transfer can be avoided.

However, in this case, number of destinations of the coherence maintenance messages increases especially when an update type protocol is used. If there are considerable number of destinations, it will take a long time to send a message if they are sent one after another(sequentially).

To cope with this problem, the hierarchical bit-map directory scheme transfers messages for different destinations simultaneously (i.e. multicast) using a tree structured multicasting paths (multicasting tree).

Figure 5 illustrates the concept of hierarchical bit-map directory scheme. Leaves of the tree correspond to the clusters of JUMP-1 and a message is first sent from

the root of the tree. Each node can multicast a message to its multiple branches at a time. To specify the destination leaves precisely, an n-bit bit-map is required for each node when an n-ary tree is used. Thus, for m height n-ary tree, total of $\sum_{k=1}^{m} n^k$ bits are required for each entry. Although the required amount of memory is larger than that of the full-map directory scheme in which n^m bits are required (since there are n^m leaves), the message can be multicast using the tree structured path in the hierarchical bit-map directory scheme.

3.1 Reduction of the directory length

Since JUMP-1 is a massively parallel processor, the $\sum_{k=1}^{m} n^k$ bits directory entry (for m height n-ary tree) is not feasible. Two ways of reducing the size of the directory can be considered.

1. The bit-maps are stored and maintained at the node. When a message comes from the upper level, a bit-map stored at the node is accessed according to the address (tag) of the message.

2. The bit-maps for a message are attached to the message header and referred at the appropriate nodes of the tree.

While the former scheme effectively reduces the size of directories stored in the entire system, the overhead caused by the accesses of the bit-maps will prevent quick cache coherent management. Thus, the latter scheme is mainly adopted for JUMP-1. However, in this method, a large size of header is required if bit-maps for all level of hierarchy are attached.

In order to reduce the size of bit-maps, a bit-map is provided not for each node, but for each level of the tree. Each node multicasts packets to its children either according to the bit-map for the level or all children of the node (thus, broadcasting).

Here, the LPRA which is the original scheme proposed in [10][11], and two novel schemes, the SM and LARP are introduced.

LPRA scheme: When the multicast is started, the message is sent from the source cluster to the root

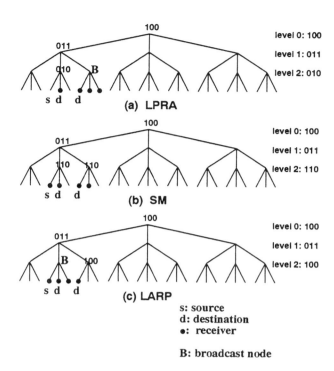

Figure 6: Hierarchical bit-map directory schemes

starts at the level 2 to the subtree which includes the source node. In this case, the broadcast is done for local nodes of the source node, while the bit-map is used for remote nodes. This scheme is advantageous when the mapping can make the best use of the locality of communication.

4 Fat tree on the RDT

In the original hierarchical bit-map directory scheme, messages are transferred on a tree structured communication paths. However, if a simple n-ary tree is used, following problems will bottleneck the performance.

- If multiple nodes multicast simultaneously, congestion may be caused around the root node of the tree.

- Depending on the location in the tree, some messages should have transferred through the very root of the tree just to move to neighboring node.

- It takes a large latency if a sender node at a leaf sends a message upstream to the root of tree.

of the multicast tree. In the LPRA (Local Precise Remote Approximate) scheme, the bit-map is used only nodes which is the root of the subtree including the source node. For nodes in other subtrees, the message is broadcast to all children. Figure 6(a) shows an example of this scheme for the 3-ary tree. In this figure, 's' is the source cluster, and 'd' is the destination to which packet is sent. • indicates clusters which receive data. It is desirable if the number of clusters which have • but without 'd' is small as possible. Using this scheme, the bit-map is used for local nodes of the source node, while the message is broadcast in the remote subtree marked **B**. This scheme is advantageous when a node send a message to a remote node group.

SM scheme: In the SM(Single Map) scheme, all nodes at a level use a unique bit-map as shown in Figure 6(b), and thus, no broadcast is made (unless a bit-map is all-1). This scheme is advantageous when the number of destination is not so large.

LARP scheme: The LARP is a complimentary scheme of the LPRA. In this scheme, broadcast is done at nodes which are the root of the subtree including the source node while a unique bit-map is used in other subtrees like the SM scheme. Note that the broadcast is started from children nodes whose parent starts the multicasting. In the example shown in Figure 6(c), since the multicast starts at the level 1 (The top node (level 0) only sends a message to a child.), the broadcast (marked **B**)

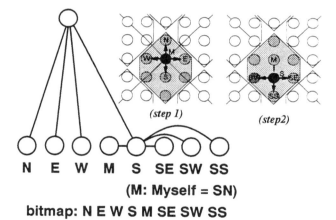

Figure 7: The 8-ary tree and bit-map pattern for multicast on the RDT

However, since the RDT involves a fat tree of toruses with multiple root nodes, these problems can be avoided. The pattern of message transfers for emulating a fat tree is shown in Figure 7. Two steps are required: (1) each node transfers a message to four neighbors. (2) a neighbor transfers the message to three neighbors. (South direction in this figure). As shown in Figure 7(b), the eight nodes which received the data does not duplicated with node which received other nodes (marked X in Figure 7(a)) on the same rank-i torus. Thus, if all nodes with rank-i toruses executes this pattern, the message is transferred to all nodes with rank-(i-1) toruses. By the iteration of this data transfer from the maximum rank to the rank-0, 8-ary tree in which is formed on the RDT. In this case, a rank in the RDT directly corresponds to the level of the tree. Moreover, in the RDT, the upper rank torus can be used within a step of message routing. Thus, the message can be directly transferred from the sender node to the root node without using the tree structure.

Figure 7 shows the 8-ary tree involved in the RDT, and bit-map pattern for the hierarchical bit-map directory scheme. When schemes described in the previous section are applied on this 8-ary tree on the RDT, followings are advantageous:

- On the RDT, there are a number of nodes with the maximum rank torus, and all of them are used as a root node. Thus, the message multicast is almost directly started from the root node without going upstream on the tree.

- Since there are many upper rank toruses in the RDT, the tree is a kind of "fat-tree" which provides many root nodes in each rank. Therefore, the congestion of root nodes is relaxed even if many source nodes multicast their data simultaneously and independently.

- In the RDT, nodes which receive the message through the tree whose root rank is 'i' are located around the source node. For larger 'i', the number of such nodes becomes large, thus the area which a message is multicast becomes wide. We call such an area "territory" of a multicast. Figure 8 shows territories of a multicast from rank-0 and rank-1. Since the territory of is always formed surrounding a source node, message multicast to local nodes are performed from a lower rank (thus, with only a small territory).

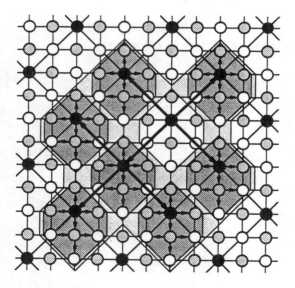

Figure 8: Territory of a multicast

5 Discussions

5.1 Comparison with other schemes

Three famous directory schemes for managing cache have been tried in multiprocessors with distributed shared memory. In the full map directory scheme which is used in Stanford DASH[2], each entry of the directory holds a bit map corresponding to each node of the multiprocessor. This scheme can be used only in a small system since the required bit map for each entry is equal to the number of processors.

In the limited pointer scheme[1] adopted in MIT Alewife[6], each entry holds a limited number of pointers each of which indicates the processor which has the copy of the line. Usually, the number of the pointers is limited in two or three. If the number of the copy is beyond to the number of pointers, the information is broadcast to every node (broadcast) or copies exceed the number of the pointers are invalidated (eviction).

In another scheme, the chained directory scheme[3][7] which will be used in Stanford FLASH[5], a chained list is used for holding node identifiers.

Both the limited pointer and chained directory will work efficiently even in massively parallel processors if the number of nodes which hold the copy of the same cache line is small. In the limited pointer, if the number of copies is large, it requires a large number of pointers. or unnecessary broadcast or invalidation is required. In the chained directory, it takes a long time to access the chained directory if the number of nodes which hold the copy is large. Through the simulation study, it appears that the number of nodes which receive the invalidate messages are one or two in most cases[12]. In this case. both schemes will work efficiently.

However, this simulation is done under the following conditions: (1) the entry of the directory is associated with the cache line, and (2) only invalidation type protocols are used. In JUMP-1, the entry of the directory is not associated with the cache line but with the page. This strategy much reduces the required memory for a directory of a large size of shared memory space that is essential for massively parallel processors. Moreover. in JUMP-1, update type protocols which are advantageous in most scientific calculations can be used with invalidation type protocols. In this case. as shown in our initial estimations, the number of copies is further increased.

Figure 9 shows the distribution of the number of destinations of invalidation/update messages when an invalidation type protocol and an update type protocol are used respectively. The bold line shows results when an entry is associated with a cache line, and the doted line shows ones with a page. Here, MP3D with 1024 molecules and water with 64 molecules from the SPLASH[8] parallel programs for shared memory is executed with 256 processors and 32 processors respectively. This result is generated from an address trace[9] when the size of a cache line is 32 byte and the size of a page is 4 Kbytes. The X axis corresponds to the number of destination processors in logarithmic scale. and the Y axis corresponds to the occurrence times.

While the bold line (entries are associated to cache

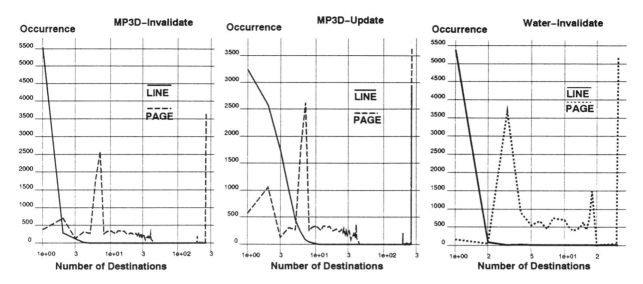

Figure 9: Distribution of number of destinations

lines) in the invalidation protocol demonstrates the small number of copies in the traditional conditions, other lines show that the number of copies exceeds three. When an entry is associated with the page, two peaks are shown. One is at 4-8 nodes and another is at all nodes in both protocols.

When 6 pointers are required in the limited pointers, the length of each directory entry becomes $6 \times 14 = 84bit$ when $2^{14} = 16K$ nodes are used. For the same size, the hierarchical directory schemes treated here require only $4(levels) \times 8(number\ of\ branch) = 32bits$. When the size of a cache line is 32 byte and the size of a page is 4 Kbyte, the total amount of required memory can be reduced to $1/128$ at maximum. Although every cache of a page is not used in practice, usually the usage ratio is quite high because of the locality of reference. This demonstrates that the required memory is much reduced in the hierarchical directory schemes. Although the chained directory scheme also can support a small cost of memory, the access time for following the directory may degrade the performance.

5.2 The number of unnecessary messages

The major disadvantage of hierarchical directory schemes is that it requires unnecessary message multicast since it only use a single multicast or broadcasting in each level of the hierarchy. If destination nodes can be mapped inside the territory, the number of unnecessary messages can be drastically reduced. However, it depends not only on the mapping strategy but also on the characteristics of application programs. Here, we estimate the number of unnecessary messages with a simple probabilistic models as the first evaluation.

The destinations are determined by a random numbers which follow a given distribution. In the following graphs, average value calculated from 10,000 trials based on the random number generation library of Sun

OS4.1.3 are shown.

Figure 10 shows number of receiving clusters when the destinations (D) follow a normal distribution of variance 1 (here, unit is a link on the base torus and the X and Y projections of the distance from the source cluster to the receiving cluster follow the distribution independently). This result represents the case which the application program has a strong locality of communication and well mapped. In this case, the number of receiving clusters is 80 to 180 even when the number of destinations are 32, thus unnecessary messages are not so many. The LARP is advantageous when the number of destinations is large.

Figure 11 shows when the distribution (D) is 5 and other parameters are unchanged. This case represents that there are a lot of destination nodes which are not local of the source node. The number of receiving clusters is considerably large, thus, many unnecessary messages are generated. The SM is advantageous when the number of receivers is small, and the LARP is advantageous when it is large. This tendency is also among other graphs, and the number of clusters when the performance of the LARP overcomes that of the SM is about 10. Although the LPRA cannot reduce the number of clusters compared with other schemes in these estimation, it will be useful when there are group of destination nodes distant from a source node.

As described earlier, when a directory entry is associated with a page, the number of destinations is likely to be at least 4. Figure 12 shows the number of average receivers when the number of destination is four and the variance is changed. the SM always yield least receivers, and the number is not more than around 120 even when the variance is large.

These results suggest that the hierarchical bit-map directory scheme requires a lot of unnecessary messages when the destination processes cannot be mapped into local processors of the source nodes. However, if processes which communicates each other can be locally

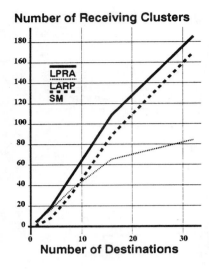

Figure 10: Number of receiving nodes vs. number of destinations (D =1)

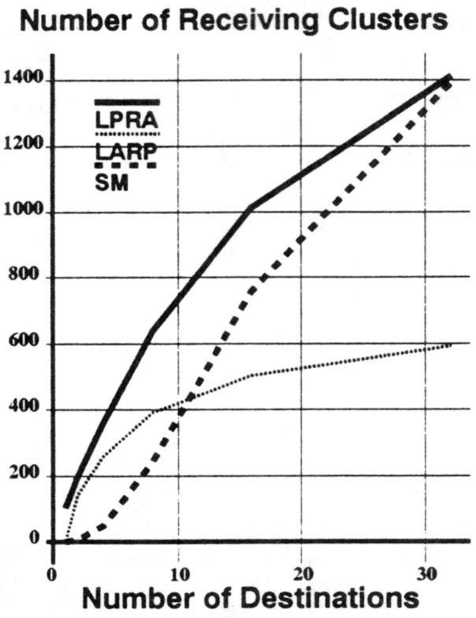

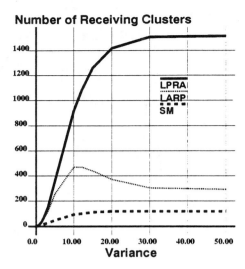

Figure 11: Number of receiving nodes vs. number of destinations (D =5)

Figure 12: Number of receiving clusters vs Variance

mapped. the number of unnecessary messages can be not so large with using the SM, LPRA and the LARP appropriately.

6 The RDT router chip

An LSI router chip which supports all hierarchical bit-map directory schemes discussed here has been implemented for JUMP-1. The structure of this RDT router chip is shown in Figure 13. The core of the chip is a 10×11 crossbar which exchanges packets from/to 10 18-bits-width links, that is, four for the rank-0 torus, four for the upper rank torus, and two for the MBPs which manage the distributed shared memory of JUMP-1. In JUMP-1, two RDT router chips are used in the bit-sliced mode to form 36 bits width for each link.

All packets are transferred between router chips synchronized with a unique 60MHz clock. In order to maximize the utilization of a link, packets are bi-directionally transferred. Maximum packets length is 16flits (36 bits-width 16flits-length) so as to carry a line of the cache. 3 flits header which carries the bit-map of the hierarchical bit-map are attached to all packets, but the length of the body is variable.

The virtual cut-through flow control is adopted to cope with frequent multicasting. Two packet buffers each of which can hold the maximum size packet form two virtual channels. In the RDT, the deadlock is avoided with the modified e-cube routing method using two virtual channels [13]. Using this technique, every 1-to-1 packet transfer or multicast is performed without deadlock. Since the route of the multicast is fixed, the FIFO assumption is ensured.

Two bits in the packet header are used for selection of three hierarchical bit-map directory schemes, the LPRA, SM and LARP. Thus, three schemes are se-

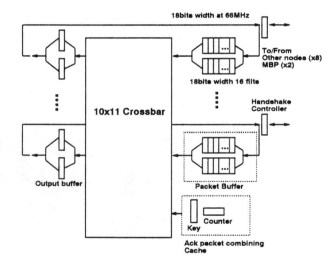

Figure 13: The structure of the RDT router

lectable for each packet, and the mixture of packet with different schemes are allowed. Moreover, the packet is forced to transfer to the MBP at any node of each hierarchy that multicast is started. In this case. the MBP re-makes the new bit-map for each hierarchy, and start multicast again. Although this scheme requires the latency to re-make the bit-map, unnecessary packets multicast can be reduced.

$0.5\mu m$ Hitachi BiCMOS SOG which provides 125K gates in maximum is utilized. Lines are directly driven with ECL interface of this chip. Using the dual port RAM, packet buffers allow to push and pull a flit of the packet simultaneously. The required number of gates are 90522 gates. Random logics require 50000 gates in total while areas corresponding to about 4000 gates are required for the dual-port RAM. The crossbar body and arbiter, which are simple but high performance is required, are designed in the logic level, while the complicated controllers are described in the VHDL.

7 Conclusion

Hierarchical bit-map directory schemes on the Recursive Diagonal Torus are proposed and discussed. Through a simple estimation, a small memory requirement of these schemes is demonstrated when a directory entry is associated with a page or update protocols are used on massively parallel processors. The problem of these schemes is congestion of the network with unnecessary multicast packets. Since the evaluation shown here is based on a simple probabilistic model, precise evaluations under practical conditions are required. However, it is difficult with simulations as a large number of processors must be simulated with practical application programs and mapping strategies.

A high speed RDT router which supports all schemes discussed here is available. Using these chips, JUMP-1 will start its operation on the next spring. A precise evaluation of these schemes will be done on this prototype.

Acknowledgment

The authors would like to express their sincere gratitude to the members of the Joint-University project for their valuable advice and discussion. The authors also express their thanks to Professor Takuya Terasawa, Mr. Osamu Gotoh, Mr. Kazuhito Kanda, and Mr. Hiromitsu Ueda of Tokyo Engineering University for supporting the environment of the multiprocessor instruction level simulator MILL.

A part of this research was supported by the Grant-in-Aid for Scientific Research on Priority Areas, #04235130, from the Ministry of Education, Science and Culture.

References

[1] Agarwal A., Simoni R., Hennessy J., and Horowitz M. An evaluation of directory schemes for cache coherence. In *Proc. of the 15th Annual International Symposium on Computer Architecture*, 1988.

[2] D.Lenoski, J.Laudon, K.Gharachorloo, W.-D.Weber, A.Gupta, J.Hennessy, M.Horowitz, and M.S.Lam. The stanford dash multiprocessor. *IEEE Computer*, 1992.

[3] James D.V., Laundrie A. T., Gjessing S., and Sohi G. S. Distributed-directory scheme: Scalable coherent interface. *IEEE Computer*, 1990.

[4] K. Hiraki, Hideharu Amano, Morihiro Kuga, Toshinori Sueyoshi, Tomohiro Kudoh, Hiroshi Nakashima, Hironori Nakajo, Hideo Matsuda, Takashi Matsumoto, and Shin ichiro Mori. Overview of the jump-1, an mpp prototype for general-purpose parallel computations. In *Proc. of the International Symposium on Parallel Architectures, Algorithms and Networks (ISPAN'94)*, 1994.

[5] J.Kuskin, D.Ofelt, M.Heinrich, J.Heinlein, R.Simoni, K.Gharachorloo, J.Chapin, D.Nakahira, J.Baxter, M.Horowitz, A.Gupta, M.Rosenblum, and J.Hennessy. The stanford flash multiprocessor. In *Proc. of the 21st Annual International Symposium on Computer Architecture*, 1994.

[6] K.Kurihara, D.Chaiken, and A.Agarwal. Latency tolerance through multithreading in large-scale multiprocessors. In *Proc. of International Symposium on Shared Memory Multiprocessing (ISSMM)*, 1991.

[7] Thapar M. and Delagi B. Distributed-directory scheme: Stanford distributed-directory protocol. *IEEE Computer*, 1990.

[8] J.P. Singh, W. Weber, and A. Gupta. Splash: Stanford parallel applications for shared-memory. In *Tech. Report. Computer System Laboratory. Stanford University*, 1992.

[9] T. Terasawa and H. Amano. Performance evaluation of the mixed-protocol caches with instruction level multiprocessor simulator. In *Proc. of IASTED International Conference of MODELLING AND SIMULATION*, 1994.

[10] T.Matsumoto and K.Hiraki. A shared-memory architecture for massively parallel computer systems. In *IEICE Japan SIG Reports. Vol. 92. No. 173. CPSY 92-26 (in Japanese)*, 1992.

[11] T.Matsumoto and K.Hiraki. Distributed shared-memory architecture using memory-based processors. In *Proc. of Joint Symp. on Parallel Processing'93 (in Japanese)*, 1993.

[12] W.D.Weber and A.Gupta. Analysis of cache invalidation patterns in microprocessors. In *Proc. of ASPLOS III*, 1989.

[13] Y. Yang and H. Amano. Message transfer algorithms on the recursive diagonal torus. In *Proc. of the International Symposium on Parallel Architectures, Algorithms and Networks (ISPAN'94)*, 1994.

[14] Y. Yang, H. Amano, H. Shibamura, and T.Sueyoshi. Recursive diagonal torus: An interconnection network for massively parallel computers. In *Proc. of 1993 IEEE Symposium on Parallel and Distributed Processing*, 1993.

The Quest for a Zero Overhead Shared Memory Parallel Machine*

Gautam Shah   Aman Singla   Umakishore Ramachandran

College of Computing

Georgia Institute of Technology

Atlanta, GA 30332-0280.

{gautam, aman, rama}@cc.gatech.edu

Abstract – *In this paper we present a new approach to benchmark the performance of shared memory systems. This approach focuses on recognizing how far off the performance of a given memory system is from a realistic ideal parallel machine. We define such a realistic machine model, called the z-machine, that accounts for the inherent communication costs in an application by tracking the data flow in the application. The z-machine is incorporated into an execution-driven simulation framework and is used as a reference for benchmarking different memory systems. The components of the overheads in these memory systems are identified and quantified for four applications. Using the z-machine performance as the standard to strive for we discuss the implications of the performance results and suggest architectural trends to pursue for realizing a zero overhead shared memory machine.*

1 Introduction

Realization of scalable shared memory machines is the quest of several ongoing research projects both in industry and academia. A collection of nodes interconnected via some kind of network, with each node having a piece of the globally shared memory is the common hardware model assumed. The cost of communication experienced on any realization of such a model is an important factor limiting the performance and scalability of a system[1]. Several techniques including relaxed memory consistency models, coherent caches, explicit communication primitives, and multithreading have been proposed to reduce and/or tolerate these communication overheads that arise due to shared memory accesses that traverse the network. In general the goal of all such techniques is to make the parallel machine appear as a zero overhead machine from the point of view of an application. It is usually recognized that no one technique is universally applicable for reducing or tolerating the communication overheads in all

situations [8].

There have been several recent studies in separating the overheads seen in the execution of an application on a parallel architecture [4, 13]. These studies shed important light on categorizing the sources of overhead, and the relative advantage of a particular technique in reducing a particular overhead category. For example, Sivasubramaniam et al. [13] break-down the overheads into algorithmic (i.e. inherent in the application such as serial component), and interaction (i.e. due to the communication and system overheads seen by the application when mapped onto a given architecture). All the techniques for latency reduction and tolerance attempt to shave off this interaction overhead. However, we would like to be able to quantify the amount of communication delay that is *inevitable* in the execution of the application on any given architecture. Any communication that is seen over and above this "realistic ideal" is an overhead, and we refer to it as *communication overhead*. Armed with this knowledge, we can then benchmark the overheads introduced by any given memory system which is meant to aid the communication in an application. Clearly, if an application has dynamic communication pattern we have to resort to execution-driven simulation to determine the communication that is inevitable in the application.

The quest for a machine model that has zero communication overhead from the point of view of an application is the goal of this work. PRAM [16] has been used quite successfully as a vehicle for parallel algorithm design. In [13], it is shown how PRAM could be used as a vehicle for determining the algorithmic overhead in an application as well by using the PRAM in an execution-driven framework. Unfortunately, the PRAM model assigns unit cost for all memory accesses and does not take into account even the realistic communication costs. Thus it is difficult to realize hardware models which behave like a PRAM and to have faith in PRAM as a performance evaluation tool.

We develop a *base machine model* that achieves this objective of quantifying the inherent communication in an application (Section 2). Then, we present an implementation of this base machine model in the SPASM simulation

---

*This work has been funded in part by NSF grants MIPS-9058430 and MIPS-9200005, and an equipment grant from DEC.

[1]We use the term system to denote an algorithm-architecture pair.

framework (Section 3). Four different shared memory systems (Section 4) are evaluated to identify the overheads they introduce over the communication required in the base machine (Section 5).

The primary contribution of this work is the base machine model that allows benchmarking the overheads introduced by different memory systems. Another contribution is the breakdown of overheads seen in different memory systems in the context of four different applications. The framework also leads to a better understanding of the application demands, and the merits and demerits of various features of the different memory systems. Such an understanding can help in deriving architectural mechanisms in shared memory systems that strive to perform like the base machine.

2 Base Machine Model

2.1 Communication in an Application

Before we identify the overheads that may be introduced by a shared memory system, we must understand the communication requirements of an application. Given a specific mapping of an application onto an architecture, its communication requirements are clearly defined. This is precisely the data movements warranted by the producer-consumer relationships between the parallel threads of the application. In order to keep track of this data flow relationship various memory systems tag on additional communication costs. There are a couple of ways we can tackle the issue of reducing communication costs. The first is by reducing the overheads introduced by the memory systems. The other is by providing ways to overlap computation with communication. The goodness of a memory system depends on its ability to keep the additional costs low and to maximize the computation-communication overlap. We therefore need a model that can give a measure of this goodness.

Given the producer-consumer relationships, applications use synchronization operations as firewalls to ensure that the program is data-race free. The inherent cost to achieve process coordination, depends on the support available for synchronization. Any additional cost that the memory system associates with the synchronization points is baggage added by the memory system to achieve its goals and thus should be looked upon as overhead.

Some of the communication requirements of an application can be determined statically based on the control flow, while the others can only be determined dynamically. Depending on the latency for remote communication on the interconnection network, the requirements will translate to a certain *lower bound* on the communication cost inherent in the application. Our goal in striving for a base machine model, is to be able to quantify this communication cost.

Figure 1 illustrates what we mean by inherent communication cost and overheads. Consider the case in which

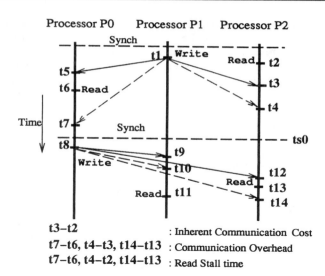

$t3-t2$: Inherent Communication Cost
$t7-t6, t4-t3, t14-t13$: Communication Overhead
$t7-t6, t4-t2, t14-t13$: Read Stall time

Figure 1: Inherent Communication Costs and Overheads

processor $P1$ writes a value at time $t1$. The reads corresponding to this write are issued by processors $P0$ and $P2$ at times $t2$ and $t6$ respectively. If we take into account the communication latencies only, the value written by $P1$ will propagate to processors $P0$ and $P2$ at times $t5$ and $t3$ respectively. Thus processor $P2$ experiences an inherent communication cost of $(t3 - t2)$, whereas there is no inherent communication cost associated with the read by $P0$ at time $t6$. Note that in the case of $P0$, there is inherent communication involved but because in this case the communication is overlapped with computation, the associated costs are absent. The inherent communication cost is dependent on task scheduling and load imbalance (which may arise due to either differing processor speeds and/or the work distribution across processors).

Any communication costs that arise in addition to this intrinsic requirement of the application is the overhead resulting from a specific shared memory system. For instance in Figure 1, although we said processor $P0$ has no inherent communication costs because of the read at $t6$, due to various architectural factors, the write by $P1$ actually propagates to $P0$ at time $t7$ and the $P0$ sees a cost of $(t7 - t6)$. In this case the observed penalty is only due to overheads. Latency and contention on the network, the memory consistency model, the cache coherence strategy, and the parameters of the cache subsystem combine to give rise to these overheads. These overheads manifest in different ways depending on the memory system and may be categorized as *read stall* time, *write stall* time, *buffer flush* time. Read stall time is the wait time seen by a typical processor for read-misses; write stall time is the wait time seen by a typical processor for write misses; and buffer flush time is the wait time seen by a typical processor at synchronization points for outstanding writes to complete. In Figure 1 the overhead for the read by processor $P0$ manifests itself as read stall time. Notice that a read stall does not mean that all of it is because of overheads - some of it may be due to the inherent communication requirements of the application, as in the case for processor $P2$. It should also be clear that if the write by processor $P1$ at time $t1$ stalls the processor from doing useful work in order to provide certain memory consistency guarantees, the time taken for $P0$ to reach synchronization point $ts0$ would be correspondingly longer.

Synchronization events have two cost components. One is for process synchronization to provide guarantees that a task has reached a certain point in execution and/or to coordinate with other tasks. This cost is inherent in the application [2]. The other cost comes about because the architectural choices made require that certain properties hold true at synchronization points. For instance, in the case of weaker memory models, synchronization is the point at which guarantees are made about the propagation of writes. This latter factor which manifests itself as buffer flush time is part of the communication overheads that were introduced by the architecture.

Thus, there is a necessity for a realistic machine model that can predict tighter lower bounds that includes the inherent communication costs to help us understand how close a given system is to the realistic ideal. Such a model can then be used to benchmark different design parameters of a memory system. The knowledge of a realistic ideal that includes communication costs can also serve as a starting point in the design of a zero overhead machine.

2.2 Description of the Model

We want the model to be cognizant of the computation and communication speeds of the underlying architecture. The communication speed helps determine the inherent communication cost of executing a given application on the architecture commensurate with the application's communication pattern; and the compute speed gives us an estimate of how much of this communication is overlapped with computation based on the timing relationships between the producers and consumers of data. Architectural artifacts such as finite caches introduce extra communication caused by replacements (due to the working set not fitting in the cache or due to poor replacement policies) and we treat them as part of the overheads introduced by the memory system. Hence, we do not want the model to be aware of the architectural limits such as finite-sized caches, or limitations of the interconnection network topology. We thus define a *zero overhead* machine (*z-machine*) as follows: it is a machine in which the only communication cost is that necessitated by the pure data flow in the application. In the z-machine, the producer of a datum knows exactly who its consumers are, and ships the datum immediately to these remote consumers. The producer does not wait for the data to reach the consumer

and can immediately proceed with its computation. We abstract out the details of the underlying communication subsystem to the point that the datum is available at a consumer after a latency L. The latency is determined only by communication speed of the interconnection network. That is, there is no contention in the z-machine for the data propagation. While this assumption has the obvious benefit of keeping the model simple, it is also realistic since we do not expect the inherent communication requirements in an application to exceed the capabilities of the network. If the consumer accesses the data at least L units of time after it is produced, then the communication is entirely overlapped with computation at both ends, and thus hidden from the application. Clearly, no real memory system can do better than the z-machine in terms of hiding the communication cost of an application. At the same time, it should also be clear that since the z-machine uses the real compute and communication times of the underlying architecture it gives a "realistic" lower bound on the true communication cost in the application. This inherent communication in an application manifests itself as read-stall times on the z-machine.

We should note that our base machine model, the z-machine, is different from base machines or idealized machines usually used in other cache studies [1, 9]. The base model in the other studies typically refers to a machine to which enhancements are applied. Many of those studies consider a sequentially consistent machine using the same cache protocol as the base model. Some other studies idealize the machine by removing some costs (like the protocol processing for coherence messages). While these models may serve the purpose in the respective studies, our model is intended for a different purpose - that of capturing the inherent communication costs and overheads. With the other models it may be possible to see benefits of the protocols in consideration. However, the extent of improvements that further enhancements may be able to obtain, is not always clear.

2.3 Overheads due to Memory Systems

Any real memory system adds overheads not present in the z-machine. For instance, the coherence protocol and the memory consistency model add to these costs in several ways. Since there is no way of pre-determining the consumers of a data item being written to, the memory system has to decide 'when' and 'where' the data should be sent. In update-based protocols, the data is sent to all processors that have currently cached this item. This increases contention on the network (since some of these might be useless updates), and manifest as write-stall times in the execution of the application. In invalidation-based protocols, the 'where' question is handled by not sending the data item to any processor (but rather by sending invalidation messages). In this case all consumer processors incur an access penalty (which manifests as read-stall times in the execution of the application) in its entirety for the remote access even if the data was produced well ahead of

---

[2]These inherent costs may certainly be changed by restructuring the application, but we are concerned only with overheads given a specific mapping of the application.

the time when it is actually requested by the consumers. Also, the z-machine can be thought of as having an infinite cache at each node. Thus real memory systems incur additional communication due to capacity and conflict misses. Further, if the memory system uses a store buffer to implement a weaker memory model, there could be additional overheads such as buffer-flush times in the execution of the application. Thus by considering each design choice in isolation with the z-machine we can determine the overhead introduced by each hardware artifact on the application performance.

By gradually developing the actual cache subsystem on top of the base machine (and hence introduce the limitations), we can isolate the effects of each of the factors individually.

3 Implementation of the z-machine

The z-machine is not physically realizable. However, we want to incorporate it into an execution-driven simulator so that we can benchmark application performance on different memory systems with reference to the z-machine. For this purpose, we have simulated the z-machine within the SPASM framework [13, 14] an execution-driven parallel architecture simulator. SPASM provides a framework to trap into the simulator on every shared memory read and write. We provide routines that are invoked on these traps based on the machine model we are simulating. The SPASM preprocessor also augments the application code with cycle counting instructions and uses a process oriented simulation package to schedule the events. Further, the simulation kernel provides a choice of network topologies. The simulated shared memory architecture is CC-NUMA using a directory-based cache coherence strategy. The cache line size is assumed to be exactly 4 bytes in the z-machine so that the only communication that occurs is due to true sharing in the application. The z-machine assumes that each producer is an oracle, meaning that the producer knows the set of consumers for a data item. In general, due to dynamic nature of applications we may not be able to determine the recipients of each write. Thus, we simulate the oracle by sending updates to all the processors on writes. The latency for these updates is directly available from the link bandwidth of the network, and is accounted for in the simulation. In order to correctly associate a read with a particular write, a counter is associated with each directory entry. Upon a write, the counter associated with the corresponding memory block is incremented. The counter is decremented after a period of L (the link latency), at which time the updates should be available at all the caches. In order to ensure that a read returns the value of the corresponding write, a read returns a value only if the counter is zero. Otherwise, the read stalls until the counter becomes zero.

The applications we are interested in are data-race free. Thus there are synchronization firewalls that separate producer-consumer relationships for shared data access in the application. The z-machine simulation has to take care of the synchronization ordering in the application. Synchronization is normally used for process control flow; but in memory systems that use weaker memory models it is also used to make guarantees about the program data flow. We take care of this separation in the simulation of the z-machine as follows. The synchronization events simulate only the process control flow aspects. Not flushing the buffers at synchronization points implies that we cannot guarantee a correct value to be be available at other processors after the synchronization point. However, the counter mechanism that we described earlier is used to delay consumers if a produced value has not been propagated. This mechanism is sufficient to ensure that consumers receive the correct value, and we thus free the synchronization events from providing data flow guarantees required by the weak memory models. It should be noted that the notion of memory consistency implemented in the z-machine is the weakest possible consistency commensurate with the data access pattern of the application.

4 Memory Systems

As described earlier, the base hardware is a CC-NUMA machine. Each node has a piece of the shared memory with its associated full-mapped directory information, a private cache, and a write buffer (not unlike the Dash multiprocessor [10]). In our work we consider the following four memory systems (RCinv, RCupd, RCadapt, RCcomp) that are built on top of the base hardware by specifying a particular coherence protocol along with a memory model.

RCinv: The memory system uses the release consistent (RC) memory model [6] and a Berkeley-style write-invalidate protocol. In this system, a processor write that misses in the cache is simply recorded in the write-buffer and the processor continues execution without stalling. The write is completed when ownership for the block is obtained from the directory controller, at which point the write request is retired from the buffer. In addition to the read stalls, a processor may incur write stalls if the write-buffer is full upon a write-miss, and incurs a buffer flush penalty if the buffer is non-empty at a release point.

RCupd: This memory system uses the RC memory model, a simple write-update protocol similar to the one used in the Firefly multiprocessor [15]. From the point of view of the processor, writes are handled exactly as in RCinv. However, we expect a higher write stall time for this memory system compared to RCinv due to the larger number of messages warranted by update schemes. To reduce the number of messages on the network we assume an additional *merge buffer* at each node that combines writes to the same cache line. While it has been shown that the merge buffer is effective in reducing the number of messages [5], it does introduce additional stall time for flushing the merge buffer at synchronization points for guaranteeing the correctness of the protocol.

RCcomp: This memory system also uses the RC memory model, and a merge buffer as in RCupd. However the cache protocol used is slightly different. Simple update protocols are incapable of accommodating the changing sharing pattern in an application, and thus are expected to perform poorly. The redundant updates incurred in such protocols increase both the stall times at the sender as well as potentially slowing down other unrelated memory accesses due to the contention on the network. The RCcomp memory system uses a competitive update protocol to alleviate this problem. A processor self-invalidates a line that has been updated *threshold* times without an intervening read by that processor. By decreasing the number of messages this memory system is expected to have lower write stall and buffer flush times when compared to RCupd.

RCadapt: This memory system is also based on the RC memory model but uses an adaptive protocol that switches between invalidation and update based on changes in the sharing pattern of the application. The protocol used in this memory system was developed for software management of coherent caches through the use of explicit communication primitives [11]. The directory controller keeps state information for sending updates to the active set of sharers through the *selective-write* primitive. When the selective-write primitive is used the corresponding memory block enters a special state. The presence bits in the directory represent active set the of sharers of that phase. Whenever a processor attempts to read a memory block with an already established sharing pattern (which implies the memory block is in a special state) it is taken as an indication of a change in the sharing pattern with the application entering a new phase. Therefore the directory controller re-initializes (by invalidating the current set of sharers) through an appropriate state transition upon such reads. In this study, we treat all writes to shared data as selective-writes. This memory system is expected to have read stall times comparable to update based schemes, and write stall times comparable to invalidate schemes.

5   Performance Results

In this section, we use four different applications to evaluate the memory systems presented in the previous section. In most memory systems studies, a sequentially consistent invalidation-based protocol is used as the frame of reference for benchmarking the performance of a given memory system. While this is a reasonable approach for benchmarking a given memory system, it fails to give an idea of how far off its performance is from a realistic ideal. The objective of our performance study is to give a break down of the overheads introduced by each memory system relative to such a realistic ideal which is represented by the z-machine. Therefore, we limit our discussions to executions observed on a 16 node simulated machine. For all the memory systems, we assume the following: infinite cache size; a cache block size of 32 bytes (as mentioned earlier, in the case of the z-machine the cache block is 4

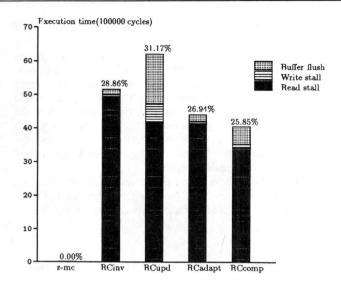

Figure 2: Cholesky

bytes to discount overheads and effects caused by larger line sizes); a mesh interconnect with a link latency of 1.6 CPU cycles per byte; store buffer of size 4 entries; and a merge buffer of one cache block. The applications we studied include Cholesky and Barnes-Hut from SPLASH suite [12], Integer Sort from the NAS parallel benchmark suite [3], and Maxflow [2].

Cholesky performs a factorization of a sparse positive definite matrix. The sparse nature of the matrix results in an algorithm with a data dependent access pattern. Sets of columns having similar non-zero structure are combined into supernodes. Supernodes are added to a central work queue if a set of criteria are satisfied. Processors get tasks from this central work queue. Each of these tasks is used to modify subsequent supernodes and if the criteria of the supernode being changed are satisfied then that node is also added to the work queue. Communication is involved in fetching all the required columns to a processor working on a given task and to obtain tasks from the central work queue. The communication pattern is totally dynamic based on the access to the work queue. In the study we consider a 1086x1086 matrix with 30,824 floating point non-zeros in the matrix and 110,461 in the factor with 506 supernodes. Barnes-Hut is an N-body simulation application. The application simulates over time the movement of these bodies due to the gravitational forces exerted on one another, given some set of initial conditions. The implementation statically assigns a set of bodies to each processor and goes through three phases for each time step. The producer-consumer relationship is well-defined and gradually changes over time. The problem size considered in this study used 128 bodies over 50 time steps [3]. The integer sort kernel uses a parallel bucket

---

[3]The problem included an artificial boost to affect the sharing pattern every 10 time steps. This boost simulates the effects of more time steps.

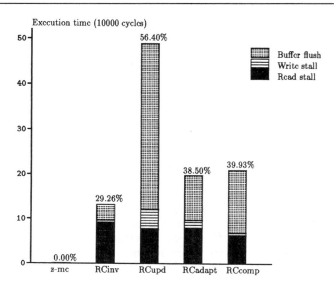

Figure 3: IS

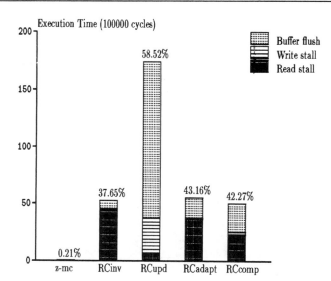

Figure 4: Maxflow

sort to rank a list of integers. The communication pattern is well defined statically. The problem size considered is 32K integers with 1K buckets. The Maxflow application finds the maximum flow from the distinguished source to the sink, in a directed graph with edge capacities. In the implementation [2], each processor accesses a local work queue for tasks to perform. These may in turn generate new tasks which are added to this local work queue. Each task involves read and write accesses to shared data. The local queues of all processors interact via a global queue for load balancing. Thus the producer-consumer relationship for data in this application is quite dynamic and random. The amount of computation required in the application is small and most of the time is spent in data movement. The input graph we consider has 200 vertices and 400 bidirectional edges.

As we mentioned earlier (see Section 2), by definition the z-machine will not have any write stall or buffer flush times. The inherent communication cost in the application will manifest as read stall times due to the timing relationships for the memory accesses from the producers and consumers of a data item. From our simulation results we observe that this cost is virtually zero for all the applications (see Figures 2,3,4 and 5). This indicates that all the inherent communication cost can be overlapped with the computation. Table 1 gives the time spent by each application on the network in the z-machine, most of which is hidden by the computation. This is a surprising result since the z-machine does take into account the two important parameters of any parallel machine, namely, the compute and the communication speeds in accounting for the inherent communication costs. In other words, the performance on the z-machine for these applications matches what would be observed on a PRAM. Given this result, any communication cost observed on the memory systems is an overhead.

Application	Number of Writes	% of Total Execution Time that these writes represent	Observed Costs (in cycles)
Cholesky	103915	1.477	54.6
IS	6353	3.779	0.0
Maxflow	38209	1.788	7257.8
Nbody	4542	0.002	0.0

Table 1: Inherent communication and observed costs on the z-machine

Figures 2,3,4 and 5 show the performance of these applications on the memory systems discussed in the previous section. The y-axis shows the execution time. For each memory system we have a bar which shows the amount of time due to each of the overheads. The total percentage of the overall execution time that these overheads represent is shown on the top each bar. The figures show that the overheads range from 6% to 37% for RCinv, 3% to 58% for RCupd, 3% to 42% for RCcomp, and 3% to 43% for RCadapt for the applications considered. As expected, the dominant component of the overheads for RCinv is the read stall time, and is significantly higher than those observed for the other three memory systems. The difference between the read stall observed on RCupd and the z-machine gives the read stall due to cold misses for any memory system[4] (since there are no capacity of conflict misses with the infinite cache assumption). Significant difference in the read stall times between RCinv and RCupd implies data reuse. This is true for Barnes-Hut and Maxflow applications (see Figures 5,4), and not true for Cholesky and IS (see Figures 2,3). The RCcomp and RCadapt can exploit data reuse only when the applica-

---

[4]Note that the observed read stall time due to cold misses could be slightly higher in an update protocol due to increased contention owing to update traffic.

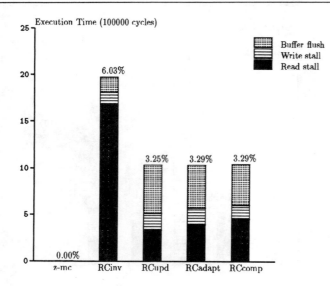

Figure 5: Barnes-Hut

tion has well established producer-consumer relationships over a period of time. This can be seen in the read stall times observed for Barnes-Hut in Figure 5. On the other hand, in Maxflow (see Figure 4) the producer-consumer relationship is more random making the read stall times for RCcomp and RCadapt to be similar to that of RCinv.

As expected, the write stall times for RCinv are significantly lower when compared to the other three. The write stall time is directly related to the number of messages that a protocol incurs on the network. Due to the dynamic nature of RCadapt and RCcomp (in switching between invalidation and update), these two memory systems incur lesser number of messages than RCupd. Decreasing the number of messages frees up the store buffer sooner, and also reduces the contention on the network.

The buffer flush time is a function of the store buffer size, the merge buffer size, and the frequency of synchronization in the application. The use of merge buffer results in a significant increase of buffer flush time for RCupd, RCcomp, and RCadapt compared to RCinv. Contention on the network plays a major role on the buffer flush time. This is evident when we compare RCupd with RCadapt and RCcomp.

6 Architectural Implications

The first observation is that since the z-machine has close to zero read stall times, the corresponding times observed on RCinv are unwarranted. The observations implies that it is possible to overlap most of the communication with the computation in an application. Figure 1 illustrates the above situation. The write by processor $P0$ at $t8$ is propagated well before the intended read by processor $P1$ at $t11$ even including the overheads. Our goal should be to reduce the unwarranted costs without bur-

dening the application programmer. One approach to reach the goal is moving towards an update based protocol, where this overhead component is usually low and is due to cold misses. However update based protocols do not handle all the situations – our results showed large read stall times even in the update protocol for Cholesky. Thus, another approach is to employ effective prefetching strategies. Applications in which there is considerable cold miss penalty (for e.g. Cholesky) prefetching and/or multithreading are more promising options.

We know that the write stall and buffer flush times are pure overheads from the point of view of an application. In order to realize a zero overhead machine we have to drive these two components to zero. The excessive message traffic of the update protocols is the main culprit for these components. To keep the traffic low (and thus the overheads), it is important for the protocol to adapt to changing sharing patterns in the applications (i.e. lean towards protocols such as RCadapt and RCcomp). Such an adaptive protocol could bring these two overhead components close to what is observed for RCinv. In our quest for a zero overhead machine, we now have to identify architectural enhancements that will drive this overhead further down. Write stall time is is dependent on two parameters: the store buffer size and the relative speed of the network with respect to the processor. Improving either of these two parameters will help to lower the write stall time.

Increasing the write buffer size could potentially increase the buffer flush time. The buffer flush time is a result of the RC memory model that links the data flow in the program with the synchronization operations in the program. As the performance on the z-machine indicates, there is an advantage in decoupling the two, i. e., use synchronization only for control flow and use a different mechanism for data flow. The motivation for doing this is to eliminate the buffer flush time. One approach would be associating data with synchronization [7] in order to carry out smart self-invalidations when needed at the consumer instead of stalling at the producer.

7 Concluding Remarks

The goal of several parallel computing research projects is to realize scalable shared memory machines. Essentially, this goal translates to making a parallel machine appear as a zero overhead machine from the point of view of an application, In striving towards this goal, we first need a frame of reference for the inherent communication cost in an application. We developed a realistic machine model the z-machine which would help to serve as such a frame of reference. We have incorporated the z-machine in an execution-driven simulator so that the inherent communication cost of an application can be quantified. An important result is that the performance on the z-machine for the applications used in this study matches what would be observed on a PRAM. Using the performance on the z-machine as a realistic ideal to strive for, we benchmarked four different memory systems. We presented a breakdown of the overheads seen in these memory systems and derived architectural implications to drive down these overheads.

There are several open issues to be explored including the effect of finite caches on the overheads, the use of other architectural enhancements such as multithreading and prefetching to lower the overheads, and the design of primitives that better exploit the synchronization information in the applications.

Acknowledgment: The selective-write primitive used in the adaptive protocol was co-designed with Anand Sivasubramaniam and Dr. H. Venkateswaran. We would also like to thank the members of our architecture group meetings, particularly Dr. H. Venkateswaran, Vibby Gottemukkala, Anand Sivasubramaniam and Ivan Yanasak, with whom we had many helpful discussions.

References

[1] F. Dahlgren aand M. Dubois and P. Stenstrom. Combined performance gains of simple cache protocol extensions. In *The 21st annual international symposium on computer architecture*, pages 187–197, April 1994.

[2] R. J. Anderson and J. C. Setubal. On the parallel implementation of goldberg's maximum flow algorithm. In *4th Annual ACM Symposium on Parallel Algorithms and Architectures*, pages 168–77, June 1992.

[3] D. Bailey et al. The NAS Parallel Benchmarks. *International Journal of Supercomputer Applications*, 5(3):63–73, 1991.

[4] M. E. Crovella and T. J. LeBlanc. Parallel Performance Prediction Using Lost Cycles Analysis. In *Proceedings of Supercomputing '94*, November 1994.

[5] F. Dahlgren and P. Stenstrom. Reducing the write traffic for a hybrid cache protocol. In *1994 International Conference on Parallel Processing*, volume 1, pages 166–173, August 1994.

[6] K. Gharachorloo, D. Lenoski, J. Laudon, P. Gibbons, A. Gupta, and J. L. Hennessy. Memory consistency and event ordering in scalable shared-memory multiprocessors. In *Proceedings of the 17th Annual International Symposium on Computer Architecture*, pages 15–26, May 1990.

[7] J. R. Goodman, M. K. Vernon, and P. J. Woest. Efficient synchronization primitives for large-scale cache-coherent multiprocessors. In *Third International Conference on Architectural Support for Programming Languages and Operating Systems*, pages 64–75, April 1989.

[8] A. Gupta, J. Hennessy, K. Gharachorloo, T. Mowry, and W-D. Weber. Comparative evaluation of latency reducing and tolerating techniques. In *Proceedings of the 18th Annual International Symposium on Computer Architecture*, pages 254–263, May 1991.

[9] M. Heinrich, J. Kuskin, D. Ofelt, and et al. The performance impact of flexibility in the Stanford FLASH multiprocessor. In *6th International Conference on Architectural Support for Programming Languages and Operating Systems*, pages 274–285, October 1994.

[10] D. Lenoski, J. Laudon, T. Joe, D. Nakahira, L. Stevens, A. Gupta, and J. Hennessy. The DASH prototype: Logic overhead and performance. *Transactions on parallel and distributed systems*, 4(1):41–61, January 1993.

[11] U. Ramachandran, G. Shah, A. Sivasubramaniam, A. Singla, and I. Yanasak. Architectural mechanisms for explicit communication in shared memory multiprocessors. Technical Report GIT-CC-94/59, College of Computing, Georgia Institute of Technology, December 1994.

[12] J. P. Singh, W.-D. Weber, and A. Gupta. SPLASH: Stanford parallel applications for shared-memory. *Computer Architecture News*, 20(1):5–44, March 1992.

[13] A. Sivasubramaniam, A. Singla, U. Ramachandran, and H. Venkateswaran. An Approach to Scalability Study of Shared Memory Parallel Systems. In *Proceedings of the ACM SIGMETRICS 1994 Conference on Measurement and Modeling of Computer Systems*, pages 171–180, May 1994.

[14] A. Sivasubramaniam, A. Singla, U. Ramachandran, and H. Venkateswaran. A Simulation-based Scalability Study of Parallel Systems. *Journal of Parallel and Distributed Computing*, 22(3):411–426, September 1994.

[15] C. P. Thacker and L. C. Stewart. Firefly: A Multiprocessor Workstation. In *Proceedings of the First International Conference on Architectural Support for Programming Languages and Operating Systems*, pages 164–172, October 1987.

[16] J. C. Wyllie. *The Complexity of Parallel Computations*. PhD thesis, Department of Computer Science, Cornell University, Ithaca, NY, 1979.

A CIRCULATING ACTIVE BARRIER SYNCHRONIZATION MECHANISM

Donald Johnson, David Lilja[*], & John Riedl

Department of Computer Science, [*]Department of Electrical Engineering

University of Minnesota

djohnson@d.umn.edu, 320 Heller Hall, Duluth, MN 55812

lilja@ee.umn.edu, riedl@cs.umn.edu, 200 Union Street, SE, Minneapolis, MN 55455

Abstract -- *A new low-cost, high-performance hardware mechanism for synchronizing multiple processing elements (PEs) at fine-grained programmed barriers is described. This mechanism is significantly less complex than other mechanisms with equivalent performance, using only a single conductor (i.e. -- a wire or copper run on a printed-circuit board) between PEs. This serial ring transports packets consisting of an active barrier identifier followed by a count of PEs which must yet check-in at that barrier. Since the ring itself has no sequential logic within the circulating active barrier (CAB) hardware, the only latency introduced is a single gate delay per PE and the conductor transmission time. The ring has a cluster controller (CC) which generates packets for active barriers, removes packets when not needed, and resets counters when all PEs have seen the zero-count. The expected synchronization times are shown to be under one microsecond for as many as 16384 PEs or 1024 workstations, if clustered appropriately. The ideal number of clusters for a two-dimensional hierarchy of N PEs is shown to be $\sqrt{N(D+G)/(I+G)}$, where G is the gate propagation delay, D is the inter-PE delay, and I is the inter-cluster transmission time. This mechanism allows rapid, contention-free check-in and proceed-from-barrier, and is applicable to a wide variety of systems architectures and topologies.*

Keywords: Distributed, Synchronization, Hardware, Barrier, VLIW

1.0 Introduction

Barriers are important synchronization operations in multiprocessor systems. When a processor executes a barrier instruction, it first checks-in at the barrier to indicate to the other processors that it has arrived at the specified synchronization point. It then must wait for all other processors participating in the barrier to check-in. After all participating processors have checked-in, the barrier is released so that the processors can proceed past the barrier to begin executing their next assigned tasks. This sequence of operations ensures that all processors have completed some specified sequence of tasks before any of the processors begin executing operations from subsequent tasks. Barrier synchronization is commonly used to synchronize processors at the end of the execution of a parallel loop [1], or when sequential tasks are assigned to different processing elements (PEs) [2]-[3], with correct execution requiring a synchronization barrier before any PE proceeds to its next assigned task.

For coarse-grained parallel tasks, the delay introduced by most current barrier mechanisms is relatively small compared to the task execution times. Synchronization of fine-grained tasks (i.e. those with fewer than 1000 instructions), however, requires much more stringent delay characteristics to prevent the synchronization delay from overwhelming the individual task execution times. It also precludes multi-tasking on any PE involved in a fine-grained barrier because of the high overhead of task-switching. Current barrier implementations typically use software trees [4], software or hardware-based counters [5]-[6], or hardware fan-in trees [7]-[8]. While the software trees are inexpensive, their synchronization delay is too long for fine-grained applications, especially on systems lacking a shared-bus. Counter-based methods introduce potentially high contention to access the shared counters, which can lead to unacceptably long, and unpredictable, synchronization delays. Combining networks have been proposed as a means to reduce such hot-spots [9], but their use may increase the switch cost and size by factors of between six and thirty-two [10]. Hardware fan-in trees provide fast check-in and release barrier operations, but they also are complex and expensive to implement.

The barrier mechanism proposed in this paper is very inexpensive, using only simple bit-serial hardware in each PE with only a single-conductor serial ring between PE's. Since the ring itself has no sequential logic, the only latency introduced by the barrier hardware is a single gate delay for each PE, plus the conductor's unavoidable time-of-flight delay between PE's. This configuration allows the performance of this new barrier mechanism to approach the performance of a fully-connected hardware fan-in tree.

Section 2 of the paper reviews existing barrier mechanisms, while Section 3 describes the new mechanism for both single and hierarchical clusters of processors. Its expected performance is analyzed in Section 4 for three different system configurations: a "single-box" of processors, such as the Connection Machine CM-5 or Cray T3D, a network of clustered multiprocessor systems, such as a cluster of Silicon Graphics machines, and a completely distributed network of workstations. This section also develops a concise analytical expression for determining the optimal number of PE's in each cluster to minimize the barrier synchronization time. Finally, the last section summarizes the results and conclusions.

2.0 Related Work

Mechanisms for barrier synchronization can be either primarily software-based or primarily hardware-based, although there are often requirements on both hardware and software. For example, access to a hardware mechanism is typically done via software, and some software mechanisms require specific hardware support (such as fetch-and-increment). We review existing barrier synchronization mechanisms in the following sections, and identify some of their limitations.

2.1 Software-Based Synchronization

Software-based barrier synchronization schemes can generally be used for medium-coarse execution granularity with acceptable efficiency, especially if fast shared-memory is used for synchronization. If synchronization takes place via operating system calls, only coarse-grained problems are reasonable because of the long latencies. Barrier synchronization algorithms based on a shared-memory tournament tree scheme, each with $O(logN)$ run-time and $O(N)$ memory and messages, have been proposed for different hardware primitives [5]. Performance using a bus-based Sequent Balance 21000 was demonstrated to be equivalent to performance using a counter-based barrier for up to 25 PEs, where the single-bus bottleneck became overwhelming. A scalable tournament barrier algorithm with separate arrival and wake-up trees that has a constant order of remote reference complexity was speculated to be fastest for large-scale broadcast cache coherence systems, even though it was experimentally outperformed by the dissemination barrier [4]. The dissemination barrier, which has a shorter critical path than the tournament barrier, should be better for noncoherent or directory-based systems which can have multiple simultaneous network transactions. For up to 16 local PEs, the synchronization time was below 40 μsec, while hundreds of μsecs were needed for remote PE synchronization [4].

2.2 Hardware-Based Synchronization

Hardware for barrier synchronization can involve a centralized array of check-in flags (one per barrier for each processor) with circuitry that detects "all-checked-in" to reset the flags associated with that specific barrier [7]. The highest performance would involve $O(N^2)$ wiring complexity, but $O(NlogN)$ complexity can be achieved with a hierarchy of barriers. Similar performance can be obtained using hardware proposed for static, single-program, compile-time determined barriers [2] or dynamic, multiple instruction stream barriers [11]. A dynamic barrier module may contain an associative memory that stores barrier masks, each mask containing the set of processors involved in a specific barrier. These masks are matched against current barrier requests and "GO" signals are transmitted to each processor waiting for the barrier when a match occurs (the match can also set-up the next barrier). An inexpensive hardware mechanism has been demonstrated to make applications using MIMD, SIMD, and VLIW execution

applicable to conventional processors [3]. This mechanism maps an output word into a barrier mask, returning a new mask for the next barrier, with each PE knowing all other PEs to be involved in a barrier. Performance considerations limit the number of PEs participating in a barrier to two less than the width of the address bus, with $O(N^2)$ wiring complexity. Hierarchical AND-tree barriers have also been included in commercially-available systems, such as the Cray T3D [8]. A distributed counter-based mechanism for shared-bus systems has been proposed [6] which requires only one shared-bus transaction per processor checking-in at a barrier, and nearly instantaneous detection and execution resumption when all PEs for a particular barrier have checked-in. While this mechanism provides the optimum barrier performance when using shared-bus communication, bus bandwidth constraints limit the number of PEs that can be used for fine-grained synchronization.

2.3 Fine-Grained Problems of Existing Barrier Mechanisms

Software barrier mechanisms have long latencies that make them inappropriate for fine-grained parallelism. Mechanisms using shared communications media produce high contention when many PEs check-in within a short time-span, with accompanying long latencies. In general, the best hardware solutions are not easily adapted to a wide-range of existing systems, but depend on a particular topology, or even a particular implementation. The hardware required to implement the best techniques is often complex and expensive, requiring $O(N^2)$ wiring complexity for the fastest performance. Associative memory, which is slow and expensive, may also be required. Without extremely fast context-switching [12]-[13], fine granularity precludes using barriers in a multiprogramming environment since a task-switch by one PE would delay any barrier(s) in which it is involved until that PE's execution resumes.

3.0 A New Barrier Synchronization Mechanism

The existing barrier synchronization mechanisms typically trade-off cost and complexity with performance. An ideal barrier mechanism, however, would:
1) allow for multiple barriers, each with potentially different sets of processors;
2) have user-level access to avoid context-switching into the OS kernel;
3) allow for very rapid check-in with no contention in the typical case;
4) have very rapid execution resumption when all PEs have checked-in;
5) be applicable to a wide variety of system architectures and topologies.

The proposed circulating active barrier (CAB) synchronization mechanism described in this section meets these goals by using simple and inexpensive hardware with an integrated special-purpose network. Section 3.1 describes an overview of the new mechanism, with user-level access and examples described in Section 3.2. Section 3.3 extends the basic concepts to multi-cluster systems with potentially thousands of PEs. Section 3.4 describes some implementation considerations.

3.1 Circulating Active Barrier (CAB) Overview

CAB uses a simple bit-serial network between processors. Bit-serial data transmission and manipulation is usually simpler and less expensive to implement than a parallel version, with the major problem being the time required to perform the serial operations. Bit-serial network rings, for example, typically introduce at least one clock delay for each node in the ring, restricting their use to applications which can tolerate slowness and/or few nodes on each ring. The CAB mechanism, however, takes advantage of the simplicity of serial data operations while minimizing the latency such operations introduce. An added benefit of this simplicity is minimized cost of the replicated hardware at each PE, with somewhat more complexity in the common cluster controller.

Barriers which are available for PE check-in are classified as "active" and are circulated around a ring of CAB hardware modules, each of which is attached to a PE. Each active barrier is in a packet consisting of a barrier identifier and a count of the remaining PEs to check-in at that barrier. When a PE reaches a barrier, it loads its CAB hardware with the barrier identifier that it is awaiting. The CAB hardware can activate an appropriate local signal, such as I/O wait, to block its PE until all PEs for that barrier have checked-in. CAB hardware will monitor the packets as they pass through it, detecting a match (using a bit-serial exclusive-OR) when the desired barrier reaches it. A bit-serial decrement of the count field of the packet is then performed.

It is important to realize that the comparison and decrement operations within each CAB are done with only combinational logic in the loop itself. There are, of course, sequential components within the CAB hardware which must be able to control the loop data during the next clock cycle. As a result, each node in the ring adds only the delay of a single XOR gate to the ring latency, regardless of the clock speed. In fact, if the clock time is longer than the inter-PE time, multiple CABs will be operating simultaneously on the same bit (with time differing by the propagation delays along the wire and through the XOR gate). Following the decrement operation, the CAB hardware continues to monitor the packets as they pass through it until a barrier matches with a count of zero, at which time it unblocks its PE. Note that the last PE completing check-in is immediately unblocked. A cluster controller is responsible for placing active barriers on the ring, resetting counts when all PEs have proceeded from a barrier, and removing barrier packets when no longer needed.

Figure 1 shows an example implementation of CAB with barrier packets inserted or deleted by the cluster controller, which may be attached to one of the loop node PEs or a separate PE. The CAB at each PE is able to detect a match of the desired barrier (dbar) and the packet's barrier (PBAR) and to decrement the packet's count (CNT) or detect a count of zero. Note that a zero count for a packet is the "all-checked-in" indicator. Table 1 shows the semantics of the various barrier function commands. Figure 2 shows the state transitions of the CAB for each PE and those of the cluster controller based on the flags and the barrier function.

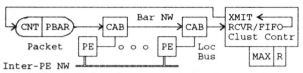

a) Single-Cluster Circulating Active Barrier (CAB)

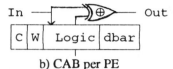

b) CAB per PE

PBAR: Circulating active barrier identifier
CNT: Number of PEs remaining to check-in at PBAR
dbar: Desired barrier # PE is awaiting (if W is true)
W flag: True when waiting for more PEs to check in
C flag: True when waiting to decrement CNT
MAX: Array of CNT resets, one for each active barrier
R flag: Array of reset flags, one for each active barrier (true when zero is circulating)
Logic: Includes clock generator, ID comparator, and decrement circuitry

Figure 1 Barrier Synchronization Hardware

Table 1 Barrier Function Semantics

#	Barrier function	Semantics of function
1	BWAIT	While W=t : block PE
2	BCHECK(n)	W=C=t, dbar=n (non-blocking)
3	BCHKW(n)	W=C=t, dbar=n, while W=t : block PE
4	PBAR=dbar&C=t	C=f,Dec CNT, if 0: W=f, unblock PE
5	PBAR=dbar&W=t &CNT=0	W=f, unblock PE
6	SETBAR(n,P)	$CNT=MAX_n=P$,PBAR=n (in cluster controller: if P=0,remove PBAR)
7	$CNT=0\&R_{PBAR}=f$	R_{PBAR}=t (in the cluster controller, cc)
8	$R_{PBAR}=t$	R_{PBAR}=f, $CNT=MAX_{PBAR}+CNT$ (in cc)

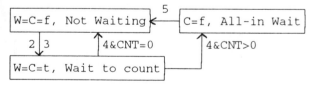

(States valid in blocking/non-blocking modes)
a) CAB States at each PE

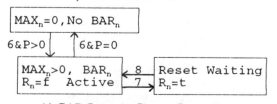

b) CAB States in Cluster Controller
Figure 2 Barrier State transitions by Function Number

Each processor can BCHECK(n) or BCHKW(n) to check-in at barrier n in a non-blocking or blocking mode, respectively. The CAB hardware waits for and decrements the counter associated with barrier n, with CNT's value originally set to the same value as MAX_n. If the counter becomes zero, the wait state is immediately removed. Otherwise the CAB hardware waits until CNT_n=0, at which point it removes the wait state. BCHECK(n) will not block the PE, thereby creating a "fuzzy barrier" [14]. The PE can do other work after checking-in, polling the W-flag or issuing a BWAIT to proceed beyond the barrier. When the cluster controller detects that a counter becomes zero, a flag, R_n, is set to wait until that barrier packet again reaches the controller. At that point, the counter is reset to the MAX value adjusted for any PEs checked-in at the new barrier. CAB modules, starting with the final one to check-in, will see the resetting packet twice, and if any are already checked-in for the next synchronization, CNT will become negative (2's complement) so that no early check-ins are lost. BWAIT or BCHKW(n) will indefinitely BLOCK the processor until its CAB W-flag is cleared, eliminating the need for spinning and reducing the start-up time when all PEs have reached the barrier to a very low value. Note that any number of barrier counters are possible, limited only by the number of PBAR bits, and that any barrier may be used by any PE. For example, this structure allows nested barriers, which could be the result of nested loops.

3.2 User-Level Access to the CAB System

The CAB hardware must be accessible at the user-level to allow rapid access, but it must be protected from unauthorized access. For example, more than one set of PEs may have independent sets of barriers on a particular cluster. Since the barrier identifiers are physically available to any PE on the cluster, the CAB hardware must protect the integrity of the system by mapping logical barrier IDs into authorized physical IDs. On some systems, external register-mapping may be used for fastest performance [15], but most systems can at least accommodate memory-mapping. The logical addresses needed and the access mode (R=read, W=write) are:

vbarw (W): BCHKW(map(vbn)) : check-in at virtual
           barrier vbn & wait until all-in
vbar (W): BCHECK(map(vbn)) : check-in at virtual
           barrier vbn (non-blocking)
bwait (R): BWAIT (may return status as data read)
lstat (R): returns the status

The mapping function in each CAB and the cluster controller may be a range of barrier numbers, with the starting barrier and the number of barriers set by the kernel, or a hardware array of individual barrier numbers available to this processor. In any case, the available barriers must be mapped into hardware using the kernel mode execution by using physical barrier identifiers supplied by a barrier server similar to that for acquiring and releasing UNIX ports. For performance considerations, however, their use must not involve context-switching into the kernel mode. The barrier server would also initiate generation and removal of circulating barriers through communication with the cluster controller, which may be attached to a PE which also has a CAB module. For example, the PE that initializes the barrier (e.g. -- the PE executing the sequential code) is responsible for communicating with the cluster controller to insert a new barrier packet for circulation. Also, this PE must send the barrier number to all participating PEs to initialize their CAB modules using the general-purpose inter-PE communication network. This is one-time overhead for establishing a barrier, which, once created, may be repeatedly used by the PEs involved since the cluster controller will reset CNT.

The example code in Figure 3 assumes that the logical addresses described above have been mapped for vbarw, vbar, bwait, and lstat. Figure 3a shows an example of a parallel loop with a barrier synchronizing the PEs at the loop's completion. Figure 3b shows an example of more generic barrier synchronization, which could be used, for instance, to emulate a VLIW (very long instruction word) [16], a taskflow [17], or a dataflow [18] machine using general-purpose PEs. Note that other interprocessor interactions (e.g. -- shared data access) are not considered here.

```
/* Cluster node has already: SETBAR(bar_num, P)
     to activate each barrier
  Each CAB has been initialized with its valid barrier(s)*/
  do {  /* start of DoAll loop for barrier synchronization
              (each PE uses different data for its loop) */
  ... /* processing prior to barrier for this PE */
  vbar = bar_num;  /* check-in at barrier bar_num in
                      non-blocking mode */
  ... /* optional processing after check-in at loop end */
  register = bwait;  /* BLOCK until all checked-in */
  }while continue;  /* loop repeats as soon as CAB
                      hardware detects all PEs have checked-in */
/* cluster node will clear barrier with:
         SETBAR(bar_num, 0) when barrier is removed*/
         a) Loop Barrier "Join" Operation


/* The barriers needn't involve the same PEs.
  Each PE knows the barrier number for each synchroniza-
  tion point. Non-overlapping barriers involving the
  same number of PEs can use the same bar_num. */
  ... /* Do part of the VLIW assigned by compiler to this
          PE (e.g. — a few instructions) */
  vbarw = bar_num1;  /* wait for other PEs to do their
                      part, then proceed */
  ... /* Do next VLIW part assigned to this PE
          (the set of PEs may differ from bar_num1) */
  vbarw = bar_num2;  /* wait for others to do their part */
         b) VLIW Emulation Using PEs
```

Figure 3 Example Code for Using CAB Mechanism

3.3 Multi-Cluster CAB Implementation

When N PEs are in a loop, the loop time is O(N), with a

very small proportionality constant because of the absence of sequential logic in the loop. For a large number of PEs, this will result in unacceptable latencies, and therefore a hierarchy of loops is desirable, although in this paper we limit our discussion to two levels of hierarchy. Extensions to additional levels are expected to be relatively straightforward. Barriers which involve only PEs within a cluster would be handled by the mechanism described in Section 3.1. If PEs accessing the same barrier reside in different clusters, an inter-cluster loop, similar to the intra-cluster loop, will be needed. In this case, the cluster controller initializes CNT to local PEs+1 and will detect when a global barrier's CNT=1 within the local cluster. It will then will check-in at the appropriate global barrier, whose count was originally the number of clusters required to check-in. When a cluster controller sees a global-loop barrier with a count of zero, it will finish global check-in by circulating the matching local barrier with a zero count. Figure 4 shows an example of a two-level hierarchy of loops, and Figure 5 shows an example of a cluster controller for such a hierarchy.

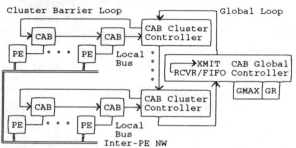

GMAX and GR are global equivalents of MAX and R of figure 1 (arrays for resetting the global barriers).

Figure 4 Two-Level Synchronization Overview

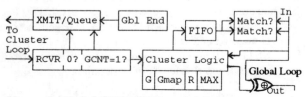

0?	Detect CNT=0 to by-pass Queue on barrier
Queue	For packet transmission in cluster
Gbl End	Insert resetting 0 at head of queue when all-in
GCNT=1?	Direct packet to cluster Queue (normal) or global check-in FIFO (if GCNT=1)
G array	True when the cluster's barrier ID is global
Gmap	Maps cluster barrier ID into global ID before placing into FIFO for global match
MAX	Array of number of PEs, one for each barrier
R flag	Array of reset flags, one per barrier in cluster
Match?	Parallel comparators to detect global ID match
Logic	To control the cluster controller (including the global-loop XOR gate)

Figure 5 Hierarchical Barrier Cluster Controller

Note that checked-in global barriers of a cluster will be simultaneously checked by parallel serial-comparators. Two queued comparators are most likely sufficient since a particular cluster is unlikely to be involved in more than two global barriers; global barriers would normally be involved only if the number of PEs for a barrier exceeds the PEs of a cluster, and all PEs of a cluster would normally be involved with a barrier before requiring inter-cluster protocols. For more than two levels of hierarchy, good performance would require additional comparators above the lowest cluster controller . As on the local loop, there are no serial components in the global loop outside of the global controller to minimize the latency.

3.4 Implementation Considerations

Since each barrier involves at least two nodes, the number of reachable barriers at a particular time is, at most, one-half the nodes on a local or global ring. If properly mapped, only nodes/2 barriers must be accommodated, with each barrier capable of counting all nodes. Since the count requires $\log_2$Nodes bits and the barrier identifier $\log_2$(Nodes/2) bits, the optimized packet size for either a cluster or an inter-cluster ring which can accommodate all possible reachable barriers would be of bit-length:

$$\log_2\text{Nodes} + \log_2(\text{Nodes}/2) = 2\log_2\text{Nodes} - 1.$$

There are circumstances which require packet lengths greater than the optimum. It may be desirable to add a packet parity bit to increase transmission reliability, especially for a network of workstations. Also, the number of active barriers could be greater than $\log_2$ of the ring nodes by increasing the number of packet bits used to represent the barrier identifier. While it is possible for a ring controller to avoid circulating barrier packets on a ring if the barrier could not be reached, such as the outer-most barrier of nested barriers, the deletion and insertion of such barriers would make the cluster controller more complex. Adding bits to packets for whatever reason would make the maximum latency between 4 - 17 % longer for 1 - 2 added bits for closely-spaced PEs, but would typically only add 1 - 2 clock cycles to the expected latency, assuming the cluster controller is holding only one bit more than one packet, which is the minimum number of bits for making a recirculate decision. The impact of adding one or two bits per packet on a network of workstations would be minimal since conductor latency, not bit-time, is the major time-factor.

It should be noted that the loop controller can reduce the latency (with increased cost) by immediately transmitting, rather than buffering, any resetting zero packet. It could also reduce the latency by prioritizing the packets based on history and/or proximity to the reset value. Note also that clock synchronization is necessary via transmission of a particular bit pattern (e.g. — all ones), with the controller ensuring that other packets could not produce that pattern.

4.0 Performance Evaluation

In Section 4.1, equations will be developed for calculating the barrier latency times for single and two-dimensional loops. Section 4.2 graphically illustrates the expected barrier latency for three different classes of systems: a) single-box multiprocessors, b) networks of single-PE workstations, and c) networks of multi-PE workstations. Section 4.3 derives an expression for finding the optimum clustering for two-dimensional systems. The following are used to represent the various parameters in this section:

D = inter-PE prop. time I = inter-cluster time
X = # global CNT bits Y = # global PBAR bits
P = period of Xmit clock N = # PEs total
G = gate-delay time Z = # clusters
C = # CNT bits N/Z = # PEs per loop
B = # PBAR bits $\lg N$ = $\log_2 N$

4.1 Latency Equations

The time from the last check-in until all PEs are notified is the release latency produced by the CAB synchronization mechanism. For barriers totally within a single cluster, the lowest time (T_L) involves one traversal of the packet around the loop, with check-in occurring just prior to the desired packet arriving at the "last-in" node. This time includes the time the packet spends in the cluster controller, $(C+B+1)P$, since the entire packet must be inside the controller before a decision can be made as to whether to recirculate or remove a barrier (which takes one clock cycle). The traversal time outside of the controller, $(D+G)N$, includes the sum of all gate delays and inter-PE propagation times. The total lowest time is then: $T_L = (D+G)N + (C+B+1)P$, or, if optimized, as described in Section 3.4 with $C=\lg N$ and $B=C-1$, the resulting time within the controller is $P\log_2 N$, yielding:

$$T_L = (D+G)N + 2P\lg N.$$

The expected latency waiting for the desired packet would be one-half the loop time. Once the desired packet is seen, it must make, at most, a complete loop traversal (with a CNT=0) before all PEs will be unblocked. Assuming the number of active barriers is small enough so that the FIFO in the cluster controller is empty, which is the typical case if the number of PEs per barrier is high, the mean expected time is then: $T_E = 1.5T_L$, or, if optimized:

$$T_E = 1.5(D+G)N + 3P\lg N.$$

The normal longest time would involve a complete loop traversal before encountering the desired barrier, leading to double the lowest time. The very worst case would involve a situation with N/2 active barriers (two PEs per barrier), which is extremely unrealistic. In this hypothetical case, the time spent outside the controller is insignificant, since the limiting factor is transmitting the backed-up packets within the controller. Assuming two loop traversals for the packet, with the resetting zero packet bypassing the controller's FIFO, the worst-case time is: $T_w = (C+B)PN$, or, if optimized: $T_w = (2\lg N - 1)PN$.

When multiple clusters are involved, the minimum latency includes times spent in the local loop, $(D+G)N/Z$, in the local cluster controller, $(C+B+2)P$, which includes one

bit-time for mapping to the global loop, in the global loop, $(I+G)Z$, and in the global controller, $(X+Y+1)P$. The minimum time is then:

$$T_L = (D+G)N/Z + (C+B+2)P + (I+G)Z + (X+Y+1)P,$$

or, if optimized with $C=\lg N/Z$, $B=C-1$, $X=\lg Z$, and $Y=X-1$, the lowest time is:

$$T_L = (D+G)N/Z + 2P\lg N + (I+G)Z+P.$$

The expected time is $1.5T_L$, which if optimized is:

$$T_E = 1.5(D+G)N/Z + 3P\lg N + 1.5(I+G)Z + 1.5P.$$

Similarly the worst-case time is: $T_w = 2(C+B)PN/Z$, or if optimized: $T_w = (2\lg N - 1)NP$.

4.2 Example Synchronization Times on Three Systems

The graphs in Figure 6 show the expected time (T_E) based on three models: a) a 1-box arrangement of PEs with each cluster on a printed circuit board for very small inter-PE distance, b) a network of closely-spaced single-PE workstations (SP-WS), and c) a network of closely-spaced multi-PE workstations (MP-WS: one cluster per workstation). The following are assumed:

1) packets are of the optimum size, $2\log_2\text{Nodes} -1$, as described in section 3.4,
2) the CAB XOR gate has a 1.5 nsec propagation delay,
3) the inter-PE time is 0.15 nsec for 1-box or MP-WS systems and 5 nsec for SP-WS systems,
4) inter-cluster time is double the inter-PE time, except MP-WS time is 5 nsec (workstations closely-spaced),
5) the serial clock period is 2 nsec,
6) only dedicated (single-task) PEs are involved, and
7) the number of active barriers is small enough so that only one packet is in a cluster controller.

Note that the best-case times (T_L) would be 67% of the plotted expected times, and the normal worst-case times would be 133% of the expected times, assuming the number of active barriers is not a limiting factor. Note also that doubling the clock period from 2 to 4 nsec increases the times only slightly (from 6-13%) unless the cluster controller holds more than one packet so that clock period becomes a limiting factor.

As can be seen from Figure 6, the clustering becomes very important as the number of PEs increases because of the O(N) loop-time outside the cluster controller. Each line represents a particular total number of PEs, so the PEs per cluster can be calculated by dividing by the number of clusters. Notice that most curves have an optimum clustering, producing a minimum synchronization time. This minimum depends on the type of system, but the slope is gentle enough so that a slightly "non-optimum" choice would not be disastrous. For example, with 1024 PEs, the minimum time occurs with 32 clusters, except for the multiprocessor workstation configuration, where the minimum occurs using 16 clusters. Note also that the major time difference is due to inter-PE time, with four times as many PEs in the MP-WS as in the SP-WS requiring approximately the same synchronization time.

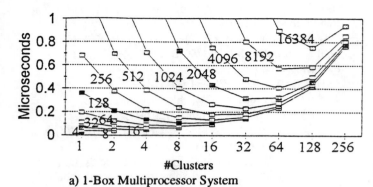

a) 1-Box Multiprocessor System

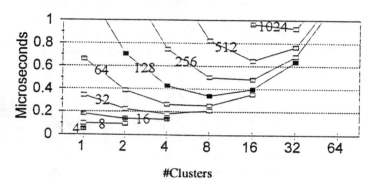

#Clusters

b) Network of Single-PE Workstations

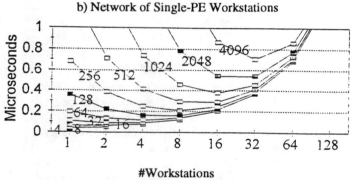

#Workstations

c) Network of Multiprocessor Workstations

Line labels are the total PEs involved in a barrier

Figure 6 Expected CAB Synchronization Time

4.3 Optimum Clustering for a Two-Dimensional CAB

Rather than generating plots in order to observe the minimum synchronization time for a given configuration, it could prove useful to analytically determine the optimum clustering based on the number of PEs. Since $T_E = 1.5(D+G)N/Z + 3PlgN + 1.5(I+G)Z + 1.5P$ when using the optimized packet sizes, the optimum number of clusters occurs when T_E's derivative with respect to Z is zero:

$$I+G - (D+G)/Z^2 = 0, \text{ or } Z = \sqrt{N(D+G)/(I+G)}.$$

For the examples used in Section 4.2, the optimum number of clusters is: $0.96\sqrt{N}$ (for 1-box system), $0.75\sqrt{N}$ (for SP-WS network), or $0.50\sqrt{N}$ (for MP-WS network). For the 1-box system, the inter-PE time (D) and inter-cluster time (I) are small compared to the gate delay (G), producing a near-unity $(D+G)/(I+G)$ ratio. For the SP-WS configuration, $D \approx 3G$ and $I \approx 7G$, resulting in a ratio near 0.5. For the MP-WS configuration, D has little influence, and $I \approx 3G$, making the ratio close to 0.25. Other configurations, with differing times, could be analyzed in a like manner.

5.0 Summary

A new fine-grained barrier synchronization mechanism has been described. This mechanism can be implemented with relatively inexpensive serial hardware while offering performance comparable to the highest-performance of other mechanisms. A contention-free ring with no clocked delays allows for multiple operations to be simultaneously performed. The mechanism allows for multiple and nested barriers and can be applied to both existing and new architectures.

As has been demonstrated, the expected synchronization times for the circulating active barrier mechanism can be under one microsecond for as many as 16384 PEs or 1024 single-processor workstations using conservative assumptions on parameters such as gate delay and electrical propagation time. This makes fine-grained problems such as VLIW emulation feasible using such networks of processors. It also provides a fast, contention-free mechanism for loop synchronization that will not adversely affect the performance of other inter-PE or PE-memory communications. When more than 16 PEs are involved in a barrier synchronization, implementing a hierarchical CAB system improves performance, with the optimum number of clusters for a two-dimensional hierarchy being $\sqrt{N(D+G)/(I+G)}$, where N is the number of PEs, G is the gate delay, D is the inter-PE delay, and I is the inter-cluster time. Future work will examine extending the mechanism beyond two levels of hierarchy and extending the performance evaluation through simulation.

Acknowledgement

This work was supported in part by the National Science Foundation under grant number MIP-9221900.

References

[1] D. Lilja, "Exploiting the Parallelism Available in Loops," *Computer*,(2/94), pp. 13-16.

[2] M. O'Keefe & H. Dietz, "Hardware Barrier Synchronization: Static Barrier MIMD (SBM)," *ICPP Proceedings*, (1990), pp. I:35-42.

[3] W. Cohen, H. Dietz, & J. Sponaugle, "Dynamic Barrier Architecture for Multi-Mode Fine-Grain Parallelism Using Conventional Processors," *ICPP Proceedings*, (1994), pp. I:93-96.

[4]   J. Mellor-Crummey & M. Scott, "Algorithms for Scalable Synchronization on Shared-Memory Multiprocessors," *ACM Trans Comp Sys,* (2/91), pp. 21-65.

[5]   B. Lubachevsky, "Synchronization Barrier and Related Tools for Shared Memory Parallel Programming," *ICPP Proceedings,* (1989), pp. II:175-179.

[6]   D. Johnson, D. Lilja, & J. Riedl, "A Distributed Hardware Mechanism for Process Synchronization on Shared-Bus Multiprocessors," *ICPP Proceedings,* (1994), pp. II:268-275.

[7]   C. Beckmann, & C. Polychronopoulos, "Fast Barrier Synchronization Hardware" *Proc. of Supercomputing '90,* (11/90), pp. 180-189.

[8]   Cray Research, *Cray T3D System Architecture Overview,* (10/93), Chapt 3: 24-27.

[9]   J. Edler, A. Gottlieb, C. Kruskai, K. McAuliffe, L. Rudolph, M. Snir, P. Teller, & J. Wilson, "Issues Related to MIMD Shared-Memory Computers: The NYU Ultracomputer Approach," *ISCA,* (1985), pp. 126-136.

[10] G. Pfister & V. Norton, "Hot-Spot and Contention and Combining in Multistage Interconnection Networks," *IEEE Trans. on Computers,* (10/85), pp. 943-948.

[11] M. O'Keefe & H. Dietz, "Hardware Barrier Synchronization: Dynamic Barrier MIMD (DBM)," *ICPP Proceedings,* (1990), pp. I:43-46.

[12] R. Alverson, D. Callahan, D. Cummings, B. Koblenz, A. Porterfield, & B. Smith, "The Tera Computer System," *International Conf. on Supercomputing,* (1990), pp. 1-6.

[13] X. Fan, "Realization of Multiprocessing on a RISC-like Architecture," *Multiprocessing and Microprogramming,* (6/92), pp. 195-206.

[14] R. Gupta, "The Fuzzy Barrier: a Mechanism for High Speed Synchronization of Processors," *Third ASPLOS,* (4/89), pp. 54-63.

[15] D. Henry & C. Joerg, "A Tightly-Coupled Processor-Network Interface," *ASPLOS V,* (10/92), pp. 111-122.

[16] M. Lam, "Software Pipelining: An Effective Scheduling Technique for VLIW Machines," *SIGPLAN 88 Conf on Prog Lang Design & Implem,* (6/88), pp. 318-328.

[17] R. Horst, "Task-Flow Architecture for WSI Parallel Processing," *Computer,* (4/92), pp. 10-18.

[18] R. Iannucci, "Toward A Dataflow/Von Neumann Hybrid Architecture," *International Symposium on Computer Architecture,* (1988), pp. 131-140.

Experimental Analysis of Some SIMD Array Memory Hierarchies*

Martin C. Herbordt[†]
Department of Electrical and Computer Engineering
University of Houston

Charles C. Weems[‡]
Department of Computer Science
University of Massachusetts

Abstract: Although SIMD arrays have been built since the 1960's, there have been few, if any, empirical studies of possible memory hierarchy designs. The underlying problems—which have included the lack of a unified architectural framework and the computational intractability of simulating large PE arrays—are addressed here through the use of *trace compilation*, a novel approach to trace-driven simulation. The results indicate the benefits of adding another level to current SIMD array memory designs. Also, surprising results were obtained about performance effects of varying cache associativity and block size. Together, they indicate that while SIMD array programs have sufficient locality to make PE caches worthwhile, the type of locality may differ fundamentally from that of serial-machine and multi-processor programs.

1 Introduction

A critical issues in computer architecture continues to be the allocation of silicon between the processors and the memory hierarchy. With SIMD arrays, as with microprocessors a few generations ago, the questions in particular are 1) whether to set aside area on the processor chip for cache or whether to depend entirely on off-chip memory, and 2) if so then how much. Once this issue has been resolved, further questions remain about the trade-off between register file and cache sizes, and about the design of the cache itself: in particular, its size, block size, and associativity. Also, as with microprocessor architecture, these issues are application dependent making the use of experimentation using non-trivial programs essential. This paper presents results from a study of memory hierarchies on SIMD arrays using real program executions.

This investigation is now particularly timely: improvements in process and packaging technologies are enabling the construction of increasingly complex processing elements (PEs) and instruction issue rates to pass 100 Mhz [4]. Together, these developments are increasing the attractiveness of SIMD arrays for a number of applications [7], and as compute servers [2] especially in heterogeneous environments [15, 10].

Previous work in the evaluation of SIMD array architectures using real program executions includes Marek's SIMD Simulator Workbench and [12] and GT-RAW, developed by Ligon and Ramachandran [11]. Neither, however, measures performance with respect to varying of the memory hierarchy concentrating instead on issues such as granularity, reconfigurability, and network support. One of the limitations of these previous studies was the cost of the simulations: the entire simulation must be repeated for every design change. Another is the range of features it is possible to examine with the SIMD Simulator Workbench, and—because of its emphasis on broader questions in the domain of reconfigurable multiprocessors—the detail at which features are simulated in GT-RAW.

The issue of simulator efficiency is largely addressed here through the use of the *trace compilation* technique by ENPASSANT, the ENvironment for Parallel System Simulation ANalysis Tools [7, 8]. Trace compilation is a novel approach to trace-driven simulation: codes are run on an abstract virtual machine to generate a coarse-grained trace, which is then refined through a series of transformations wherein greater resolution is obtained with respect to the details of the target architecture. We have found this approach to be one to two orders of magnitude faster than detailed simulation, while still retaining much of the accuracy of the model. Furthermore, abstract machine traces must only be generated for a small fraction of the possible architectural parameter combinations.

The primary memory hierarchy results obtained using ENPASSANT are as follows:

- The shapes of the memory-reference-time versus register-file-size curves indicate distinct reference locality. This is significant as SIMD arrays are still being built with flat memory hierarchies.

- The register assignment policy has little impact. This is significant for machines using dynamic register assignment as it encourages use of trivial (i.e. fast) replacement algorithms.

- The level of associativity has a major impact on cache performance. This is somewhat surprising given that the well-known result by Hill [9] has become virtually a rule of thumb.

- Increasing the block size has a *negative* effect on cache performance. This is an extremely surprising result indicating that the type of locality avail-

*This work was edited substantially to meet the page requirement. A much more detailed version is available via anonymous ftp as pub/techrept/techreport/1995/UM-CS-1995-007.ps from ftp.cs.umass.edu. This work was supported in part by an IBM Fellowship and in part by the Defense Advanced Research Projects Agency under contract DACA76-89-C-0016, monitored by the U.S. Army Engineer Topographic Lab; under contract DAAL02-91-K-0047, monitored by the U.S. Army Harry Diamond Lab; and by a CII grant from the National Science Foundation (CDA-8922572).

[†]Houston, TX 77204-4793; NetAd: herbordt@uh.edu.
[‡]Amherst, MA 01003; NetAd: weems@cs.umass.edu.

able in SIMD array programs is quite unusual.

The overall significance of this work is that it is the first systematic study of the memory hierarchy of SIMD arrays, that it indicates the benefits of multi-level memories for these processors, and that the design of SIMD PE caches requires substantially different assumptions than in other architectures.

The rest of this paper is organized as follows. In the next section the design and application spaces are presented. There follows a discussion of the methodology. After that come the results for the register file and cache designs, respectively, and concluding remarks.

2 Architecture and Application Spaces

The overall design space is the class of architectures known as massively parallel SIMD arrays (MPAs). Architectures of this type consist of a controller and an array consisting of from a few thousand to a few hundred thousand processing elements (PEs). The PEs execute synchronously the code broadcast by the controller. As a consequence, a characteristic all the MPA PE memory configurations have in common is that all array instructions operate on the same register within each PE at the same time. The same is also largely true for memory locations within each PE. Local indexing is not a critical issue in the current test suite and will be examined in a later study.

The following are the memory architectures of some more recent and well-known MPAs. For simplicity we refer to explicitly controlled on-chip storage as *register file* while off-chip storage that must be brought on-chip to be operated upon is referred to as *memory*. Not included in the descriptions are the specific accumulator registers in some processors. The memory sizes indicated are per PE.

- **Abacus** [4] – The Abacus is a reconfigurable processor. In its one-bit PE configuration it has 4 bytes of on-chip and 2K bytes of off-chip storage. The on-chip storage is loaded at 40 cycles/bit.

- **ADSP-21020** [5] – The ADSP is a powerful DSP chip that can be configured to run in SIMD mode. It has 80 bytes of on-chip register file, 512K bytes of on-chip cache, plus external memory.

- **CAAPP** [15] – The CAAPP has 40 bytes of on-chip and 4K bytes of off-chip storage. The on-chip storage is loaded at 5 cycles/bit.

- **CM1/2** [14] – The CM1 and CM2 have a flat storage space: the 8K bytes of storage are all off-chip.

- **DAP** [1] – The DAP also has a flat storage space: the 128K bytes of storage are all off-chip.

- **MP1/2** [3] – The MP1 and MP2 have 160 bytes of on-chip and 16K bytes of off-chip storage. The on-chip storage is loaded at a rate of 1.14 cycles/bit.

MPA memory architectures can be partitioned into two types, flat and hierarchical. The flat memory has the advantages of simplicity of design, since the entire memory can be constructed from commodity parts,

and that the entire PE chip can be used for the data-paths. A drawback is that the average PE data access latency can be longer than in hierarchical schemes.

In most hierarchical designs, the storage can be partitioned into on-chip and off-chip components. Historically, the on-chip memory has been very small and has been loaded from the side of the array. As a result, load latencies were on the order of a hundred cycles/bit [13]. Since many important computations require substantial storage, most recent designs have either increased the amount of register file, introduced high-speed register load mechanisms, or both. We have included the ADSP-21020 as an example of the likely future direction for MPA memory architectures. It is possible, if not likely, that future MPA designs will have a cache level in the memory hierarchy in addition to the register file and main memory. EN-PASSANT has the capability of evaluating three level memory designs, as shown in Figure 1.

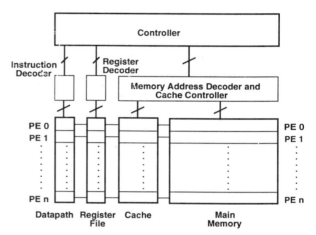

Figure 1: Shown is the MPA PE memory design space. Note that all PEs share the same controllers.

The fundamental criterion for selecting test suite programs is that they should be representative of the expected workload of the target machine. SIMD arrays, however, are not fully general purpose processors and as such general purpose benchmarks such as SPEC are not appropriate. Rather, we have taken the approach of evaluating MPAs with respect to a representative set of programs within one of the domains wherein they are most commonly used: spatially mapped applications. By spatially mapped applications we mean those applications derived from real world processes where data are mapped to the processor array in such a way as to preserve the spatial relations inherent in the application.

The six programs in the test suite are listed in Table 1; they are described in detail in [7]. Some general comments are as follows: they are all non-trivial in that they consist of from several hundred to several thousand lines of code; they are, with the exception of the IU Benchmark, in use in a research environment; and they span a wide variety of types of computations. As an example of the last point, some are dominated by gray-scale computation (which is mostly 8 bit integer), others by floating point.

Application	Description	Characteristics
ARPA IU Benchmark II	Combines several low- and inter-mediate-level vision tasks	Integrated series series of tasks. 8-bit int. ops. Region- and window-based comm.
Daumueller Line Finder	Region-based edge grouping	Region-based reductions and broadcasts.
Fast Line Finder	Region-based edge grouping	Integer ops and broadcasts.
Correspondence Matcher	Correlation-based corre-spondence	Integer ops and window-based comm.
Dutta Motion System	Depth from correspondence	FP ops
Weymouth-Overton Image Preproc.	Curve fitting filter	FP ops and window-based comm.

Table 1: List of test suite tasks.

3 Methodology

The simulation of MPAs at the machine instruction level, as is necessary for accurate evaluation, requires orders of magnitude more processing than the simulation of a serial processor. This is largely because of the tremendous number of processing elements (PEs) that need to be simulated. For example, running a modest sized program that takes milliseconds to run on a target machine (the CAAPP [15]) and seconds to run on a SUN SPARC-2 workstation can take hours, or even days, to simulate on that same workstation. Such long turnaround times make it impossible to examine a large number of design alternatives, even with a substantially faster host machine.

There are two keys to overcoming the computational intractability of the simulations: simulating at higher (less detailed) level, and reusing work done during one simulation for multiple designs. Both of these principles are employed in the commonly used method of trace-driven simulation. Trace-driven simulation is a two part process. In the first part, a trace is generated using a slightly less than comprehensive model of the target machine. We call this a *detailed* simulation, as opposed to a *complete* simulation where all components would be modeled. For example, though the instruction and register architectures are specified in a detailed simulation, typically the cache and pipeline architecture are not. This brings us to the second part of the process: the trace can now be used any number of times to drive simulators of those components that were not initially modeled. Trace-driven simulation works because the design of the components being modeled does not affect the program execution that drives the generation of the initial trace.

Trace-driven simulation, however, does not help us address the problem of computational intractability: The initial detailed simulation is still too costly. In fact, for target machines with simple PE ALUs the detailed and the complete simulation are identical. Instead, our approach is to run the initial simulation

and trace generation at an even higher level—on what we call the MPA *virtual machine*. The MPA virtual machine is the minimum configuration necessary to run a program written in a generic MPA language, in this case ICL[6]. The trace generated from the virtual machine *emulation* is then passed through a series of transformations with respect to the attributes of the target architecture until a trace emerges that closely resembles the trace that would have been generated by a detailed simulation of that target machine. We call this process of trace reconstruction *trace compilation*.

Together, virtual machine emulation and trace compilation are typically 30 to 50 times faster than detailed simulation. Also, traces need to be generated only rarely. See Figure 2 for block diagrams of the different methods just described.

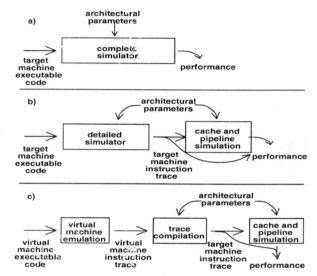

Figure 2: Block diagrams of three possible methods of evaluating a class of architectures: a) complete simulation, b) trace-driven simulation, and c) virtual machine emulation and trace compilation.

The basic problem in evaluating a potential memory hierarchy design in ENPASSANT is that the generic MPA virtual machine model has a flat memory space with locations specified only by variable name and type, rather than the physical memory and register locations of the target machine architectures. This 'gap' is the cost of being able to generate traces quickly and to evaluate the traces with respect to a large number of target machine register/cache designs without regeneration. As a consequence, the physical memory, cache, and register behavior of the program execution must be (re)constructed *a posteriori*.

4 MPA Register Architectures

In order to evaluate a register architecture it is necessary to determine the number of memory references required during the execution of a test suite program.

There are two ways that PE registers can be assigned: statically at compile time or dynamically at run time. ENPASSANT currently supports several dynamic replacement policies, including, LRU, random replacement, and a hybrid policy that combines evaluation stack allocation with random replacement. Note that this latter policy makes much more sense for SIMD PE registers than for serial processor or multiprocessor PE register assignment: SIMD PE registers are used almost exclusively to hold elements of parallel variables, (i.e. array elements).

Figure 3 contains the results of six of the benchmark programs using the hybrid replacement policy. A CAAPP-like model was used; however changing the datapath design does not change the shape of the graphs, only the scale of the Y-axis.

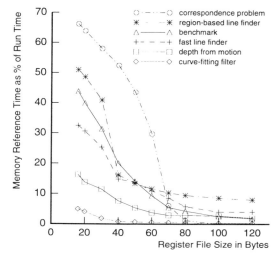

Figure 3: Effect of the register file size on the fraction of program execution time spent on memory references. A CAAPP-like model is used.

The key observation is that the curves all have distinctive 'knees'. In this respect, the shapes are similar to what one finds when doing memory analysis on serial processors. These knee shapes in Figure 3 are significant because they record a pattern of locality in variable references: in particular they signify a pattern of locality similar to that found in the memory references of serial programs and that analogous architectural techniques will also be useful for MPAs. We now compare the LRU and hybrid replacement policies. We observe in Figure 4 that LRU has slightly better performance than the hybrid policy. This is to be expected since LRU uses significant run-time information to decide which variable should be swapped out. Another result, however, is that the effect of the register replacement policy on memory performance is relatively small, especially at the design target past the knee of the curve.

The significance of the result lies in its implication on the reconstruction of compile-time register assignment. The hybrid policy, blindly assigning registers to the evaluation stack and randomly assigning the rest, should give worse performance than the policy of a reasonable compiler. LRU, as it uses dynamic information about locality of references, should give better performance. The policies, as a pair, are thus likely to provide bounds on the possible effect of compile-time register assignment on register file performance. Since the spread is relatively small it follows that either of these policies can be used to obtain a first

order approximation of compile-time register assignment, without the need the need to recompile and regenerate the traces for each data point.

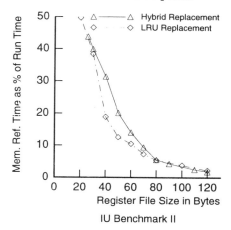

Figure 4: Effect of dynamic register replacement policy on the total cost of memory references.

5 MPA Cache Architectures

Evaluating cache architectures from a memory reference trace is straightforward and uses the same basic technique that was used for reconstructing dynamic register assignment. The cache evaluator in ENPAS-SANT allows cache architectures to be evaluated with respect to cache size, line size, and associativity.

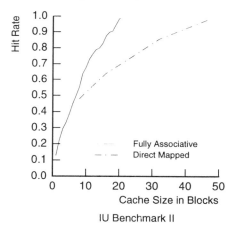

Figure 5: Comparision of fully associative and direct mapped cache performance.

In Figure 5 we compare associative and direct-mapped caches by showing the effect of associativity on the relationship between the cache size and the hit rate. The benchmark programs were run on the CAAPP-like model with a virtualization factor of 4. The register file size was 30 bytes. Although these results are not sufficient for use in final design decisions—the effect of the cache design within the entire memory architecture must be examined for that to be possible—they are useful in showing what those effects are likely to be. To make caching cost-effective, high hit rates are essential. In particular, we see that whereas an associative cache having a size of 20 blocks is sufficient to consistently achieve hit rates over 95%, a direct-mapped cache must be twice that size to achieve the same performance. The exceptions are the programs that require only a small amount of memory [7].

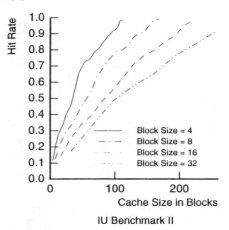

Figure 6: Effect of varying the cache block size on the relationship between the hit rate and the overall cache size. The cache is fully associative.

In Figure 6 we evaluate the effect of block size on cache performance. All programs were again run on a CAAPP-like model with a virtualization factor of 4. Here we see a result that is completely counter to what has almost universally been found in similar analyses of serial processor codes: the hit rate *decreases monotonically* as the block size increases. The explanation is that there is much less spatial locality in the memory references in the MPA test suite programs than in serial programs. One reason for this has been alluded to earlier: much of the spatial locality in serial machine programs comes from stepping through arrays. In MPAs we are already processing arrays two dimensions at a time.

6   Conclusion

In this paper we have presented the results from a study of SIMD array memory hierarchies using EN-PASSANT, an evaluation system designed specifically to take advantage of the properties of MPAs to perform efficient simulations. Using the trace compilation technique, we were able to generate hundreds of data points in a short time on a modest workstation.

As far as we are aware, these are the first published results of working set sizes and locality effects for SIMD array programs. The first-order working sets for the test suite programs running with a virtualization factor of 1 (as indicated by the knees in the graphs in Figure 3) are between 25 and 80 bytes. Further locality with higher virtualization factors is shown in Figures 5 and 6 clearly indicating the advantage of using a multilevel memory hierarchy.

However, although these results indicate high temporal locality, spatial locality was much lower than expected. Cache associativity was found to be much more important for MPAs than for serial or multiprocessors: direct mapped caches must be more than

double the size of fully associative caches to achieve similar performance. Also, in an even more surprising result, we found that the caches with the smallest block sizes yielded the best performance. The likely explanation for these phenomena is the extremely fine granularity of MPAs: parallel variables (typically images, arrays, etc.) are distributed over the processor array such that only a few elements are deposited in each PE. So although variables are used in working sets (as in common serial processor programs) yielding temporal locality, a major causes of spatial locality—stepping through arrays—is largely non-existant.

Much more work remains to be done. Ongoing projects include expanding the application domain, examining locality of programs that use extensive local indexing, and developing practical schemes for implementing MPA caches.

References

[1] Active Memory Technology, Inc. *AMT DAP Series: Tech. Overview*. AMT, Inc., Irvine, CA 92714, 1988.

[2] Bell, G. Scalable, parallel computers: Alternatives, issues, and challenges. *IJPProg. 22*, 1 (1994), 3–46.

[3] Blank, T. The MasPar MP-1 architecture. In *Proc. 35th IEEE Comp. Conf.* (1990), pp. 20–24.

[4] Bolotski, M., Simon, T., Vieri, C., Amirtharajah, R., and Knight, Jr., T. F. A 1024 processor 8 ns simd array. In *Proc. 16th Conf. on Adv. Res. in VLSI* (1995).

[5] Brewer, J. E., Miller, L. G., Gilbert, I. H., Melia, J. F., and Garde, D. A single-chip digital signal processing subsystem. In *Proc. 1994 IC on WSI* (1994), pp. 1–8.

[6] Burrill, J. H. *The Class Library for the IUA: Tutorial*. AAI, Inc., Amherst, MA 01003, 1992.

[7] Herbordt, M. C. *The Evaluation of Massively Parallel Array Architectures*. PhD thesis, Dept. of CS, U. of Mass. (also TR95-07), 1994.

[8] Herbordt, M. C., and Weems, C. C. Enpassant: An environment for evaluating massively parallel array architectures for spatially mapped applications. *Int. J. of PR and AI 9*, 2 (1995).

[9] Hill, M. D. A case for direct mapped caches. *Computer 21*, 12 (1988), 25–40.

[10] Klietz, A. E., Malevsky, A. B., and Chin-Purcell, K. A case study in metacomputing: Distributed simulations of mixing in turbulent convection. In *Proc. Work. on Het. Proc.* (1990), pp. 416–423.

[11] Ligon III, W. B., and Ramachandran, U. An empirical methodology for exploring reconfigurable architectures. *JPDC 19* (1993), 323–337.

[12] Marek, T. C. *Comparative Evaluation of Processor Architectures for Massively Parallel Systems*. PhD thesis, North Carolina State University, 1992.

[13] Potter, J. L., Ed. *The Massively Parallel Processor*. The MIT Press, Cambridge, MA, 1985.

[14] Tucker, L. W. Architecture and applications of the Connection Machine. *Computer 21*, 8 (1988), 26–38.

[15] Weems, C. C., Levitan, S. P., Hanson, A. R., Riseman, E. M., Nash, J. G., and Shu, D. B. The Image Understanding Architecture. *IJCV 2*, 3 (1989).

Profile-guided Multi-heuristic Branch Prediction

Pohua Chang
Utpal Banerjee
Intel Architecture Lab
M/S RN6-18
2200 Mission College Blvd.
Santa Clara, CA. 95052-8119
pohua@gomez.sc.intel.com
banerjee@csrd.uiuc.edu

Abstract

Superscalar and VLIW processors must dispatch a large number of instructions to the function units every cycle to keep the machine saturated. Because there are usually many branch instructions in scalar programs, speculative execution is limited by branch prediction. To improve branch prediction, researchers have studied various branch prediction algorithms over the years. In this paper, we study a multi-heuristic branch prediction algorithm that is based on profiling and compile-time selection of branch prediction heuristic for each branch instruction. We study a two-heuristic prediction algorithm: One of the heuristics is a modified Two-level Adaptive algorithm [7], the other heuristic is the AVG heuristic, which makes prediction based on the average number of Not-Taken branches that are seen before a Taken branch in the branch history. We show that the two-heuristic branch prediction algorithm is better than the modified Yeh's algorithm.

1. Introduction

Branch prediction is a critical factor in pipelined processor design [1]. To improve branch prediction, researchers have studied various branch prediction algorithms [2, 3, 4, 5, 6, 7, 8, 9]. In this paper, we study a multi-heuristic branch prediction algorithm that is based on profiling and compile-time selection of branch prediction heuristic for each branch instruction. A multi-heuristic branch predictior is constructed by combining two or more single-heuristic branch predictors [8, 9]. On top of the single-heuristic branch predictors, there is a meta-predictor that chooses which single-heuristic branch predictor that is to be used for each branch instruction. There are two strategies. The first strategy is to use a hardware meta-predictor for selecting which single-heuristic branch predictor is to be used for each dynamic branch instruction [8]. The second strategy is to provide a way for the compiler to tell the hardware which single-heuristic branch predictor is to be used for each branch instruction [9]. The compiler can profile the program to collect run-time information, based on which, selection decisions can be made. In this study, we focus on the compiler selection strategy.

A two-heuristic branch predictor is proposed and studied. We show that it has superior performance over single-heuristic branch predictors.

The organization of this paper is as follows: Section 2 describes our method. Section 3 presents experimental results. Section 4 offers some concluding remarks.

2. Prediction Method

In this section, we will first describe some known dynamic branch prediction algorithms [4, 5, 6, 7], and then describe our two-heuristic branch predictor.

2.1. Two-Bit Counter Predictor

In the Two-bit Counter prediction scheme [4], an array of two-bit saturating counters is used to make branch predictions. The lower bits of the branch instruction address is used to select a two-bit saturating counter. If the value of the selected counter is greater or equal to 2, the branch is predicted to be Taken; otherwise, it is predicted to be Not-Taken. For a Taken branch, the value of the corresponding two-bit saturating counter is set to min(3, old value + 1). For a Not-Taken branch, the value of the corresponding two-bit saturating counter is set to max(0, old value - 1).

2.2. Global Two-Level Branch Predictor

As in the Two-bit Counter prediction scheme, two-level branch prediction schemes also use 2-bit counters to record branch history. However, the selection of the associated counter for a branch is based on both the lower bits of its instruction address and the history of recently executed branches.

The table of 2-bit counters is set-associative; each set is called a Pattern History Table (PHT). For a given branch, the lower-bits of its instruction address is used to select the associated PHT.

An n-bit shift register, called a branch history register (BHR), records the directions of the n most recently executed branch. For a taken branch, a 1 is shifted into the BHR. For a not-taken branch, a 0 is shifted into the BHR. The most significant bit of the history register, once shifted out, is discarded. The value of the BHR is then used to select the associated counter within the associated PHT.

An alternative two-level branch prediction scheme is the Per-set scheme. In the Per-set scheme, branches are partitioned into different sets. For each set of branches, an n-bit shift register is used to record the directions of the n most recently executed branches from that set. The value of the associated shift register is used to select the associated entry in the PHT.

2.3. Modified Per-set Branch Predictor

We have studied a new variation of the Per-set Two-Level Branch Predictor is studied. A global history register is added to the Per-set prediction scheme. The selection of the associated PHT is based on the value of the global history register instead of the branch instruction address.

For this prediction scheme (SAs.GH), we assume the following configurations:
(1) branches are partitioned into 128 sets based on their instruction addresses.
(2) the global history register has 2 bits (4 PHTs).
(3) each PHT has 512 counters; thus, each shift register has 8 bits.

2.4. AVG Predictor

A *pattern* is a finite sequence of zeroes followed by a 1. The *length* of a pattern p is the number of bits in p. A pattern is uniquely determined by its length.

Consider a given conditional branch instruction in a given program. During the execution of the program, this instruction will be executed a number of times. At any given time, the branch is taken if the condition of the instruction (a boolean expression) has the value TRUE or 1, and not taken if the value is FALSE or 0. This way we associate a sequence of zeroes and ones with the branch instruction. This sequence can be broken up into a sequence of patterns (as defined above):

$$P1, P2, ..., Pn, ...$$

The problem is to predict Pn for a given n, when we know

$$Pn-1, Pn-2, ..., Pn-k$$

where k >=1 is suitably chosen.

Let Ln denote the length of the pattern Pn for n>=1. We assume that Ln is a function of Ln-1, Ln-2,..., Ln-k:

$$Ln = f(Ln-1, Ln-2, ..., Ln-k).$$

The problem then is to find a suitable function that will closely approximate the value of Ln in practice.

AVG predictor: Take
$$Ln = floor((Ln-1 + Ln-2 + ... + Ln-k) / k).$$

In this study, we chose k to be 8.

Because the AVG predictor is more expensive to implement, we assume that we implement a small number of fully associative entries, e.g., 20 entries. The compiler needs to carefully choose the best candidates to be placed in the AVG predictor.

2.5. Combined Predictor

Our compiler can insert a probe for each branch instruction in a program. Each probe is a call to a branch predictor simulator, simulating both the SAs.GH and the AVG prediction algorithms. Running the instrumented program will produce a file that specifies, for each branch instruction, the difference between the number of prediction misses by SAs.GH and the number of prediction misses by AVG. From that data, we know the branch instructions that should be predicted by the AVG predictor. Using the simulation data, in the second compilation pass, the compiler marks each branch instruction as a SAs.GH branch or a AVG branch. The hardware will place the branches in the appropriate predictor automatically.

3. Experimental Results

We use the SPEC CINT92 programs as the benchmark programs.

Program	Profile Input
008.espresso	bca.in
022.li	li-input.lsp6
023.eqntott	int_pri_3.eqn
026.compress	cccp.c
072.sc	load1
085.gcc	1cexp.i

Because we are interested only in conditional branches, we only count conditional branch prediction statistics in the following presentation. Therefore, the miss ratios are higher than if unconditional branches are also accounted for.

We conduct two experiments: (1) use the same set of input data to measure the branch prediction results, (2) use a different set of input data to measure the branch prediction results. The purpose is to see the difference between using the same set of data to profile and to simulate branch performance, and using different sets, which is more realistic.

In the following figures, we show the branch miss prediction ratios on the Y-axis. The X-axis shows the number of entries that we have in the AVG predictor. We simulate up to 20 entries. This means that at most 20 branches can be predicted by the AVG predictor.

Same Input/008.espresso:

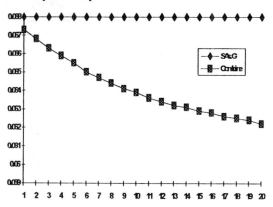

Same Input/022.li:

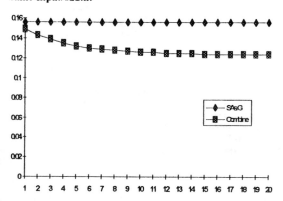

Same Input/023.eqntott:

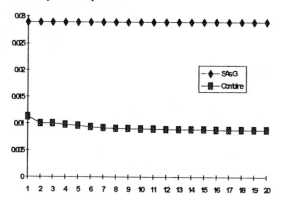

Same Input/026.compress:

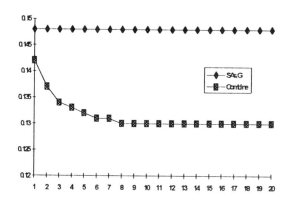

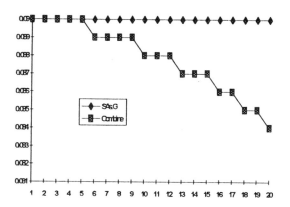

Same Input/072.sc:

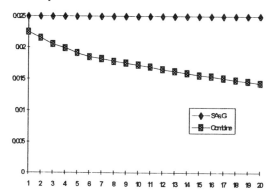

Different Input/022.li:

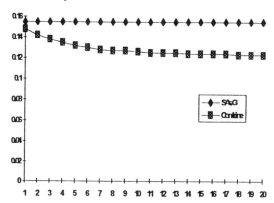

Same Input/085.gcc:

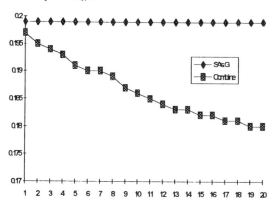

Different Input/023.eqntott:

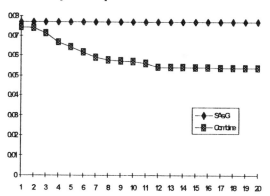

Different Input/008.espresso:

Different Input/026.compress:

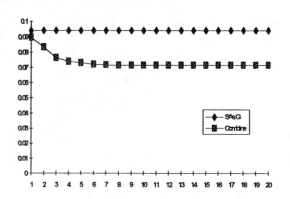

Different Input/072.sc:

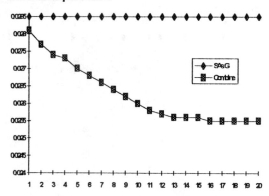

Different Input/085.gcc:

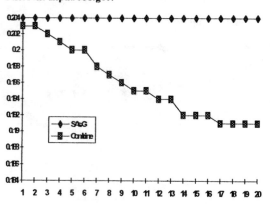

The next tables shows the percentages of branch prediction misses that are eliminated by the combined approach over the SAs.GH approach.

Program	Same Input	Different Input
008.espresso	0.5%	0.5%
022.li	3.5%	3%
023.eqntott	2%	2%

026.compres	2%	2%
072.sc	1%	0.3%
085.gcc	2%	1.3%

For high-performance processors, a 1% reduction in branch miss prediction ratio results in substantial speedup. Therefore, we have shown that the combined SAs.GH+AVG multi-heuristic predictor is far more superior than SAs.GH predictor.

4. Conclusion

We proposed a multi-heuristic branch predictor that is based on profiling feedback to make selection of single-heuristic branch prediction heuristics. We have shown by simulation that it is superior to single-heuristic branch predictors.

References

[1]   P. M. Kogge, *The Architecture of Pipelined Computers*, pp. 237-243, McGraw-Hill, 1981.

[2]   J. A. DeRosa and H. M. Levy, "An Evaluation of Branch Architectures," Proceedings of the 14th International Symposium on Computer Architecture, May 1989.

[3]   S. McFarling and J. L. Hennessy, "Reducing the cost of branches," Proceedings of the 13th International Symposium on Computer Architecture, pp.396-404, June 1986.

[4]   J. E. Smith, "A Study of Branch Prediction Strategies," Proceedings of the 8th International Symposium on Computer Architecture, pp. 135-148, June 1981.

[5]   T.-Y. Yeh and Y. N. Patt, "Alternative Implementations of Two-level Adaptive Branch Prediction," Proceedings of the 19th Annual International Symposium on Computer Architecture, pp. 124-135, May 1992.

[6]   T.-Y. Yeh and Y. N. Patt, "Two-level Adaptive Branch Prediction," Proceedings of the 24th ACM/IEEE International Symposium on Microarchitecture, pp.51-61, November 1991.

[7]   T.-Y. Yeh and Y. N. Patt, "A Comparison of Dynamic Branch Predictors that use Two Levels of Branch History," Proceedings of the 20th Annual International Symposium on Computer Architecture, pp.257-266, May 1993.

[8]   S. McFarling, "Combining Branch Predictors," WRL Technical Note TN-36, Digital Equipment Corporation, June 1993.

[9]   P.-Y. Chang, E. Hao, T.-Y. Yeh, and Y. N. Patt, "Branch Classification: a New Mechanism for Improving Branch Predictor Performance," Proceedings of the 27th ACM/IEEE International Symposium on Microarchitecture, pp.22-31, November 1994.

TABLE OF CONTENTS- FULL PROCEEDINGS

(R): Regular Papers
(C): Concise Papers

Session 1A: Practical Multiprocessors

Chair: Prof. J. Torrellas, University of Illinois

Session 2A: Networks

Chair: Prof. C. S. Ragavendra, Washington State University